This stimulating new volume on vision extends well beyond the trad........ ..... vision research and places the subject in a much broader philosophical context. The emphasis throughout is to integrate and illuminate the visual process. The first three Parts of the volume provide authoritative overviews on computational vision and neural networks, on the neurophysiology of visual cortex processing, and on eye-movement research. Each of these parts illustrates how different research perspectives may jointly solve fundamental problems related to the efficiency of visual perception, to the relationship between vision and eye movements and to the neurophysiological 'codes' underlying our visual perceptions. In the fourth Part, leading vision scientists introduce the reader to some major philosophical problems in vision research such as the nature of 'ultimate' codes for perceptual events, the duality of psycho-physics, the bases of visual recognition and the paradigmatic foundations of computer-vision research.

This volume will be of interest to all neuroscientists, cognitive scientists, neurophysiologists, psychologists and to those working on neural networks, AI and computer vision.

Other titles of interest

Vision: Coding and Efficiency (1990) edited by Colin Blakemore

An authoritative and comprehensive collection of essays on all aspects of vision, which covers experimental and computational approaches.

Images and Understanding (1990) edited by Horace Barlow, Colin Blakemore and Miranda Weston-Smith

An imaginative and stimulating account by scientists and artists which surveys the communication of images in all their forms, as ideas, pictures, words, music and dance.

Night Vision: Basic, Clinical and Applied Aspects (1990) edited by R. F. Hess, L. T. Sharpe and K. Nordby

A detailed, current account of the mechanism of low-light vision which includes full scientific, physiological and clinical background.

Representations of Vision

Representations of Vision
Trends and Tacit Assumptions
in Vision Research

A Collection of Essays based on the
13th European Conference on Visual Perception
organised by Andrei Gorea
and held in Paris at the Cité des Sciences et de l'Industrie
in September 1990

Editor-in-Chief
ANDREI GOREA

Associate Editors
YVES FREGNAC
ZOI KAPOULA
JOHN FINDLAY

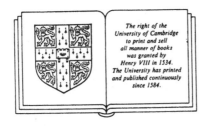

The right of the
University of Cambridge
to print and sell
all manner of books
was granted by
Henry VIII in 1534.
The University has printed
and published continuously
since 1584.

CAMBRIDGE UNIVERSITY PRESS

Cambridge
New York   Port Chester
Melbourne   Sydney

CAMBRIDGE UNIVERSITY PRESS
Cambridge, New York, Melbourne, Madrid, Cape Town, Singapore, São Paulo, Delhi

Cambridge University Press
The Edinburgh Building, Cambridge CB2 8RU, UK

Published in the United States of America by Cambridge University Press, New York

www.cambridge.org
Information on this title: www.cambridge.org/9780521115056

First published 1991
This digitally printed version 2009

A catalogue record for this publication is available from the British Library

ISBN 978-0-521-41228-5 hardback
ISBN 978-0-521-11505-6 paperback

CONTENTS

	page
Preface	ix
Acknowledgements	xi

Part 1: Visual Pyramids and Neural Networks — 1
E. Bienenstock & A. Gorea

Pyramids and multiscale representations — 3
E. H. Adelson, E. P. Simoncelli & W. T. Freeman

Multidimensional pyramids in vision and video — 17
A. B. Watson

Self-similar oriented wavelet pyramids: conjectures about neural non-orthogonality — 27
J. G. Daugman

Issues of representation in neural networks — 47
E. Bienenstock & R. Doursat

Part 2: Visual Cortical Processing: From Perception to Memory — 69
Y. Frégnac & A. Gorea

Recent progress toward an understanding of experience-dependent visual cortical plasticity at the molecular level — 73
M. F. Bear

Synchronous neuronal oscillations in cat visual cortex: functional implications — 83
C. M. Gray, A. K. Engel, P. Konig & W. Singer

How many cycles make an oscillation? — 97
Y. Frégnac

Elements of form perception in monkey prestriate cortex — 111
E. Peterhans & R. von der Heydt

Manipulating perceptual decisions by microstimulation of extrastriate visual cortex — 125
W. T. Newsome, C. D. Salzman, C. M. Murasugi & K. H. Britten

Primal long-term memory in the primate temporal cortex: linkage between visual perception and memory — 141
Y. Miyashita, N. Masui & S. Higuchi

Part 3: Eye-movements and Vision — 153
J. Findlay & Z. Kapoula

The effect of cortical lesions on visuo-motor processing in man — 155
C. Kennard

Binocular eye-movements and depth perception — 165
H. Coolewijn, J. van der Steen & L. J. van Rijn

The parsing of optic flow by the primate oculomotor system — 185
F. A. Miles, U. Schwarz & C. Busettini

Current views on the visuo-motor interface of the saccadic system — 201
J. A. M. Van Gisbergen, A. J. Van Opstal & A. W. H. Minken

Part 4: Tacit Assumptions in Vision Research   217
A. Gorea & M. Imbert

Thoughts on the specific nerve energy   219
A. Gorea

The duality of psycho-physics   231
S. Klein

Some tacit assumptions in visual psychophysics   251
C. W. Tyler

Hidden assumptions in seeing shape from shading and apparent motion   279
S. Antis

What's up in top-down processing?   295
P. Cavanagh

Tacit assumptions in the computational study of vision   305
S. Ullman

Vison tells you more than "what is where"   319
H. B. Barlow

Some strategic questions in visual perception   331
B. Julesz

PREFACE

Representation and Vision

This volume is based on the four invited symposia of the *13th European Conference on Visual Perception* (*ECVP 90*), held at La Cité des Sciences et de l'Industrie, Paris at the beginning of September 1990. The choice of the four symposia was based on the belief of their organizers (Elie Bienenstock, John Findlay, Yves Frégnac, Zoi Kapoula, Michel Imbert and myself) that they address representative trends in vision research. Necessarily, some major lines of interest in this field (such as visual attentional processes, visual learning and memory) have been totally or partially omitted but being exhaustive was not our purpose. The reader may, however, get an idea of the full scope of vision research by referring to Bela Julesz's chapter in this book and to the abstracts of all the communications presented at the *ECVP 90* and published in a special issue of *Perception* (1990, **19**, pp. 323-418).

The choice of the four symposia was also based on our belief that time is ripe for some fields in visual research to enter in closer communication and even merge. We therefore wanted to encourage the discussion of those experimental and theoretical problems the solution of which might require an attack on more than one front. This explains why we emphasized the link between computational and connectionist approaches in early vision (**Part 1**: *Visual Pyramids and Neural Networks*), between cortical processes underlying visual perception as such and visual memory (**Part 2**: *Cortical Visual Processing: From Perception to Memory*) and between visual perception and eye-movements (**Part 3**: *Eye-movements and Vision*). Finally, a most exciting endeavour was to encourage visual scientists to take up the challenge and reveal their philosophical attitudes related to their own experimental work but also vis-à-vis enduring questions such as the mind-body problem (**Part 4**: *Tacit Assumptions in Vision Research*).

There is another, quite different way to take up the reading of this book. It consists in looking at the different approaches to vision research that the book presents as distinct *representations* of what vision is. It might be trivial but still worth mentioning that, for the professional but also for the layman, vision -- and for that matter, the object of vision research -- is the set of representations explicitly or implicitly formulated by vision scientists. So is science... which, we all know, is more than just science...

The triviality of the above statement consists in its equivalence to the statement that "vision is what we (professionals) believe it is", although not everybody would agree with this definition. The interesting aspect of the statement is related to the fact that *representation* is, perhaps, the most notorious term and concept in vision research independently of how it is approached. Notorious and crucial, indeed, since equally fundamental for computer vision (Marr) as for Gestalt psychology and "ecological vision" (Gibson), for visual psychophysics and neurophysiology (as general techniques) which perpetually strive to define (usually in the first place) and to reveal (usually in the second place) bottom-up and top-down visual *functional entities*.

It might thus be an interesting and instructive endeavour to take up the reading of this book while keeping in mind that "pyramids" and "neural networks" (as theoretical approaches), spike frequencies and resonant oscillations (as dependent variables), continuous re-mapping of the sensory coordinates as a function of eye-movements (as a fundamental problem), etc., are nothing but representations in both of the above mentioned senses: they reflect, on the one hand, our choices of relevant vision problems together with the related approaches meant to solve them and, on the other hand, our personal beliefs (representations) on how vision *represents* reality.

   It is my hope that the last part of this volume will help the reader in this enterprise. The contributors to **Part 4** (*Tacit Assumptions in Vision Research* -- all visual scientists and not professional philosophers) were invited to venture to discuss vision and also higher-order associative processes, from the perspective of their current research.

   While most of the contributors to this volume only occasionally address the problem of representation as such (the chapters by Elie Bienenstock and by Patrick Cavanagh are exceptions to this rule), this is a book on representation in and of vision. What is science if not a representation of its object together with the capacity of manipulating experiment and theory to render this representation self-consistent? Revealing the nature of self-consistent representations is a scientific matter. It may also be a source of pleasure for the reader whom I invite to meditate on what representation is and to try to figure out how it is manipulated by the contributors to this book.

                                                            Andrei Gorea

ACKNOWLEDGEMENTS

The organization of the *13th European Conference on Visual Perception* on which the present volume is based was made possible by the following sponsors:

CITE DES SCIENCES ET DE L'INDUSTRIE
COMMISSION OF THE EUROPEAN COMMUNITIES
MINISTERE DE LA RECHERCHE ET DE LA TECHNOLOGIE
VILLE DE PARIS
FONDATION DU CREDIT NATIONAL
MINISTERE DE LA DEFENSE
DIRECTION DES RECHERCHES, ETUDES ET TECHNIQUES
AIR FRANCE
UNIVERSITE RENE-DESCARTES
INTERNATIONAL BRAIN RESEARCH ORGANIZATION
UNESCO
US AIR FORCE
INSTITUT NATIONAL DE LA SANTE ET DE LA RECHERCHE MEDICALE
CENTRE COMMUN D'ETUDES DE TELEDIFFUSION ET TELECOMMUNICATIONS
CENTRE NATIONAL DE LA RECHERCHE SCIENTIFIQUE
ASSOCIATION NATIONALE POUR L'AMELIORATION DE LA VUE
ESSILOR
MINISTERE DE L'EDUCATION NATIONALE, DE LA JEUNESSE ET DES SPORTS
FONDATION NATURALIA ET BIOLOGIA
THOMSON CSF
SOCIETE FRANCAISE D'OPTIQUE PHYSIOLOGIQUE

Particular thanks to

Catherine MARLOT

and to

LABORATOIRE DE PSYCHOLOGIE EXPERIMENTALE, PARIS

PART 1

Visual pyramids and neural networks

Elie Bienenstock and Andrei Gorea

Modelling early vision as a parallel, multiple-scale process has been increasingly popular since the pioneering work of Marr et al. (1979) and of Burt and Adelson (1983). In the same time, the parallel distributed processing approach has pervaded our field of research, although its domain of application was initially - and still is - meant to cover chiefly higher-level visual processes, such as form recognition, shape from shading, etc. It is our belief that the two approaches need not stay separate; they could indeed fuse into a more comprehensive one where the extraction of visual primitives would be incorporated into largely distributed neural networks. Current work in early vision modelling suggests that the time may be right for such a fusion.

The first two chapters in this Part deal essentially with the classical problems related to the data structures meant to represent multi-scale information in a natural way. E.H. Adelson, E.P. Simonicelli and W.T. Freeman discuss a number of types of pyramids in terms of their efficiency for image coding and emphasize the superiority of those based on *quadrature mirror filters* (QMFs) over more classical pyramids based on Gaussian and Lapalacian image filtering. QMF representations of the image are localized in space and spatial frequency, are self-similar and orthogonal. They are likely to be more useful for general vision applications including models of early vision although they involve some difficulties with mixed orientations. Another class of pyramids, based on *steerable filters*, may solve such problems offering, in addition, excellent properties for image enhancement. As emphasized by A.B. Watson, the pyramid approach to early vision can be generalized to other visual dimensions such as color and motion and may thus provide a general conceptual model of the representation employed at early stages in human vision.

While the efficiency of multiple-scaling coding is directly dependent on its orthogonality, J. Daugman's contribution draws attention on the fact that the receptive fields of visual neurons responding over a given region of visual space *are not mutually orthogonal* and on the difficulty of obtaining the correct coefficients on a non-orthogonal set of expansion functions. He presents a general three-layered "relaxation neural network" that can compute complete image representations with the constraint of orthogonality removed. The relaxation network permits the exploration of many degrees-of-freedom, such as the trade-off between orientation sampling and positional sampling.

The key-word for all the pyramidal approaches, as well as for a large class of neural networks is *rate-coding*: information is encoded in the rate of activity of cells, i.e. the intensity of firing over a brief period. While this might be (at least partly) true for early sensory processing, E. Bienenstock and R. Doursat argue that rate-coded representations may be inappropriate at higher processing stages because they lack inner structure. They are mere lists of features, and the computations performed by neural networks amount to *associating* these features with each other and/or with labels. Yet shape recognition and scene interpretation require (a) explicit representations of what in the object or scene relates to what, and how, and (b) computations involving *matching of such structured relational representations* with each other. E. Bienenstock and R. Doursat suggest that structured representations may be achieved in living brains by exploiting the fine temporal structure of neural activity, i.e. by encoding information in the *co-activity* or correlations between cells.

Marr D., Ullman S. & Poggio T. (1979) Band pass channels, zero-crossings, and early visual information processing. *J. Opt. Soc. Am.* **69**, 914-916.

Burt P. & Adelson E.H. (1983) The Laplacian pyramid as a compact image code. *IEEE Transactions on Communications* **COM-31**, 532-540.

Pyramids and multiscale representations

Edward H. Adelson†, Eero P. Simoncelli‡ and William T. Freeman

MIT Media Laboratory,
†Brain and Cognitive Sciences Department,
‡Electrical Engineering and Computer Science Department
Cambridge, Massachusetts 02139

Introduction

Images are composed of features of many sizes, and there is no particular scale or spatial frequency that has a special status in natural scenes. Therefore a visual system, whether natural or artificial, should offer a certain uniformity in the representation and processing of visual information over multiple scales.

Primate visual systems achieve a multiscale character in two ways. First, in the retina, there is a continuous variation in the sizes of the receptive fields of ganglion cells, with the size increasing roughly in proportion to distance from the fovea (and spatial resolution descreasing accordingly); a similar scaling is reflected in cortex. And second, for a given patch of the visual field there are numerous cells in striate cortex which are tuned for different bands of spatial frequency. The decomposition of each part of the image into a set of spatial frequency tuned responses seems to be critical to vision systems in nature, and has been found to be very useful in many artificial settings as well.

One approach to understanding the issues that a natural vision system must face is to build artificial systems and discover the power and limits of different representational schemes. This paper will present a brief overview of some of the multiscale representations that we have explored in our laboratory, and will describe some of the lessons we have learned about representing and using multiscale image information. In this limited space it is impossible to present much detail, so readers may wish to consult the original sources for further information.

Pyramids

A pyramid is a multiscale representation that is constructed with a recursive method that leads naturally to self-similarity. The first basic idea is shown in figure 1, which shows a "Gaussian" pyramid (Burt, 1981; Burt and Adelson, 1983). The original test image is convolved with a low-pass filter and subsampled by a factor of two; the filter-subsample operation is repeated recursively to produce the sequence of images shown. Such a pyramid can be useful for operations that require access to information about low frequencies. The pyramid is also highly efficient to compute.

This research was supported by grants from NSF (IRI 871-939-4), and DARPA (Rome Airforce F30602-89-C-0022).

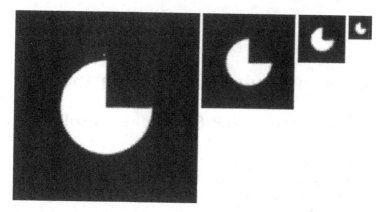

Figure 1: A 4-level gaussian pyramid constructed on a test image.

Figure 2 shows a "Laplacian" pyramid of the same image, in which a bandpass filter is used rather than a low-pass filter. A Laplacian pyramid is a complete representation of an image, in the sense that one can perfectly reconstruct the original image given the coefficients in the pyramid. The reconstruction process is straightforward: one simply interpolates ("expands") each image up to the full size of the original image using the correct interpolation filter, and then sums all of the interpolated images.

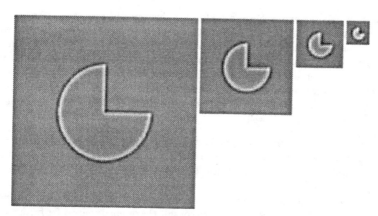

Figure 2: A 4-level laplacian pyramid constructed on a test image.

The hierarchical filtering procedures lead to equivalent filters which are illustrated in figure 3. The equivalent filters used in building the Gaussian pyramid are shown in figure 3(a), while the equivalent filters used in building the Laplacian pyramid are shown in figure 3(b).

Completeness is a valuable property for a representation in early vision, not because a visual system needs to literally reconstruct the image from its representation, but rather because completeness guarantees that no information has been lost, i.e. that if two images are different

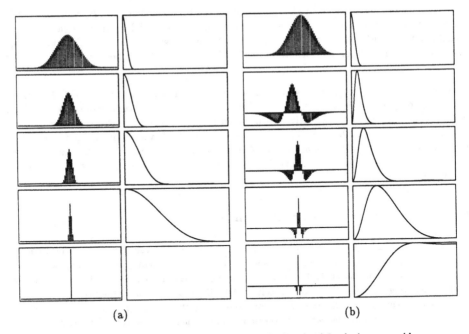

Figure 3: (a) Five example basis functions of a four level Laplacian pyramid, along with their Fourier transforms. (b) The corresponding inverse (sampling) functions of the pyramid, with their Fourier transforms. The transforms are plotted on a linear scale over the range from 0 to π.

then their representations are different.

A complete representation can also be used as a method of storing image information, and the Laplacian pyramid offers an efficient means of storage (Burt and Adelson, 1983). The image encoding procedure is as follows: first the Laplacian pyramid is constructed; then the coefficients in each level are quantized into a fairly small number of bins; then entropy coding techniques such as Huffman coding are applied to the quantized values. Because the Laplacian pyramid values tend to cluster around zero, and because the higher frequencies coefficients tend to have low variance, this technique allows a 256x256 8-bit greyscale image to be stored with about 1.5 bits per pixel with little degradation in image quality.

What can the research on image data compression tell us about multiscale representations in biological system? A biological system is not trying to store an image for reconstruction and display, and cannot use digital techniques such as Huffman coding to gain efficiency. Nonetheless there are important lessons to be learned. The first task of most image coding schemes is to find a representation that is robust and is well-matched to scene statistics. The quantization step noted above leads to random perturbations in the pyramid coefficients, and these perturbations translate into local contrast errors in the bandpassed images. In spite of the random perturbations, it is possible to reconstruct the image with little degradation due to the robust nature of the pyramid. Moreover we find that, from the standpoint of scene statistics, one can allow the perturbations to be quite large in the high frequency bands, while one needs more accuracy in the medium and low frequency bands. For a biological representation this means that it is possible to get away with noisy neurons without losing very much image information, and that the representation of the high frequencies can be particularly tolerant of error.

The efficiency of pyramid representation has relevance in other domains as well. For example, many computational operations such as coarse-to-fine motion processing or stereo matching can be accomplished very efficiently in a pyramid structure. Computational advantages will also be found in such applications as texture analysis, orientation analysis, and pattern matching. A pyramid typically uses as few coefficients as are possible at a given scale, and this reduces both the storage requirements and the number of operations that must be performed in a given task.

Gabor Functions and Orientation Tuning

Orientation tuning is one of the most salient aspects of the cells found in striate cortex, and so it would be useful to understand how to build and use oriented multiscale image representations. Two-dimensional Gabor functions have been the most popular idealized receptive field models (e.g. Granlund, 1978; Marcelja, 1980; Daugman, 1985). One difficulty with the Gabor transform, at least in its original formulation, is that it is highly non-orthogonal. To understand what this means, we have to discuss some general properties of linear transforms.

A linear transform expresses a given (discrete) signal, $f(n)$, as a sum of a set of basis functions, $b_i(n)$:

$$f(n) = \sum_i c_i b_i(n)$$

In the familiar case of the Fourier transform, the $b_i(n)$'s are sinusoids. The c_i's are the coefficients indicating the amount of each basis function that must be added in order to synthesize the original signal.

The value of each coefficient c_i can be determined by taking a weighted sum of the pixels in the input signal, i.e. by taking a dot product of the input and a sampling function which represents a "receptive field." That is, for the ith coefficient there is a sampling function $s_i(n)$ such that

$$c_i = \sum_n f(n)s_i(n)$$

In the case of an orthogonal transform, such as the Fourier transform, the sampling functions $s_i(n)$ and the basis functions $b_i(n)$ are identical, so that one detemines the coefficient of a given sinusoid by computing the dot product of the image with that same sinusoid. But in the case of non-orthogonal transforms, the sampling and basis functions can be quite different.

The Gabor transform is invertible because its basis functions are linearly independent; howevever it is not orthogonal and the sampling functions are quite different from the basis functions. Figure 4 shows the Gabor basis functions, along with their Fourier transforms, and also the inverse functions. The inverse set is quite poorly behaved and not at all like one expects to find in a biological system. If one wants to use the Gabor functions as a basis set with which to build images, then one must derive the coefficients by applying the inverse functions, i.e. one would have to use a visual system with these bizarre receptive fields. Or, if one builds an image representation by applying the receptive fields comprising the Gabor set, then the resulting coefficients implicitly represent the image as a sum of the unpleasant inverse functions.

The original Gabor transform has some additional difficulties, one being that it is not self-similar since all the Gabor functions are windowed by a Gaussian of the same width. Many of the investigators who have used Gabor functions in their work have devised self-similar, pyramid-like, approaches.

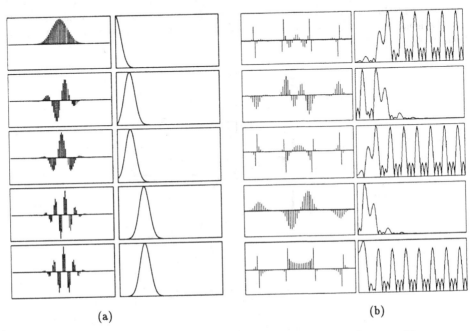

(a)   (b)

Figure 4: (a) Five of the sixteen basis functions of a Gabor filter set, with their corresponding Fourier transforms. (b) The inverse (sampling) functions of the Gabor filter set. The transforms are plotted on a linear scale over the range from 0 to π.

Quadrature Mirror Filter Pyramids and Wavelets

Is it possible to construct a representation that has many of the aspects of a self-similar Gabor transform, and yet which is orthogonal? The answer is yes, as we will now discuss.

Quadrature mirror filters (QMF's) are a class of band-pass filters that were described in the speech domain by Croisier, Esteban and Galand (Croisier et. al., 1976; Esteban and Galand, 1977), and have more recently been applied to the decomposition of images (Vetterli, 1984;

Woods and O'Neil, 1986; Adelson *et al.*, 1987). Although the filters were originally developed using signal processing concepts, they can be easily understood in terms of orthogonal linear transforms (Simoncelli, 1988; Simoncelli and Adelson, 1990b). There has been considerable theoretical and applied work on QMF's in recent years, much of which is reviewed in a book edited by Woods (Woods, 1990). In addition, it has been shown that QMF pyramids are a discrete orthogonal form of wavelets (Mallat, 1989), and for image representation the terms "wavelets" and "QMF's" are sometimes used interchangeably.

Figure 5 shows a self-similar set of QMF's derived from a basic one that has 9 coefficients. These filters can be used as a self-similar basis set for an orthogonal pyramid, where the sampling density of each level is one-half that of the previous level. The result is a pyramid which is "critically sampled," i.e. the number of coefficients is equal to the number of pixels in the original image. The filters are not perfect, in that the reconstructed image will differ from the original very slightly.

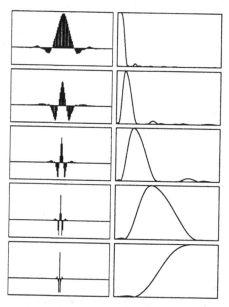

Figure 5: Five of basis functions of a 9-tap QMF/wavelet pyramid transform, along with their Discrete Time Fourier Transforms.

In addition to offering a self-similar orthogonal basis set, the QMF's shown in figure 5 are compact in both space and spatial frequency. Since they are (approximately) orthogonal, the sampling functions are identical to the basis functions. Filters like these form a promising set for application to many problems in image processing. Before one can apply them to images, however, one must extend the one-dimensional functions into two dimensions.

The most straightforward method of two-dimensional generalization involves the separable application of band-splitting QMF's. The QMF's shown in figure 5 come in pairs, which split the frequency band into high-pass and low-pass components. In two dimensions, such filters can be applied separably in the x- and y- dimensions to produce four filters, which may be labeled low-low, low-high, high-low, and high-high. The low-high and high-low bands contain oriented information about vertical and horizontal components of the image. The low-low band contains low-passed information which can be further decomposed in the next level of the pyramid. The high-high band contains a mixture of left and right diagonal information. Figure 6 depicts this decomposition in the frequency domain. The separable decomposition retains the orthogonality of the one-dimensional transform.

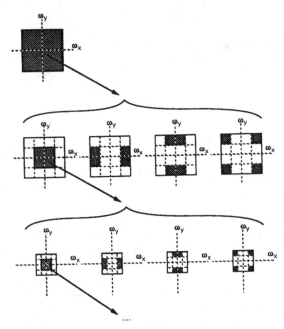

Figure 6: Idealized diagram of the partition of the frequency plane resulting from a 4-level pyramid cascade of separable 2-band filters. The top plot represents the frequency spectrum of the original image, with axes ranging from $-\pi$ to π. This is divided into four subbands at the next level. On each subsequent level, the lowpass subband (outlined in bold) is subdivided further.

A QMF pyramid built using filters of the sort described above turns out to be extremely good for image data compression. Indeed, "subband coders" based on such pyramids are among the best techniques known for efficient coding and are being widely investigated for application in image archiving and digital video transmission.

In spite of its simplicity and its success in image coding, the separable approach has some problems from the standpoint of vision systems. The high-high filter is not nicely oriented, since it contains equal contributions from the two diagonal orientations. This problem is not easily remedied; i.e. it is not easy to split the diagonal band into two oriented parts. Therefore one must seek other approaches in order to achieve orientation specificity in all of the bands.

Simoncelli and Adelson (Adelson *et al.*, 1987, Simoncelli and Adelson, 1990a) have described

 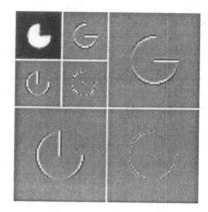

Figure 7: Construction of a 4-level QMF pyramid built on a test image.

a method of constructing QMF pyramids on a hexagonal lattice, in which the basis functions are all of the same shape and are all well-tuned in orientation and spatial frequency. An idealized frequency-domain decomposition is shown in figure 9. The actual frequency tuning of a set of such filters is shown in 10. These filters, like those of the separable pyramid, change scale by octaves from one level to the next.

The hexagonal QMF pyramid demonstrates that it is possible to capture many desirable properties in a single representation. It is an orthogonal wavelet transform; it is complete and is critically sampled, utilizing the same functions for basis functions and sampling functions. The transform is self-similar: the basis functions are all of the same shape, but appear at various sizes, positions, and rotations. The basis functions are smoothly overlapping and are well localized in both space and spatial frequency; they also bear a certain resemblance to the sort of functions that are used in modeling biological visual systems. One can use the hex pyramid for some tasks in early vision (Simoncelli and Adelson, 1990a). The hex pyramid is also a very good structure for image data compression, possibly better than the separable QMF pyramid mentioned above.

An alternate structure for building hexagonal pyramids, but with non-overlapping filters, has been described in (Crettez and Simon, 1982) and in (Watson and Ahumada, 1989). The resulting filters display an unusual blocky structure. Although we have not made a direct comparison, published results (Watson, 1990) suggest that image coding with these alternate hex pyramids requires data rate that is 2 to 4 times higher than with our hex pyramid.

Steerable Pyramids

Although the hex QMF pyramid of figure 9 manages to achieve a great many desirable properties, it does have its limitations. QMF's violate the Nyquist criterion for sampling, and are able to provide successful image representation because aliasing from adjacent bands has opposite sign and therefore cancels during reconstruction. However, there is still aliasing within

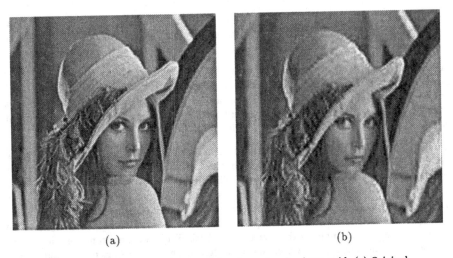

(a)   (b)

Figure 8: Data compression example using a four-level pyramid. (a) Original "Lena" image at 256 × 256 pixels. (b) Compressed using 9-tap separable QMF bank. The pyramid data was compressed to a total of 16384 bits (i.e. total first-order entropy was 0.25 bit/pixel)

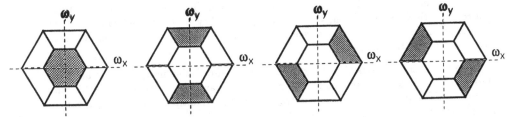

Figure 9: The low-pass and the three oriented high-pass bands of the hexagonal pyramid. Note that the high-pass subbands are not mixtures of different orientations, as in the separable decomposition. This may improve performance for coding and image analysis applications

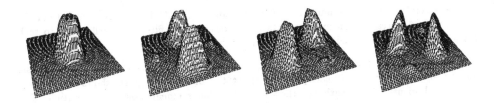

Figure 10: The power spectra for a "4-ring" set of hexagonal QMF filters.

a given band and this expresses itself as a lack of positional invariance in the representation. The problem can be improved if the QMF's have sharp-cut frequency responses, but then the filters lose spatial localization and produce significant "ringing" artifacts.

In fact, the very properties that tend to characterize good filters for data compression (sharp frequency cut-offs with flat responses in between) cause problems for image analysis and early vision applications. For example, in orientation analysis it is necessary that filter responses be very smooth and without flat regions, in order that the population response to different orientations should vary continuously as the orientation of the stimulus is varied.

An interesting class of filters that are well suited to orientation analysis are "steerable filters" (Freeman and Adelson, 1990). A small bank of these filters (say, 4), tuned for different orientations, can be used to analyze an image. Then if one wishes to know the response of a filter of arbitrary orientation, one can compute it as a linear combination of the responses of the original filters. Thus one can derive information about a continuum of possible oriented filters by the application of only a few. The concept can also be extended to allow steerability in phase as well as orientation: by applying the correct set of basis filters, with appropriate orientations and phases, one can synthesize the response of a filter at an arbitrary orientation and phase. One can also extract local energy measures, find the direction of maximal orientation strength, and so on, all from the same basic set of measurements.

Figure 11 (a) shows a bank of steerable filters. These particular filters were designed with one more criterion in mind, namely, that they should be useful for constructing a steerable pyramid decomposition. Indeed, these particular filters were designed to allow the construction of a self-inverting decomposition, which is to say that they were designed so that the basis functions and the sampling functions would be identical. A pyramid can be constructed from either the even or the odd phase filters.

Since the steerable filters are not orthogonal, the self-inverting property must be enforced through other means. We use a highly overcomplete set, and design the filters to "tile" in the frequency domain; i.e. the summed spectral power of the multiple bands and orientations is forced to be flat.

Figure 11 shows a steerable pyramid decomposition of a test image (which has been pre-filtered for reasons that will not be discussed here). The original image is shown in Fig. 11(b); the various levels and orientations are shown in (c), (d), and (e). Because the filters are smooth and their outputs are oversampled, the responses shown here are also quite smooth and well-behaved.

The steerable pyramid is much less efficient than the hex QMF pyramid, both from the standpoint of representation and computation. However, the filters are well-suited to such tasks orientation analysis, edge detection, and image enhancement.

Figure 12(a) shows a picture of Einstein, and figure 12(b) shows an orientation analysis applied to the same image. At each point the orientation of the line segment shows the direction of maximal orientation strength, and the length of the line segment shows the magnitude. The orientation and strength were calculated from the outputs of a pair of even and odd phase steerable filters, similar to the odd phase filters of Figure 11(a) (Freeman and Adelson, 1990) (cf. Knutsson and Granlund, 1983).

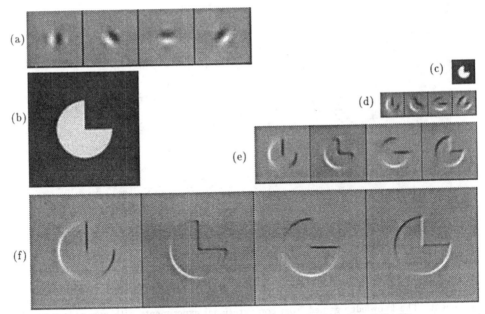

Figure 11: (a) Basis filters for steerable pyramid. Combinations of the filter at these four orientations span the space of all rotations of the filter. (b) Test image. (c) Low-pass image at top of multi-scale pyramid representation of (b). (d) - (f) Steerable, bandpass coefficients in pyramid representation. A linear combination of the transform coefficients will synthesize the response of the analyzing filter at any angle.

Figure 13 shows an example of image enhancement with steerable filters. The original image is a digital cardiac angiogram–an X-ray of a heart. The orientation was analyzed, and then a filter was applied along the direction of maximal orientation in order to enhance the oriented information. A local gain control was then applied to normalize contrast. The result is shown in Fig. 13. Linear structures, which are the ones of greatest interest here, are greatly enhanced in visibility.

Conclusions

Multiscale image representations are useful in a wide variety of vision tasks, and pyramids offer a highly convenient approach to the computation and utilization of multiscale processing. Research in pyramid image representation has revealed some of the strengths and limitations of various kinds of representaitons. Laplacian pyramids are complete and are fairly efficient for image coding, and are useful for front-end processing in various aspects of early vision. Improved coding efficiency can be achieved with QMF pyramids, which are built with orthogonal basis functions; QMF pyramids lead to discrete orthogonal wavelet transforms. By adopting a sampling structure based on a hexagonal lattice it is possible to build QMF pyramids in

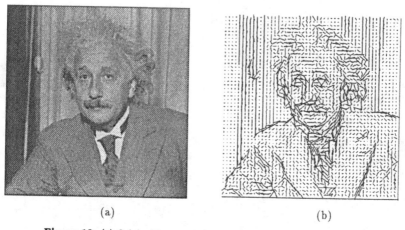

(a)   (b)

Figure 12: (a) Original image of Einstein. (b) Orientation map of (a).

which all of the basis functions are well-tuned in orientation and spatial frequency. We have recently explored new form of pyramid based on steerable filters, which is less efficient for coding but is well-suited to such tasks as orientation analysis, edge-detection, and image enhancement. The knowledge gained from computational experiments with pyramids may be helpful in understanding the representational issues faced by biological visual systems.

References

E. H. Adelson, E. Simoncelli, and R. Hingorani (1987): Orthogonal pyramid transforms for image coding. In *Proceedings of SPIE*, pp. 50–58, Cambridge, MA.

P. J. Burt (1981): Fast filter transforms for image processing. *Computer Graphics and Image Processing*, 16:20–51.

P. J. Burt and E. H. Adelson (1983): The Laplacian pyramid as a compact image code. *IEEE Trans. Communications*, COM-31(4):532–540.

J. P. Crettez and J. C. Simon (1982): A model for cell receptive fields in the visual striate cortex. *Computer Graphics and Image Processing*, 20:299–318.

A. Croisier, D. Esteban, and C. Galand (1976): Perfect channel splitting by use of interpolation/decimation/tree decomposition techniques. In *International Conference on Information Sciences and Systems*, pp. 443–446, Patras.

R. E. Crochiere and L. R. Rabiner (1983): *Multirate Digital Signal Processing. Signal Processing Series*, Prentice-Hall, Englewood Cliffs, NJ.

J. G. Daugman (1985): Uncertainty relation for resolution in space, spatial frequency, and orientation optimized by two-dimensional visual cortical filters. *J. Opt. Soc. Am. A*, 2(7):1160–1169.

D. Esteban and C. Galand (1977): Application of quadrature mirror filters to split band voice coding schemes. In *Proceedings ICASSP*, pp. 191–195.

W. T. Freeman and E. H. Adelson (1990): Steerable filters for early vision, image analysis, and wavelet decomposition. In *Proc. Intl. Conf. Computer Vision*.

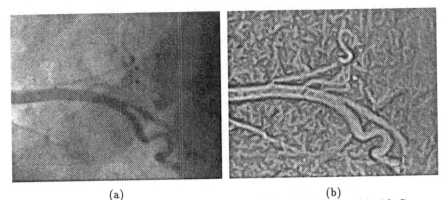

(a)   (b)

Figure 13: (a) Digital cardiac angiogram. (b) Result of filtering (a) with G_2 oriented along the local direction of dominant orientation, after local contrast enhancement (division by the image's blurred absolute value). The oriented vascular structures of (a) are enhanced. (We thank Paul Granfors of G. E. Medical Systems (Milwaukee) for providing the digital angiogram.)

G. H. Granlund (1978): In search of a general picture processing operator. *Comp. Graphics, Image Proc.*, 8:155–173.

H. Knutsson and G. H. Granlund (1983): Texture analysis using two-dimensional quadrature filters. In *IEEE Computer Society Workshop on Computer Architecture for Pattern Analysis and Image Database Management*, pp. 206–213.

S. G. Mallat (1989): Multifrequency channel decompositions of images and wavelet models. *IEEE Trans. ASSP*, ASSP-37(12):2091–2110.

S. Marcelja (1980): Mathematical description of the response of simple cortical cells. *J. Opt. Soc. Am.*, 70:1297–1300.

E. P. Simoncelli (1988): *Orthogonal sub-band image transforms.* Master's thesis, Massachusetts Institute of Technology, Department of Electrical Engineering and Computer Science, Cambridge, MA.

E. P. Simoncelli and E. H. Adelson (1990a): Non-separable extensions of quadrature mirror filters to multiple dimensions. In *Proceedings of the IEEE: Special Issue on Multidimensional Signal Processing*.

E. P. Simoncelli and E. H. Adelson (1990b): Subband transforms. In J. W. Woods, editor, *Subband Image Coding*, chapter 4, Kluwer Academic Publishers, Norwell, MA.

M. Vetterli (1984): Multi-dimensional sub-band coding: some theory and algorithms. *Signal Processing*, 6(2):97–112.

A. B. Watson (1989): Perceptual-components architecture for digital video. *J. Opt. Soc. Am. A*, 7:1943–1954.

A. B. Watson and A. J. Ahumada (1989): A hexagonal orthogonal-oriented pyramid as a model of image representation in visual cortex. *IEEE Trans. Biomedical Engineering*, 36(1):97–106.

J. W. Woods and S. D. O'Neil (1986): Subband coding of images. *IEEE Trans. ASSP*, ASSP-34(5):1278–1288.

J. W. Woods, editor (1990): *Subband Image Coding*. Kluwer Academic Publishers, Norwell, MA.

Multidimensional pyramids in vision and video

Andrew B. Watson

AeroSpace Human Factors Research Division
NASA Ames Research Center, MS 262-2
Moffett Field, CA 94035-1000

INTRODUCTION

The family tree of the image pyramid concept has two great taproots, one in the study of biological vision, the other in computer graphics, image processing, and computer vision. Physiological and psychophysical ideas about multiple spatial frequency channels may be traced back at least to the work of Robson, Campbell, Nachmias, and Blakemore beginning during the mid sixties (Blakemore & Nachmias, 1971; Blakemore, Nachmias & Sutton, 1970; Campbell & Robson, 1968), while the earliest pyramid image data structures were introduced by Tanimoto and Pavlidis in the mid seventies (Tanimoto & Pavlidis, 1975). In the 1980's, these roots grew sufficiently close together that some papers, such as that of Burt and Adelson on the Laplacian Pyramid (Burt & Adelson, 1983), could not clearly be identified as belonging to one or the other root. Today, no computer vision paper that employs a pyramid structure is complete without a scholarly allusion to biological analogs, and no computer model of pyramid processing in human vision is complete without a concluding remark that "this would no doubt make an excellent front end to a computer vision system." This confluence of biological and electronic technology is natural, given that in both systems we confront the problem of efficiently representing information that may exist at many spatial scales.

The earliest pyramid structures provided a way of explicitly representing spatial scale, as well as location in an image. The concept may be generalized, however, to create a multidimensional pyramid which also makes explicit orientation, color, and motion. In this paper I want to pursue the biological/technological connection a little further, to describe how a multidimensional pyramid model of early human vision may provide a basis for the design of digital video systems.

The current battles raging about High Definition Television (HDTV) are being fought largely over how many rows and columns of pixels there should be in some new analog standard (Lechner, 1990). Lurking in the background of these petty squabbles, however, is a much more revolutionary advance, namely digital TV. Digital TV will be important not only in commercial broadcast, but also in a myriad of other consumer, scientific, industrial and military applications (Jaworski, 1990; Watson, 1990a). Whatever its detailed form, digital TV involves breaking the analog signal down into discrete pieces, wrapping these pieces into small packets, and sending them over a network. An important technical question is: into which pieces shall we break apart the signal? I will argue that the pieces should match the pieces into which human vision analyzes the visual signal. I call this a Perceptual Components Architecture for digital video (Watson, 1990b). I will begin by reviewing briefly the manner in which early human vision, to the extent that we understand it, partitions the visual signal.

MULTIDIMENSIONAL PYRAMID MODEL OF EARLY VISION

We begin with what I will call the *image stream*. This is roughly the stimulus to one eye, and it is a function distributed over space (x,y) time (t) and wavelength (λ) (Fig. 1).

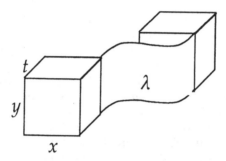

Fig. 1. The image stream. This is a distribution of light over space, time, and wavelength.

This continuous image stream is first filtered and sampled, in space and wavelength, by the receptors, of which there are three types, nominally R, G, and B. The filtering is a result of the spatial and chromatic apertures of the receptors (Packer, Hendrickson & Curcio, 1989; Schnapf, Kraft & Baylor, 1987). The resulting signal may be thought of as three three-dimensional arrays, one each for R, G and B (Fig. 2). This is a slight mis-statement, since the stream is not sampled in time.

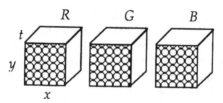

Fig. 2. The image stream sampled in space and wavelength. Each cube is one color channel.

These three color signals are then linearly transformed into three opponent channels: an achromatic signal (A), a red-green signal (RG), and a yellow-blue signal (YB) (Fig. 3) (Guth, Massof & Benzschawel, 1980 ; Hurvich & Jameson, 1957; Ingling, 1977). Roughly, these are simple sums and differences of the receptor signals (Larimer, Krantz & Cicerone, 1974, 1975).

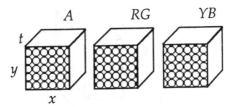

Fig. 3. Image stream after opponent color transformation.

The three remaining partitions are best viewed in the frequency domain, so we imagine that each 3D array is Fourier transformed and its axes are now spatial frequency (u,v) and temporal frequency (w). Note that I am not proposing that the visual system performs this transformation, we are simply converting in these pictures to a representation which is more informative and illuminating.

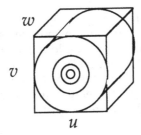

Fig. 4. Division of the achromatic channel of the image stream into multiple bands of spatial frequency.

The next partition of the image stream is by spatial frequency. The concentric rings in Fig. 4 show bands of 1 octave in spatial frequency, approximately the bandwidths of visual neurons (Campbell, Cooper & Enroth-Cugell, 1969; De Valois, Albrecht & Thorell, 1982). The center-most band contains the lowest frequencies, which correspond to the coarsest detail in the image, while the outermost ring corresponds to the highest spatial frequency, and hence the finest detail. This partition has the same general form as the earliest pyramid image data structures which separate imagery into several bands of resolution (Burt & Adelson, 1983; Crowley, 1984; Tanimoto & Pavlidis, 1975). We only show the frequency partition for the achromatic channel, but in our model it is applied independently to all three opponent color channels. There is very little evidence in the physiological or psychophysical literature on this important point (Bradley, Switkes & De Valois, 1988; Lennie, Krauskopf & Sclar, 1990; Quick & Lucas, 1979).

Next the stream is partitioned by orientation. The spatial coordinates in the frequency diagram of Fig. 4 may be regarded either as cartesian coordinates of horizontal and vertical spatial frequency (u and v) or as polar coordinates of radial spatial frequency and orientation.

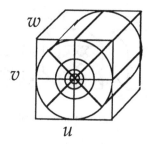

Fig. 5. Partition of the image stream by orientation.

The radial planes in Fig. 5 partition the signal into orientation bands of 45 degrees, approximately that found in visual neurons(De Valois, Yund & Hepler, 1982). Thus a particular band of spatial frequency and orientation is represented by a sector of one of the cylinders. This partition is implemented by the simple cells of primary visual cortex.

The final partition is by temporal frequency, as illustrated in Fig. 6. In this figure we have rotated the cube 90 degrees in order to better expose the temporal frequency axis, which now runs left to right, with its origin in the middle. The psychophysical and physiological evidence to date (Kulikowski & Tolhurst, 1973; Tolhurst, 1973; Watson & Robson, 1981), while equivocal, suggests a partition into just two temporal bands, low (L) and high (H). These may be associated with the magnocellular and parvocellular streams of retina, lateral geniculate nucleus, and visual cortex in primate (Merigan, 1989).

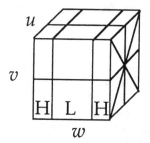

Fig. 6. Partition of the achromatic channel of the image stream by temporal frequency.

In keeping with the theme of this symposium I should point out that this partition of the temporal frequency axis is a pyramid in the time dimension. Here we have envisioned only two level to the pyramid, low and high, but the principle is the same. And indeed, it may be worthwhile to consider more than two layers in the future.

When a real signal, like the image stream, is transformed into the frequency domain, it has conjugate symmetry, which means that the value at any point is identical to the complex conjugate of the point on the opposite side of the origin. This means that we only need to consider one half of the frequency cube, since the other half is redundant. It does not matter which half we chose, but here we will ignore the negative temporal frequencies, and just consider the high and low positive temporal frequency bands. If we look at the high band, we see that there are, in this example, eight sectors of orientation. These may be grouped in two different ways.

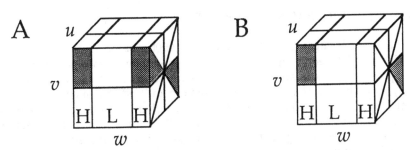

Fig. 7. Grouping of spatiotemporal components. A) Separable components tuned for orientation. B) Inseparable spatiotemporal components tuned for direction.

In the first method, shown in Fig. 7A, we group each pair of sectors diametrically opposed about the w axis. This corresponds to an image component that is separable in space and time, specifically, a flickering spatial pattern of a particular orientation and spatial frequency. Fig 7A shows one such separable component. There would then be four such oriented components in the high temporal frequency band. This would be a partition tuned for *orientation*. A neuron tuned for orientation, but not for direction, would implement this sort of partition.

The second method, shown in Fig. 7B, is to take each of the eight (positive frequency) sectors as a separate component (along with its negative frequency conjugate). Each of these sectors corresponds to a spatial pattern in *motion* (Watson and Ahumada, 1983, 1985). The angle of the sector indicates the direction of motion. These comprise components selective for *direction*, rather than orientation. This is the sort of component extracted by a direction-selective visual neuron. The shaded region in Fig. 7B shows one such motion component, moving downwards and to the right at an angle of -22.5°.

The psychophysical evidence suggests that higher temporal frequencies are coded by direction-selective mechanisms (Watson, Thompson, Murphy & Nachmias, 1980) Also, the cortical layers fed by the magnocellular stream are highly direction selective (Hawken, Parker & Lund, 1988). Together, these facts lead us to adopt the scheme of Fig. 7B, that of inseparable, directional motion components, as our model of human visual coding of the higher temporal frequency band. We therefore use the terms *static* and *motion* bands for the low and high temporal frequency partitions.

This completes our review of the partition of the image stream by the early visual system. Each band in the final partition corresponds to a distinct population on visual neurons. Each neuron may thus be considered a filter, selecting one portion of the image stream. The motion filters implied by the partition of Fig. 7B are equivalent to the filters designed by Watson and Ahumada (1983, 1985) to model direction-selective cortical neurons. To summarize, our model of signal processing in early human vision divides the image stream into three opponent color channels, each further subdivided by spatial frequency, orientation, and temporal frequency. The temporal frequency partition is in terms of static and moving components. In the following section, we use this model as a scheme for digital coding of moving color imagery.

PERCEPTUAL COMPONENTS ARCHITECTURE

The perceptual components architecture (Watson, 1990b), based upon the signal partition described above, is pictured in Fig. 8. The frequency axes have again been rotated so that temporal frequency now corresponds to elevation. Only the positive (temporal) half of the frequency space is pictured, and only achromatic and red/green channels are shown. Each spheroid represents one distinct band, and has a characteristic spatial frequency, temporal frequency, and orientation (static channels) or direction (motion channels). The image sequence is filtered to segregate these bands, and is subsampled in each band in proportion to three-dimensional bandwidth. The coefficients in each band are quantized in a manner that matches human sensitivity to that band. In general, this means that quantization is coarser at higher spatial and temporal frequencies.

Why do we go to all this trouble? What is advantageous about this particular style of breaking up the signal? There are two answers. The first is that the coefficients in each band are much less correlated over space and time, than are the image pixels. This allows them to be coded with fewer bits.

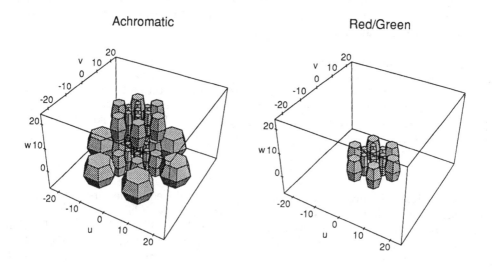

Fig. 8. Perceptual Components Architecture for digital video viewed in the frequency domain. Only achromatic and red/green channels are shown.

The second advantage is that this partition allows us to easily exploit the various blindnesses of human vision. First, the varying sensitivity to various spatial frequencies is easily implemented, since the different frequencies are neatly bundled in separate bands. Second, note that the upper tier of the "cake" in Fig. 8 lacks an outer ring. This is because we are insensitive to high spatial frequencies in motion, so those outer bands can be discarded. Those outer bands contain many coefficients, so their removal yields a large savings in bit rate. If we turn to the chromatic bands in Fig. 8, then we see further large savings. Note that the upper motion tier is completely absent, corresponding to the low temporal sensitivity of the motion system. And finally the static channel has only low to medium frequencies, corresponding to the low spatial resolution of color vision.

Together, all of these savings suggest an efficient code for the image stream. The scheme also has many other potential advantages to digital TV, such extensibility and device independence. Extension of the system to higher resolution is easily achieved by addition of extra bands. Device independence arises from the use of components specified in cycles/degree, and by divorcing the signal representation from the number of lines and pixels on the screen, or indeed their geometrical layout.

IMPLEMENTATION

To test the practicality of the above ideas we have implemented portions of the Perceptual Components Architecture. Using spatial filters with one octave bandwidth and 45 degree orientation bandwidth, we have coded still images with high quality at rates of less than 1 bit/pixel (Watson, 1987). Fig. 8 shows an example coded image. Using the opponent color transform, we have also coded full color images at less than 1 bit/pixel (Watson, 1990b).

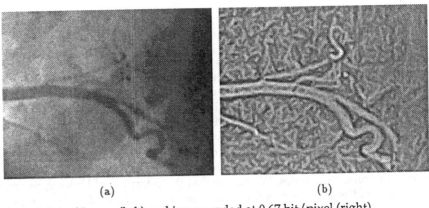

(a)   (b)

Fig. 9. Original image (left) and image coded at 0.67 bit/pixel (right).

We have recently extended the implementation to the time domain. We have used the same spatial filters as in Fig. 9, in conjunction with a band of four temporal filters shown in Fig. 10. These correspond to the low and two high temporal frequency bands (-Hi and +Hi) in Fig. 6, along with an additional fourth band (vHi) containing very high Nyquist frequency signals. We anticipate discarding this uppermost fourth band. The motion filters created with the +Hi and -Hi bands are essentially the same as filters designed elsewhere as models of direction-selective visual neurons (Watson & Ahumada, 1983, 1985).

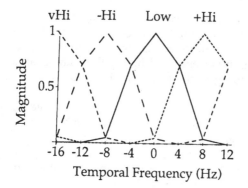

Fig. 10. Four temporal filters used to implement the Perceptual Components Architecture.

The spatial and temporal filters and subsampling scheme have been designed to allow essentially exact reconstruction when fine quantization is used (some very small aliasing artifact remains). The implementation verifies this exact reconstruction. With additional quantization, bit rates are reduced. We cannot yet report what bit rates are required for high visual quality in full-motion color video, but experiments in this direction are planned.

CONCLUSIONS

I have tried to show how the early stages of visual processing may be viewed as a multidimensional pyramid that makes explicit the dimensions of location, color, spatial frequency, orientation, and direction of motion. I have exploited this multidimensional pyramid in the design of a generic digital video code that is matched to the perceptual capabilities of human observers. This code leads naturally to the desirable attributes of efficiency, flexibility, error-tolerance, device-independence, and extensibility. We have implemented portions of this scheme and verified the mathematical details of the architecture. Experiments are underway to explore its efficiency, and to compare it to more conventional techniques. We expect that the interplay of biology, psychology, mathematics and engineering in the area of image sensing and representation will continue to provide interesting and useful ideas.

REFERENCES

Blakemore, C. B. & Nachmias, J. (1971): The orientation specificity of two visual aftereffects. *J. Physiol., Lond* 213, 157-174.

Blakemore, C. B., Nachmias, J. & Sutton, P. (1970): The perceived spatial frequency shift: evidence for frequency-selective neurones in the human brain. *J. Physiol., Lond* 210, 727-750.

Bradley, A., Switkes, E. & De Valois, K. (1988): Orientation and spatial frequency selectivity of adaptation to color and luminance gratings. *Vision Research* 28, 841-856.

Burt, P. J. & Adelson, E. H. (1983): The Laplacian pyramid as a compact image code. *IEEE Transaction on Communications* COM-31, 532-540.

Campbell, F. W., Cooper, G. F. & Enroth-Cugell, C. (1969): The spatial selectivity of the visual cells of the cat. *J. Physiol., Lond* 203, 223-235.

Campbell, F. W. & Robson, J. G. (1968): Application of fourier analysis to the visibility of gratings. *J. Physiol., Lond.* 197 , 551-566 .

Crowley, J. L. (1984). A multiresolution representation for shape. In *Multiresolution Image Processing and Analysis*, ed. A. Rosenfeld, pp. 169-190. New York: Springer-Verlag.

De Valois, R. L., Albrecht, D. G. & Thorell, L. G. (1982): Spatial Frequency Selectivity of Cells in Macaque Visual Cortex. *Vision Research* 22, 545-559.

De Valois, R. L., Yund, E. W. & Hepler, H. (1982): The orientation and direction selectivity of cells in macaque visual cortex. *Vision Research* 22, 531-544.

Guth, S. L., Massof, R. W. & Benzschawel, T. (1980): Vector model for normal and dichromatic color vision. *J. Opt. Soc. Am.* 70, 197-211 .

Hawken, M. J., Parker, A. J. & Lund, J. S. (1988): Laminar organization and contrast sensitivity of direction-selective cells in the striate cortex of the old world monkey. *J. Neuroscience* 8, 3541-3548.

Hurvich, L. M. & Jameson, D. (1957): An opponent-process theory of color vision. *Psych. Review* 64, 384-404.

Ingling, C. R. (1977): The spectral sensitivity of the opponent-colors channels. *Vision Research* 17, 1083-1090.

Jaworski, A. (1990): Earth Observing System (EOS) Data and Information System (DIS) software interface standards. *AIAA/NASA Second International Symposium on Space Information Systems*, Pasadena, CA, AIAA-90-5075.

Kulikowski, J. J. & Tolhurst, D. J. (1973): Psychophysical evidence for sustained and transient mechanisms in human vision. *J. Physiol., Lond.* 232 , 149-163 .

Larimer, J., Krantz, D. H. & Cicerone, C. M. (1974): Opponent process additivity- I: red/green equilibria. *Vision Research* 14, 1127-1140.

Larimer, J., Krantz, D. H. & Cicerone, C. M. (1975): Opponent process additivity- II: yellow/blue equilibria and non-linear models. *Vision Research* 15, 723-731.

Lechner, B. J. (1990): HDTV Systems. *Society for Information Display, 1990 Seminar Lecture Notes* 1, M-3.2-3.40.

Lennie, P., Krauskopf, J. & Sclar, G. (1990): Chromatic mechanisms in striate cortex of Macaque. *J. Neuroscience* 10, 649-669.

Merigan, W. H. (1989): Chromatic and achromatic vision of Macaques: Role of the P pathway. *J. Neuroscience* 9, 776-783.

Packer, O., Hendrickson, A. & Curcio, C. A. (1989): Photoreceptor topography of the retina in the adult pigtail macaque (Macaca nemestrina). *J. Comp. Neurol.* 288, 165-183.

Quick, R. F. & Lucas, R. N. (1979): Orientation selectivity in detection of chromatic gratings. *Optics Letters* 4, 306-308.

Schnapf, J. L., Kraft, T. W. & Baylor, D. A. (1987): Spectral sensitivity of human cone photoreceptors. *Nature* 325, 439-441.

Tanimoto, S. & Pavlidis, T. (1975): A Hierarchical data structure for picture processing. *Computer Graphics and Image Processing* 4, 104-119.

Tolhurst, D. J. (1973): Separate channels for the analysis of the shape and the movement of a moving visual stimulus. *J. Physiol., Lond.* 231, 385-402.

Watson, A. B. (1987): Efficiency of an image code based on human vision. *J. Opt. Soc. Amer. A* 4, 2401-2417.

Watson, A. B. (1990a): Digital visual communications using a perceptual components architecture. *AIAA/NASA Second International Symposium on Space Information Systems*, Pasadena, CA, AIAA-90-5075.

Watson, A. B. (1990b): Perceptual-components architecture for digital video. *J. opt. Soc. Amer. A* 7, 1943-1954.

Watson, A. B. & Ahumada, A. J., Jr. (1983). A look at motion in the frequency domain. In *Motion: Perception and representation*, ed. J. K. Tsotsos, pp. 1-10. New York: Association for Computing Machinery.

Watson, A. B. & Ahumada, A. J., Jr. (1985): Model of human visual-motion sensing. *J. Opt. Soc. Amer. A* 2, 322-342.

Watson, A. B. & Robson, J. G. (1981): Discrimination at threshold: labelled detectors in human vision. *Vision Research* 21, 1115-1122.

Watson, A. B., Thompson, P. G., Murphy, B. J. & Nachmias, J. (1980): Summation and discrimination of gratings moving in opposite directions. *Vision Research* 20, 341-347.

Self-similar oriented wavelet pyramids: conjectures about neural non-orthogonality

John G. Daugman

Cambridge University, Downing Street, Cambridge CB2 3EJ

1   Introduction

Pyramid methods in image processing are based upon multiresolution hierarchies. The key concept is that image structure is scale-specific, and that non-redundant information can be extracted across geometric sequences of sampling scales. The hierarchical nature of pyramid algorithms – in which the results of coarse analysis are used as inputs for finer analysis – lends itself to logarithmic speed-up in processing time: typically, image processing operations are executed in time proportional to the log of the image diameter (number of pixels). Originally developed by Peter Burt and Azriel Rosenfeld in the late 1970's, pyramid methods are now in widespread use in machine vision and image processing (for a review of methods, see Rosenfeld (1984), *Multiresolution Image Processing and Analysis*). By contrast, the relevance of image pyramids for visual neuroscience is less clear. Although certainly visual neural receptive fields exist in many sizes, there is little evidence for a hierarchical organization of scale in which receptive fields of one size provide a filtered image to those of the next size. Instead, at least through striate cortex, cells of many different sizes seem to receive image input in parallel.

Structurally related in some ways to image processing pyramids, artificial neural networks have re-emerged recently in models of learning, development, associative memory, categorization, speech processing, and visual perception. In such models, traditional approaches based upon rules and symbol manipulation are replaced by very different distributed processes such as stochastic annealing, adaptive weight evolution, cooperation, competition, resonance, and related concepts from dynamical systems theory. A common feature of many neural network models (excluding the Hopfield-class) is their hierarchical organization into multiple layers. While the seminal "perceptron" of Minsky and Papert was based on just two layers, the "neo-cognitron" of Fukushima, for example, is a nine-layered feedforward hierarchy. Much current attention focuses on three-layered schemes with feedback adjustment of internal weights, as in the back-propagation Rummelhart network, incorporating Widrow-Hoff weight adjustment by partial derivatives of a cost function, or the double-ended Hecht-Nielson "counter-propagation" network. In general, as is true of pyramid algorithms for image processing, neural network models incorporate some features that mime biological neural organization; but not in significant detail, nor to the degree suggested by their name.

As part of the ECVP-90 symposium *Image Pyramids and Neural Networks*, this paper considers multiresolution representations with some of the properties both of pyramids and of neural networks. We begin by examining some theoretical problems which have always been latent, but not addressed, in the classical concept of encoding image information by neural receptive field structures. Next, a general relaxation network method of non-orthogonal image coding is demonstrated using self-similar, oriented "wavelets." Finally, a recent practical application is discussed that illustrates multiresolution image analysis using this set of neural filters to encode a biometric signature, for automatic visual personal identification with extremely high confidence.

2 Completeness and Non-Orthogonality in Image Codes

Completeness of a candidate image code is a issue lying latent within most theories about biological visual mechanisms. The issue can be algebraically formulated as the question of whether all image structure (up to a resolution limit) can be represented by weighted sums of the proposed code's primitives. A given set of candidate code primitives $\{G_i(x,y)\}$ constitutes a complete linear code if, for any image $I(x,y)$, there exists some set of coefficients $\{a_i\}$ (representing the relative strength, or presense, of each of the code primitives within the image) such that

$$\sum_i a_i G_i(x,y) = I(x,y). \tag{1}$$

In vision models based on linear concepts such as receptive field profiles, the code's coefficients $\{a_i\}$ are assumed to be the firing rates of the linear neurons. Their spatial response selectivities, or receptive field profiles, correspond to the $\{G_i(x,y)\}$ code primitives. Although the question of completeness has generally been ignored in theories of biological vision, the issue is analogous theoretically to the dimensionality of color vision. In the absence of completeness, many different visual stimuli within resolution could not be uniquely represented and would appear indistinguishable. Examples of such spatial analogs to chromatic metamers, as counterexamples to Marr's (1982) theory of spatial coding by multi-scale Laplacian zero-crossings, may be found in Daugman (1988a).

Let us assume that the ensemble of primitives constituting a candidate visual code is complete, so that any and all kinds of image structure up to the resolution limit can be uniquely represented by linear combinations of them. If the code primitives happen to be mutually non-orthogonal (defined later), then in general it is very difficult to determine the correct set of coefficients $\{a_i\}$ which will correctly represent a specific image $I(x,y)$, even though the existence of such a set of coefficients $\{a_i\}$ is guaranteed by virtue of the completeness property of the ensemble of $\{G_i(x,y)\}$. In particular, the required set $\{a_i\}$ is *not* provided by the standard receptive field operation of taking the inner product of $\{G_i(x,y)\}$ with the image $I(x,y)$:

$$a_i \neq \int\int G_i(x,y) I(x,y) dx dy \tag{2}$$

Complete codes can be composed of either orthogonal or non-orthogonal sets of primitives, but orthogonality of the code primitives does not ensure completeness. Rather, completeness requires equality between the number of independent degrees of freedom in the code, and the dimension of the signal (number of independent samples, or pixels). Each primitive in the code constitutes an independent degree of freedom, provided that it cannot be represented by any linear combination of other primitives. Whether or not the code primitives are orthogonal, the completeness of the image code in a given area of visual space, up to the appropriate resolution limit, is determined by whether or not the number of linearly independent degrees of freedom in the code is at least as great as the dimension of the incoming sampled signal.

3   Ubiquitous Non-Orthogonality

Interestingly, a very common property of both biological sensory and motor control systems is their operation in non-orthogonal coordinates. For example, as pointed out by Pellionisz (1987, 1988), the three semi-circular canals of the vestibular system, functioning as an accelerometer, lie in planes inclined to each other at roughly 120°. Thus there is no axis of bodily acceleration which projects onto only one of the three vestibular canal planes. Other examples of non-orthogonal sensory and motor control systems include the non-orthogonal attachments of the eye muscles, the skeletal musculature generally, and the mutually non-orthogonal action spectra of retinal color pigments. These properties presumably have more functional significance than mere redundancy; Pellionisz (1987, 1988) has proposed that they actually reveal an underlying *tensor geometry* embedded pervasively in the nervous systems of higher animals.

Non-orthogonality among the primitives of a visual code means mathematically that their inner product (the integral of the product of any two receptive field profiles) is in general non-zero. A given set of visual receptive field profiles $\{G_i(x, y)\}$ would comprise an orthogonal code if and only if

$$\forall i, j \ (i \neq j), \quad \int \int G_i(x, y) G_j(x, y) dx dy = 0. \tag{3}$$

Clearly, this is generally not the biological situation. For example, two retinal ganglion cells with overlapping receptive fields having the generic concentric center/surround structure, possibly with different sizes or offset positions, in general would not satisfy the above criterion for orthogonality. If a stimulus pattern exactly matched the receptive field profile of one of these two cells, and if the two cells were orthogonal, the matching one alone should respond. But in reality, both cells in general would respond to some degree; this reflects their non-orthogonality. Similarly, cortical simple cells responding to a given patch of visual space are in general non-orthogonal and hence have correlated responses, although if they differ in their preferred orientation, then their inner product (a measure of their non-orthogonality, given in the left-side of Eqt. [3]) will be much reduced. One easy way to achieve orthogonality in a pair-wise sense is by preserving a quadrature phase relation, or more generally, even and odd symmetry between the pair. Figure 1 illustrates a canonical example of a pair of orthogonal receptive fields, differing in orientation by 45°, with zero inner product. The orthogonality of this particular pair is ensured simply by the fact that one has even symmetry and the other odd. Paired simple cells sharing the same orientation, and maintaining a quadrature phase relation (Pollen and Ronner, 1981) will of course also be orthogonal if they are concentric and their symmetries are purely even and odd, but the criterion of ensemble orthogonality is not satisfied by this pairwise scheme.

4   Why Non-Orthogonality?

The fact that ensembles of receptive field profiles are non-orthogonal greatly complicates the interpretation of image codes based on neural firing rates. The firing rate of a linear neuron is given by its inner product with the local image structure; indeed, this concept underlies the very definition of its receptive field profile. According to the classical view, then, firing rate is interpreted as signifying the relative presense of image structure resembling the neuron's receptive field shape. But this classical view is paradoxical. Spurious and misleading image structure would be implied by the collective response of an ensemble of linear, non-orthogonal receptive fields, unless special mechanisms are invoked which correct for their non-orthogonality. Although it is not difficult to *construct* an orthogonal basis set starting from some seed function (as by the Gram-Schmitt procedure), it is much more difficult to work within some pre-determined (i.e. hard-wired, or random), non-orthogonal basis, and recover true signal structure from a set of response coefficients on these "receptive fields."

It is therefore interesting to ask why, and how, the visual system works within the constraints of non-orthogonal representations. Probably the crucial factor would be the enormous difficulty of building and preserving orthogonality among the visual code primitives. Although non-overlapping receptive fields are trivially orthogonal, the construction of an orthogonal basis involving localized, multi-scale, oriented, overlapping receptive fields is daunting, especially in two spatial dimensions. The genetic cost of accurately specifying the detailed receptive field profile of each cell, including its 2-D position, orientation, symmetry/phase, size parameters, and gain, would clearly be prohibitive. Although there are general organizational principles in visual cortex such as sequence regularity of orientation columns, these are *statistical* regularities; the microstructure, especially for size and position, is fraught with randomness and scatter (Hubel and Wiesel, 1962, 1974). In addition to the prohibitive genetic cost, the neuroembryological process of development would be vastly complicated if orthogonality were a property of visual codes. In what might dismissively be termed "Gram-Schmidt neurodevelopment," each new receptive field coming into place in a given region of retinotopic space would have to take into account the local details of all other existing receptive fields.

A final factor might be the difficulty of *preserving* orthogonality over time, even if it could be genetically specified and embryologically implemented. Any change over time of 2-D weighting profiles or positions would destroy the delicate orthogonality of the configuration. Likewise, changes in sensitivity or contrast gain (with adaptation or attention) would be disruptive.

The death of a single neuron in an orthogonal scheme would even cause certain visual stimuli to become invisible, and other patterns to become mutually indistinguishable. It is important to note, however, that redundancy would not be ensured by non-orthogonality *per se*. Redundancy, in the sense of fault-tolerance, arises from linear dependency, a situation in which

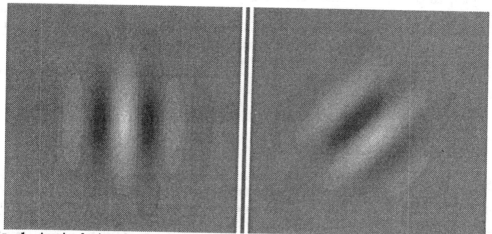

Fig. 1. A pair of oriented code primitives that are orthogonal only by virtue of symmetry. In general, as is true of most sensory and motor control systems, the overlapping and multi-scale primitives of spatial vision are very far from being orthogonal (having zero inner product). Non-orthogonality of the code primitives implies that image structure is not correctly represented by the classical concept of receptive field operation. This fact raises critical questions about the nature of visual codes.

some of the primitive elements can be constructed by some combination of the others, rendering them replaceable, if necessary. Both orthogonal and non-orthogonal schemes may be linearly *independent*, meaning that all of the primitive elements in the set are indispensable for completeness. Presumably visual codes are in fact linearly dependent, so that no neuron is indispensable. No linearly dependent system can be comprised of mutually orthogonal members. Non-orthogonality is a necessary, albeit not a sufficient, condition for fault-tolerant redundancy.

5 Decorrelation and Principal Components Analysis

Image coding is optimally efficient if the coefficients of the representation are non-redundant, in the Shannon and Weaver (1949) sense that their *mutual information* is zero. In that case, the code coefficients are *decorrelated* and they form a statistically independent ensemble, so that the conditional probability $P(x|y)$ of one coefficient's value ($X = x$), given any other coefficient's value ($Y = y$), is the same as its unconditional probability $P(x)$: $P(x|y) = P(x)$. In order for each coefficient to capture a unique property of the image that cannot be captured by any other coefficient, the expansion functions employed in the code must be linearly independent. In order for the code coefficients to have zero mutual information, i.e.,

$$H(X;Y) = \sum_{X,Y} P(x,y) \log_2 [P(x,y)/P(x)P(y)] = 0 \qquad (4)$$

the code primitives must be orthogonal so that their projections onto each other are always zero. In that case, $P(x,y) = P(x)P(y)$, and thus the logarithm above equals zero for each term in the mutual information sum.

For any given image, such a representation can always be found by its Karhunen-Loeve expansion (Karhunen, 1947; Loeve, 1948). This form of *Principal Components Analysis* (Oja (1982); Oja and Karhunen (1985); Linsker (1986)) generates a complete representation of the image in question in terms of the eigenvectors of its diagonalized covariance matrix, all of which are, of course, orthogonal. It may be shown (Pratt, 1978) that this is the *most efficient possible image code*. However, the actual eigenvectors and their eigenvalues are different for each image. It is true that if the image has stationary statistics at least locally, then the optimal Karhunen-Loeve expansion is equivalent (Pratt (1978); Bossomaier and Snyder (1986)) to a Fourier analysis in those local regions. In general, however, this approach does not lend itself to the construction of a universal image code. Rather, the primitive elements of the code would be utterly signal dependent; the neural mechanisms would be constantly re-configuring themselves for each new image.

As noted by Barlow (1961, 1972), Srinivasan *et al.* (1982), and others, the ubiquitous "center/surround" structure of a sensory neuron, in many diverse modalities, can be described as constituting a decorrelator of the incoming signal. Examples of such decorrelating mechanisms include spatial lateral inhibition in the retina, chromatic opponency among ganglion cells, temporally transient "on-off" cells, disparity-tuned cells, figure/ground motion-contrast cells, and so on. In these diverse visual subsystems, neural mechanisms respond to change; they detect epochs of the signal in which it deviates from the local average (across space, time, wavelength, depth, velocity, etc.). Srinivasan, Laughlin, and Dubs (1982) have pointed out that center/surround organization can be interpreted as a form of predictive coding. For oriented (2-D Gabor-like) receptive fields, the decorrelating effect is even more pronounced. This aspect of orientation selectivity (Daugman, 1989) will be seen by comparing Figs. 3 and 4.

6 Relaxation Network for Non-Orthogonal Visual Codes

A theoretical 'neural network' is illustrated in Fig. 2 that can correctly find the coefficients in non-orthogonal visual codes. As noted earlier, spurious and misleading image representations would result from trying to generate a visual code simply according to the classical notion of direct inner products of the image onto neural receptive field profiles, because of the mutual non-orthogonality of these code primitives. Presumably, some mechanisms to compensate for the problem of non-orthogonality exist in biological visual systems, since they are based upon mutually non-orthogonal receptive fields. The network shown in Fig. 2, which was used to generate the image representations described in subsequent sections of this paper, operates by relaxation. It asymptotically arrives at coefficient values which equate all of the mutual projections among the weighted primitives to their weighted projections onto the input signal. The result is analogous to orthogonalization of the representation, without changing or imposing constraints upon the given non-orthogonal set of visual code primitives.

For the present purposes we shall treat the original retinal image that is sampled by the photoreceptor lattice as a discrete 2-D signal $I[x, y]$, with the notational convention that square brackets [] signify a discrete rather than continuous signal. This 2-D signal is to be represented in the early visual nervous system by an ensemble of neural firing rates which comprise a set

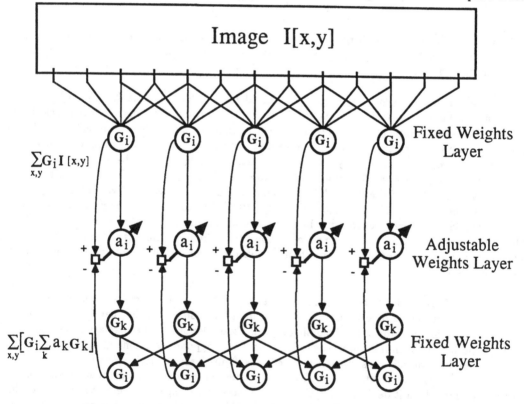

Fig. 2. A hypothetical three-layered relaxation network which can correctly find the coefficients for non-orthogonal codes, such as those observed in biological vision, among many other sensory and motor control systems. The relaxation of the network over time implements an orthogonalization of the code while allowing the underlying code primitives to remain non-orthogonal.

of expansion coefficients $\{a_i\}$ on some set of two-dimensional code primitives, $\{G_i[x,y]\}$. These could be regarded, for example, as 2-D receptive field profiles, whatever their structure may be. It can be useful to think of a given image $I[x,y]$ comprising n pixels as a single vector in an n-dimensional vector space; the elements of the vector constitute the individual pixel values which comprise the image. In the (non-biological) case of image representations based upon complete orthonormal expansions, a given orthonormal expansion set would constitute a *basis* for this vector space. Examples of such orthonormal bases include the unit basis (point-by-point image representation), and the Fourier basis (in which each basis vector is a 2-D discrete Fourier component). The unsupervised network of Fig. 2 creates a representation for any sampled image $I[x,y]$ by a set of coefficients (firing rates) on non-orthogonal code primitives (linearly *dependent* vectors), which may or may not completely span the vector space. The relaxation network implements constraint satisfaction by "relaxing" into a state of minimal energy (to be defined shortly).

Our problem is to represent $I[x,y]$ either exactly or in some optimal sense by projecting it onto some given set of code primitives $G_i[x,y]$. This requires finding projection coefficients $\{a_i\}$ such that the resultant vector $H[x,y]$

$$H[x,y] = \sum_{i=1}^{n} a_i G_i[x,y] \tag{5}$$

is either identical to $I[x,y]$ (the complete case) or generates a difference-vector $I[x,y] - H[x,y]$ of minimal length (the optimization case). If the primitives $\{G_i[x,y]\}$ comprise a complete orthogonal code, then the representation in $H[x,y]$ is exact (the difference-vector is zero) and the solution for $\{a_i\}$ is simple:

$$a_i = \frac{\sum_{x,y} (G_i[x,y] I[x,y])}{\sum_{x,y} G_i^2[x,y]} \tag{6}$$

But if the primitives do not constitute a complete orthogonal code, then in general the representation $H[x,y]$ will be inexact and the desired set of code coefficients $\{a_i\}$ must be determined by satisfying a constraint, such as minimizing the squared norm of the difference-vector:

$$E = \|I[x,y] - H[x,y]\|^2 = \sum_{x,y} (I[x,y] - H[x,y])^2. \tag{7}$$

The norm E will be minimized only when its partial derivatives with respect to all of the n coefficients $\{a_i\}$ equal zero:

$$\forall i, \quad \frac{\partial E}{\partial a_i} = -2 \sum_{x,y} (I[x,y] G_i[x,y]) + \sum_{x,y} \left[2 \left(\sum_{k=1}^{n} a_k G_k[x,y] \right) G_i[x,y] \right] = 0 \tag{8}$$

Satisfying this condition for each of the a_i then generates a system of n simultaneous equations in n unknowns, where n is equal to the total number of pixels in the image (typically on the order of 100,000):

$$\left| \sum_{x,y} \left(I[x,y] G_i[x,y] \right) = \sum_{x,y} \left[\left(\sum_{k=1}^{n} a_k G_k[x,y] \right) G_i[x,y] \right] \right| \tag{9}$$

Thus the solution which minimizes the squared norm of the difference-vector (Eqt. [7]) amounts to finding the set of coefficients $\{a_i\}$ such that the inner product of each code primitive $G_i[x,y]$

with the entire linear combination of code primitives $\sum a_k G_k[x,y]$, is the same as its own inner product with the original image $I[x,y]$. It should be noted that in the case when the $\{G_i[x,y]\}$ comprise a complete orthogonal code, then the inner products in the right-hand side of Eqt. [9] are non-zero only for $k = i$, and so each of the n equations then has only a single unknown and it is immediately apparent that the minimal-difference-vector solution for each a_i is identical to that given earlier in Eqt. [6] as the familiar orthogonal case.

The relaxation network shown in Fig. 2 converges to the correct values of $\{a_i\}$ without supervision by gradient descent along the n-dimensional $E(a_i)$ surface, which expresses the quadratic energy function's dependency on all of the $\{a_i\}$ coefficients. A common feature of theoretical neural network schemes is the combination of some adjustable (or adaptive) synaptic weights layers, and some layers with fixed weights. The present scheme begins with a layer of fixed connection strengths which are specified by a particular set of non-orthogonal code primitives $\{G_i[x,y]\}$. By summing the different image pixels through these weights, the output of the ith neuron in this layer is simply the inner product of the ith code primitive, $G_i[x,y]$, with the input image $I[x,y]$ in that region. The second layer contains adjustable weights for multiplying each of these outputs, according to a control signal which arises from inter-laminar interactions. The third layer is identical to the first layer and stores the same fixed set of elementary functions. The adjustable weights of the middle layer constitute the new image representation as the set of coefficients $\{a_i\}$.

The interlaminar control signal adjusts each of the weights by an amount Δ_i, given by the difference between a feed-forward signal and a feed-back signal. The feed-forward signal is the level of activity of the neuron from the first layer, and the feed-back signal is the inner product of the weighting function of the corresponding neuron in the third layer with the weighted sum of all the other neighboring neurons in that layer with which it is connected. Thus the weight adjustment is given by:

$$\Delta_i = \sum_{x,y} (G_i[x,y]I[x,y]) - \sum_{x,y} \left[G_i[x,y] \left(\sum_{k=1}^{n} a_k G_k[x,y] \right) \right] \tag{10}$$

and the iterative rule for adjusting the value of each coefficient is $a_i \Rightarrow a_i + \Delta_i$. It should be noted that the network does not require a "teacher" that generates the weight adjustment signal by comparing the current representation with a separate copy of the desired pattern. Rather, the adaptive control signal Δ_i arises only from inter-laminar network interactions.

It can be seen by inspecting Equations [8] and [10] that the weight adjustment rule is equivalent to:

$$\Delta_i = -\frac{1}{2} \frac{\partial E}{\partial a_i} \tag{11}$$

In the continuous-time case, the network becomes a dynamical system each of whose nodes is described by the differential equation:

$$\frac{da_i}{dt} = -\frac{1}{2} \frac{\partial E}{\partial a_i} \tag{12}$$

It should be noted that the minus sign implies that the weight adjustment is always in the downhill direction of the energy surface $E(a_i)$, and that the adjustment is proportional to the slope of the energy surface at this point. The equilibrium state of the network that is reached when all $\frac{da_i}{dt} = 0$ is the state in which the energy function E representing the difference-vector squared norm $\|I[x,y] - H[x,y]\|^2$ has reached its minimum. This is the point at which the partial derivative of E with respect to all of the adjustable weights is nil:

$$\forall i, \ \frac{da_i}{dt} = 0 \iff \forall i, \ \frac{\partial E}{\partial a_i} = 0. \tag{13}$$

Thus in the stable state, the middle layer of the network has weights which represent the optimal coefficients $\{a_i\}$ for the projection of the signal $I[x,y]$ onto any set of code primitives $\{G_i[x,y]\}$, which, as noted earlier, need neither comprise an orthogonal nor a complete code.

Figures 3, 4, 5, and 6 illustrate the use of this network for computing a complete and recoverable (invertible) image transform onto non-orthogonal code primitives, with significant data compression. The pixel histogram of Fig. 3 shows that a point-by-point image representation of the original image, such as one might associate with retinal photoreceptors, is highly inefficient because spatial correlations have not been extracted. The histogram plots the number of pixels in the Lena image having each grey level between 0 (black) and 255 (white). If the original image had been uncorrelated random noise, then this histogram would be flat (all grey levels being equally probable) and the entropy would be maximum, i.e. $S = 8$. At the other extreme, if all pixels had the same value, regardless of how many possible values there were, then the entropy would be zero. The entropy of a statistical ensemble is defined as

$$S = -\sum_i p_i \log_2(p_i) \tag{14}$$

given that for the ensemble of available states i with associated probabilities p_i,

$$\sum_i p_i = 1. \tag{15}$$

Clearly, the image pixel histrogram in Fig. 3 is broad and multi-modal. Its entropy is $S = 7.57$, close to the maximum possible value for 8-bit images (namely $S = 8$).

If the same image is represented instead by coefficients on 2-D Gabor elementary functions (Daugman 1980, 1984, 1985), then the complete representation has the histogram shown in Fig. 4, with obviously much reduced entropy. The actual values of these 2-D Gabor coefficients are displayed as luminance values in Fig. 5, organized as an image. Here the global coordinates are still the spatial (x,y) image domain, but the local coordinates systematically map out windowed 2-D spectral parameters for each local region of the image. The pixel brightness value at each location in Fig. 5 signifies the coefficient value associated with a 2-D Gabor elementary expansion function, which is spectrally tuned both for orientation and spatial frequency but at the same time localized to that specific image region.

The average grey value in Fig. 5 corresponds to a 2-D Gabor coefficient value of zero, while lighter values signify positive coefficients and darker values negative coefficients. It is obvious that most of the transform image representation in Fig. 5 is at the average grey value, with only sporadic significant departures to white or black. As a consequence, the entropy of this image is $S = 2.55$, or five $\log_2$ units smaller than the space-domain representation. This reduction in entropy was obvious from comparing the two histograms of Figs. 3 and 4; indeed, Fig. 4 is the pixel histogram of Fig. 5.

Proof that the compressive 2-D Gabor transform displayed in Fig. 5 still contains complete

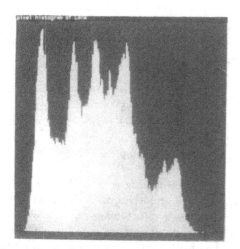

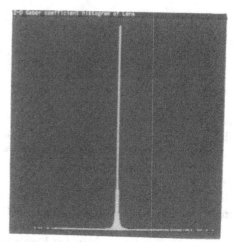

Fig. 3. Statistical complexity of a standard image when represented point-by-point, as if by retinal photoreceptors. In this pixel histogram the entropy, or mutual self-information, is 7.57 for the ensemble. A point-by-point image representation does not extract the intrinsic redundancies and correlations which distinguish structure from random noise.

Fig. 4. Histogram of the coefficients comprising a complete 2-D Gabor representation of the same image as described by Fig. 3, with the same number of degrees of freedom (65,536) and the same quantization (8-bits per coefficient). Compared with the distribution in Fig. 3, the entropy of this distribution has been reduced from $S = 7.57$ to $S = 2.55$, but it nonetheless represents a complete encoding of the image. The oriented code primitives have a significant decorrelating effect.

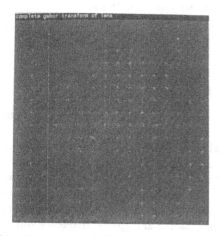

Fig. 5. Complete 2-D Gabor representation of the Lena portrait described in Figs. 3 and 4, plotting coefficient values as luminance. The coordinate system is globally still the spatial (x, y) image domain, but embedded within this are local 2-D spectral parameters, which may be interpreted as locally-windowed spatial frequency and orientation. The strategy of embedding angular and spectral variables within global retinotopic coordinates resembles the functional architecture of striate cortex.

Fig. 6. Demonstration of completeness of 2-D Gabor representations. The original Lena portrait can be recovered without loss from the 2-D Gabor coefficients shown in Fig. 5, despite the extreme sparseness of the representation.

information about the original image is provided in Fig. 6, showing lossless reconstruction of the original image from the 2-D Gabor coefficients plotted in Fig. 5. The source of the entropy reduction from 7.57 bits/pixel to 2.55 bits/pixel in the lossless 2-D Gabor transform is the fact that the 2-D Gabor elementary functions represent the image in *decorrelated* form. The fact that these code primitives are oriented has increased their efficacy as decorrelators, compared with isotropic center/surround structures.

Bastiaans (1980) provided the first method for actually computing the coefficients needed for Gabor's expansion of 1-D signals, 34 years after Gabor's classic proposal. (Gabor proved that for any signal such a representation always exists, but he was unable to provide a method for obtaining it.) Bastiaans' method is to generate a dual basis, so that the sampling functions are not the same as the code primitives. But in creating the ancillary, bi-orthogonal basis, the method requires that all terms in the signal expansion use the same Gaussian envelope, regardless of modulation frequency. This restriction is not biologically realistic. If anything, as Kulikowski and Bishop (1981), among others, have pointed out, the population of cortical simple cells tend to have a common template. This requires the Gaussian width to scale proportionally with modulation wavelength, rather than being strictly independent of it. We turn now to a use of the relaxation network of Fig. 2 for computing image representations with the restriction of fixed-size primitives removed; specifically, image representations in which all of the code primitives are *dilations, translations*, and *rotations* of a common underlying 2-D template. If there are as many degrees of freedom (linearly independent code primitives) as there are pixels in the original image, then such a representation can be complete, i.e. lossless. If there are fewer code primitives, then the network converges to a set of coefficients which offer the optimal image representation in the least-squares sense.

7   A Self-Similar Polar Pyramid

Both neurophysiological and psychophysical data suggest that spatial visual mechanisms have 2-D spectral selectivities which are organized in log-polar form. That is, their response selectivities tend to be defined in terms of an approximately constant angular bandwidth (say, ±15 degrees) corresponding roughly to the changes in preferred orientation between cortical orientation columns, and their spatial frequency bandwidths tend to be roughly constant in *octave* terms (e.g. 1 - 2 octaves) rather than in linear bandwidth terms. Moreover, the distribution of preferred spatial frequencies among cells also tends to be spaced in octave, not linear, terms (Pollen and Feldman, 1979). Taken together, these sampling rules imply a self-similar polar pyramid of spatial filters with a 2:1 aspect ratio in both the space and spectral domains (Daugman 1984, 1985). Any two of these three properties (an aspect ratio of 2:1, an orientation bandwidth of about ±15 degrees, or a spatial frequency bandwidth of about 1.5 octaves) suffice to determine the third of these properties for an ensemble of self-similar filters that uniformly cover the 2-D spectral domain. These various classes of data on tuning curves can be regarding as reflecting the inescapable conjoint 2-D/2-D uncertainty principle (Daugman, 1985) that constrains the encoding and representation of image information. Subsequent sections of this paper apply this self-similar polar pyramid to image coding and pattern recognition tasks.

7.1   Relation to Wavelet Theory

In recent years, much interest has attended mathematical developments in "Wavelet Theory," largely made possible by the work of the French mathematician Yves Meyer and his students (see references by Meyer, Morlet, Mallat, Grossmann, Daubechies, and Holschneider). Wavelets are families of localized, undulatory expansion functions all of whose members are dilations and translations of each other. Thus any 1-D wavelet $\Psi_{mn}(x)$ can be generated from any other wavelet in its family through the following generating function, parameterized by integers (m, n) for octave dilations in size (m) and uniform shifts in position (n):

$$\Psi_{mn}(x) = 2^{-m}\Psi(2^{-m}x - n) \tag{16}$$

The generating function captures the *self-similarity* property, since whatever the shape of $\Psi(x)$, all other members of the set having different sizes and locations will just be scaled versions of the same underlying shape.

We may generalize to two-dimensional wavelet functions $\Psi(x, y)$ which are anisotropic (oriented) by adding another degree of freedom to the generating function, namely, periodic rotations through some angle θ:

$$\Psi_{mpq\theta}(x, y) = 2^{-2m}\Psi(x', y') \tag{17}$$

where the substituted variables (x', y') capture dilations in size (m), 2-D translations in position (p, q), and changes in orientation (θ) as follows:

$$x' = 2^{-m}[x\cos(\theta) + y\sin(\theta)] - p \tag{18}$$

$$y' = 2^{-m}[-x\sin(\theta) + y\cos(\theta)] - q \tag{19}$$

Figure 7 displays as luminance primitives several self-similar examples of 2-D Gabor elementary functions, all of which can be generated from a single complex member by dilations, translations, and rotations, using the above 2-D wavelet generating functions. The five examples in this figure differ in size and spatial frequency by factors of 2 (hence m in Eqt.s [17]-[19] equals 0,1,2,3,4, although amplitudes here are constant), and five different orientations θ are represented. Since the underlying 2-D Gabor elementary function which enters into the 2-D wavelet generating equation (Eqt. [17]) as $\Psi(x, y)$ is complex, the implicit complex constant multiplying each 2-D elementary function specifies not only contrast but also phase. Two quadrature phases are represented among the five examples shown in Fig. 7. Can such an ensemble of self-similar oriented primitives be the basis of an image code?

Demonstrations of several different 2-D Gabor wavelet codes are provided in the 12 panels of Fig. 8. The images are reconstructed from coefficients on the 2-D Gabor wavelets, obtained despite the non-orthogonality of these elementary functions, using the relaxation network described in Section 6. The four columns represent different choices for the number of distinct orientations θ used for generating the wavelets (Eqt.s [17]-[19]), namely, from left to right, 6 orientations, 4, 3, and 2. The rows represent different quantization depths for the computed 2-D Gabor coefficients, in order to show how compact the codes can become without unacceptable loss of image information. For all schemes shown, the coefficients computed for the coarsest 2-D Gabor elementary functions have 2 bits greater quantization accuracy than those for the finest ones (following psychophysical evidence on the detectability of quantization error). Thus the top row across Fig. 8 uses 8 bits for the coefficients on the coarsest terms and 6 bits for the finest; the bottom row uses 6 bits and 4 bits, respectively. Entropy rates are displayed in bits-per-pixel (bpp) above each recovered image.

7.2   Sampling Theory and the Self-Similar Polar Pyramid

It is interesting that full representation (and recovery) of image structure is possible with these oriented codes, despite a number of properties which may seem to violate traditional intuitions

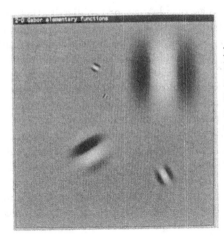

Fig. 7. Five members of a self-similar 2-D Gabor wavelet code viewed in the space domain. All members of this visual code, described in either 2-D spatial or 2-D spectral terms, can be generated by dilations, rotations, and translations of each other, or of a basic underlying complex member. (See Eqt.s [17]-[19].)

Fig. 8. Compact image representation in self-similar 2-D Gabor wavelet codes. The columns correspond to different numbers of orientations used in the code, ranging from 6 orientations (at left) to only 2 orientations (at right). The rows represent different levels of quantization accuracy for each 2-D Gabor coefficient, with the coarsest terms (ranging from 8 bits at top to 6 bits at bottom) having 2 bits more accuracy than the finest terms (6 bits to 4 bits). The resulting codes' entropies, starting with an image quantized at 8 bits/pixel, range in the encoded representations here from 2.87 bits/pixel (upper left) to 0.38 bits/pixel (lower right).

about sampling requirements. First, because of the log-polar lattice for the distribution of 2-D spectral parameters, the only spatial frequencies represented follow the geometric sequence 1, 2, 4, 8, 16,... , rather than the linear sequence 1, 2, 3, 4, 5,... familiar from Nyquist sampling theory. Second, and related to this, the orientations represented in any given scheme are always from a fixed set of angles (e.g. 0, 30, 60, 90, 120, and 150 degrees), regardless of spatial frequency. By contrast, Nyquist sampling theory might lead one to expect that the higher the spatial frequency, the larger the number of orientations that must be sampled, as in the familiar case of covering the 2-D Fourier plane with a Cartesian sampling grid. Both of these unusual properties can be understood in light of the *localization* of the code primitives, as opposed to the global expansion functions used in a complete 2-D Fourier representation. Crudely speaking, the fact that each 2-D Gabor wavelet is localized creates additional degrees of freedom. These position-related degrees of freedom compensate for the apparent undersampling of the spatial frequency and orientation parameters on the log-polar grid. But the fact that rather complete image representations are possible even with oriented code primitives so sparsely sampled in orientation, makes all the more compelling the unanswered question of why the visual cortex incorporates such dense oversampling for orientation.

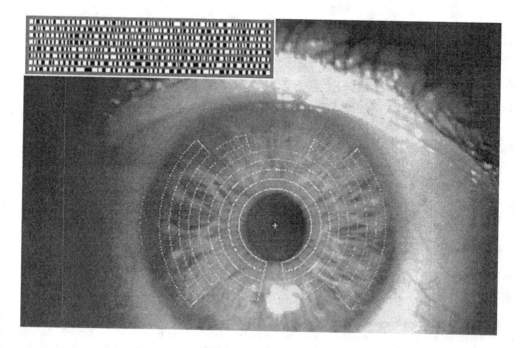

Fig. 9. Zones of analysis and bar code generated for a typical iris, by the biometric signature security system for automatic visual personal identification. The polar pyramid of self-similar, multiresolution 2-D Gabor filters extracts several hundred independent degrees of freedom from the iris texture. These can be compactly encoded as a unique biometric signature.

8   Application: Automatic Visual Personal Identification

The self-similar polar pyramid has been incorporated into a rapid personal identification system, based on a facial feature that serves as a unique "biometric signature." The purpose of the system is to establish or to confirm personal identity automatically, with extremely high confidence. The biometric feature used as a signature is the iris of the eye, which contains a unique texture that constitutes a kind of visible fingerprint (indeed, two) for each individual. The visible texture of the iris, as illustrated in Fig. 9, arises primarily from random radial furrows, or wrinkles, as well as fibrils, crypts, freckles, annular rings, striations, and a serpentine vasculature. When illuminated with well-chosen wavelengths, this rich texture can be acquired at some distance with a videocamera for image analysis and coding. Analysis of a substantial data base of iris images reveals that the variation in human iris textures extends into more than 400 independent degrees-of-freedom. A signal of this complexity, if it can be rapidly and reliably encoded, ensures the possibility of automatic individual identification and recognition with extremely high confidence. Such a system has been built, producing identifications in less than one second with confidence levels corresponding to error probabilities of typically 1 in 10^{27}.

A standard grey-scale image involves far too much data to serve usefully for efficient identification. More important than data compression, however, is the need to encode the image structure in a size-invariant and rotation-invariant fashion that reliably extracts as many textural degrees of freedom as possible. The strategy used for this pattern recognition task was a pyramid of self-similar 2-D Gabor filters defined in polar coordinates, distributed across predetermined zones of analysis in the iris. Preprocessing stages first find the iris in a video image of a person's face, and establish the inner and outer perimeters of the iris (pupil and limbus). With the iris image defined in polar coordinates $I(\rho, \phi)$ around the center of the pupil, a "bar code" of just 256 bytes is computed from both even- and odd-symmetric 2-D Gabor coefficients, at multiple scales of analysis, keeping just the most-significant-bit (MSB) in each case:

$$\text{MSB}(r, \theta) = 1 \text{ if } \text{Re} \int_{\rho} \int_{\phi} e^{-2\pi i \omega(\theta - \phi)} e^{-(r-\rho)^2/\alpha^2} e^{-(\theta-\phi)^2/\beta^2} I(\rho, \phi) \rho d\rho d\phi > 0 \qquad (20)$$

$$\text{MSB}(r, \theta) = 0 \text{ if } \text{Re} \int_{\rho} \int_{\phi} e^{-2\pi i \omega(\theta - \phi)} e^{-(r-\rho)^2/\alpha^2} e^{-(\theta-\phi)^2/\beta^2} I(\rho, \phi) \rho d\rho d\phi \leq 0 \qquad (21)$$

The multiresolution pyramid is implemented by selecting the parameters α, β, and ω according to a sequence of binary decimation, as indicated in Eqts. [17]-[19]. An example of the resulting bit stream can be seen at the top of Fig. 9. Through this multi-resolution polar coding pyramid, the original iris image, which initially contains about 300,000 bytes, is reduced to a compact and unique "signature template" of only 256 bytes.

Because each bit in the code has equal probability of being a 1 or a 0, the fraction of bits which agree, in comparing any two codes from different irises, will be binomially distributed. A Hamming distance metric (fraction of disagreeing bits) converts this recognition problem into a classic signal detection problem. Figure 10 illustrates the decision task for the familiar dual Gaussian model, in terms of distributions of Hamming distances which might arise in comparing two codes from different irises ("imposters"), and in comparing two codes from the same iris under different conditions ("authentics"). The four areas under these distributions defined on either side of the criterion indicated, correspond to the probabilities of false accept, false reject, correct accept, and correct reject.

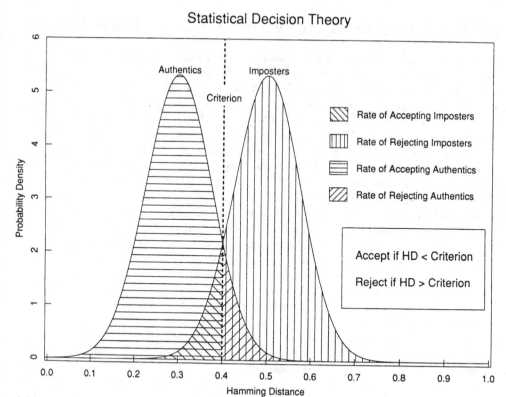

Fig. 10. The classical Signal Detection paradigm. The four areas under the two distributions, on either side of the chosen Hamming distance criterion, correspond to the probabilities of false accept, false reject, correct accept, and correct reject, of an individual claiming a particular identity.

Finally, Fig. 11 shows the actual performance histograms with the current image coding algorithms. The two histograms for "authentics" and "imposters" have zero empirical overlap, signifying the absence of any false accepts or false rejects in performance trials when using a Hamming distance criterion of 0.40. The mean of the "authentics" distribution of Hamming distances was 0.20, which generates an extraordinary confidence level of authenticity. The probability of such a measure arising by chance from the "imposters" distribution (i.e., the expected area under the solid distribution in Fig. 11, to the left of 0.20) is less than 1 in 10^{27}. Potential applications of this pattern recognition system include passport control, premises access security, automatic bank teller machines, or perhaps even in lieu of keys and ID cards generally.

Acknowledgments: Supported by N.S.F. Presidential Young Investigator Award IRI-8858819 and by AFOSR University Research Initiative F49620-87-C-0018.

Hamming Distances for Authentics and Imposters

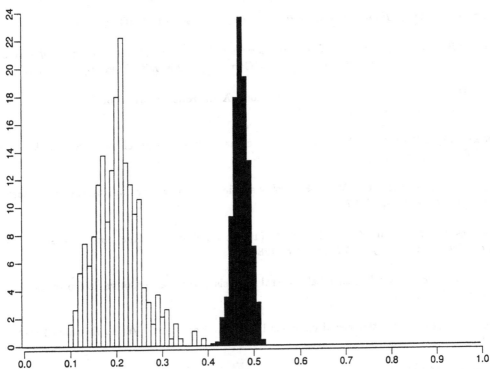

Fig. 11. Empirical distributions of Hamming distances for the automatic personal identification system based on iris texture. At a criterion of 0.40, there were no false accepts and no false rejects, as indicated by the absence of any overlap between the two distributions.

REFERENCES

Abramson, N. (1963): *Information Theory and Coding.* New York: McGraw-Hill.

Barlow, H.B. (1961): The coding of sensory messages. In *Current Problems in Animal Behavior* (Edited by W.H. Thorpe and O.L. Zangwill), Cambridge University Press, pp. 331-360.

Barlow, H.B. (1972): Single units and sensation: A neurone doctrine for perceptual psychology? *Perception* **1**, pp. 371-394.

Bastiaans, M. (1980): Gabor's expansion of a signal into Gaussian elementary signals. *Proc. I.E.E.E.* **68**, pp. 538-539.

Bossomaier, T., and Synder, A.W. (1986): Why spatial frequency processing in the visual cortex? *Vision Research* **26**, pp. 1307-1309.

Daubechies, I., Grossmann, A., and Meyer, Y. (1986): Painless non-orthogonal expansions. *Journal of Mathematical Physics* **27**, pp. 1271-1283.

Daugman, J. (1980): Two-Dimensional spectral analysis of cortical receptive field profiles. *Vision Research* **20**, pp. 847-856.

Daugman, J. (1984): Spatial visual channels in the Fourier plane. *Vision Research* **24**(9), pp. 891-910.

Daugman, J. (1985): Uncertainty relation for resolution in space, spatial frequency, and orientation optimized by two-dimensional visual cortical filters. *Journal of the Optical Society of America, A* **2**(7), pp. 1160-1169.

Daugman, J. (1988a): Pattern and motion vision without Laplacian zero-crossings. *Journal of the Optical Society of America, A* **5**(7), pp. 1142-1148.

Daugman, J. (1988b): Complete discrete 2-D Gabor transforms by neural networks for image analysis and compression. (Invited Paper) *I.E.E.E. Trans. Acoustics, Speech, and Signal Processing* **36**(7), pp. 1169-1179.

Derin, H., and Cole, W. (1986): Segmentation of textured images using Gibbs random fields. *Computer Vision, Graphics, and Image Processing* **3**, pp. 72-98.

Field, D.J. (1987): Relations between the statistics of natural images and the response properties of cortical cells. *Journal of the Optical Society of America, A* **4**(12), pp. 2379-2394.

Gabor, D. (1946): Theory of communication. *J. Inst. Electr. Eng.* **93**, pp. 429-457.

Gallagher, R. (1968): *Information Theory and Reliable Communication.* New York: Wiley.

Geman, S., and Geman, D. (1984): Stochastic relaxation, Gibbs distributions, and the Bayesian restoration of images. *I.E.E.E. Trans. Pattern Analysis and Machine Intelligence* **6**, pp. 721-741.

Grossmann, A., and Morlet, J. (1984): Decomposition of Hardy functions into square integrable wavelets of constant shape. *SIAM Journal of Mathematical Analysis* **15**, pp. 723-736.

Grossmann, A., Morlet, J., and Paul, T. (1985): Transforms associated to square integrable group representations. I. General results. *Journal of Mathematical Physics* **26**, pp. 2473-2479.

Holschneider, M. (1988): On the wavelet transformation of fractal objects. *Journal of Statistical Physics* **50**(5/6), pp. 963-993.

Hotelling, H. (1933): Analysis of a complex of statistical variables into principal components. *J. Educ. Psych.* **24**, pp. 417-441.

Hubel, D., and Wiesel, T. (1962): Receptive fields, binocular interaction, and functional architecture in the cat's visual cortex. *Journal of Neurophysiology (London)* **160**, pp. 106-154.

Hubel, D., and Wiesel, T. (1974): Sequence regularity and geometry of orientation columns in the monkey striate cortex. *Journal of Comparative Neurology* **158**, pp. 267-293.

Jacobson, L. and Wechsler, H. (1988): Joint spatial / spatial-frequency representation. *Signal Processing* **14**, pp. 37-68.

Jones, J. and Palmer, L. (1987): An evaluation of the two-dimensional Gabor filter model of simple receptive fields in cat striate cortex. *Journal of Neurophysiology* **58**(6), pp. 1233-1258.

Karhunen, H. (1947): On linear methods in probability theory. (English trans. by I. Selin.) The Rand Corp., Doc. T-131, 1960.

Kulikowski, J. and Bishop, P.O. (1981): Fourier analysis and spatial representation in the visual cortex. *Experientia* **37**, pp. 160-163.

Linsker, R. (1986): From basic network principles to neural architecture: Emergence of spatial-opponent cells, of orientation-selective cells, and of orientation columns. *Proc. Nat'l Acad. Sci. USA* Vol. **83**, pp. 7508-7512, pp. 8390-8394, pp. 8779-8793.

Loeve, M. (1948): *Fonctions Aleatories de Seconde Ordre*. Paris: Hermann.

Logan, B.F. (1977): Information in the zero-crossings of bandpass signals. *Bell System Technical Journal* **56**, pp. 487-510.

Mallat, S. (1988): A theory for multiresolution signal decomposition: The wavelet representation. *I.E.E.E. Trans. on Pattern Analysis and Machine Intelligence* **10**, in press. Also as University of Pennsylvania GRASP Lab **103**, MS-CIS-87-22.

Marcelja, S. (1980): Mathematical description of the responses of simple cortical cells. *J. Opt. Soc. Am.* **70**, pp. 1297-1300.

Marr, D. (1982): *Vision: A Computational Investigation into the Human Representation and Processing of Visual Information*. New York: Freeman.

Meyer, Y. (1986): Principe d'incertitute, bases Hilbertiennes, et algebres d'operateurs. *Seminaire Bourbaki*, Nr. 662.

Morlet, J., Arens, G., Fourgeau, I., and Giard, D. (1982): Wave propagation and sampling theory. *Geophysics* **47**, pp. 203-236.

Movshon, J. (1979): Two-dimensional spatial frequency tuning of cat striate cortical neurons. *Society for Neuroscience Abstracts*, Vol. **9**, p. 799.

Oja, E. (1982): A simplified neuron model as a Principal Component Analyzer. *J. Math. Biology* **15**, pp. 267-273.

Oja, E., and Karhunen, J. (1985): On stochastic approximation of the eigenvectors and eigenvalues of the expectation of a random matrix. *Journal of Mathematical Analysis and Applications* **106**, pp. 69-84.

Pellionisz, A., and Graff, W. (1987): Tensor network model of the "three-neuron vestibulo-ocular reflex-arc" in cat. *J. Theoretical Neurobiology* **5**, pp. 127-151.

Pellionisz, A. (1988): Tensor geometry: A language of brains and neurocomputers. Generalized coordinates in neuroscience and robotics. In: *Neuronal Computers (NATO Advanced Sciences Institutes Series (Computers and Systems Sciences)*, Vol. **41**, ed. by R. Eckmiller and C. von Malsburg), pp. 381-391. Berlin: Springer-Verlag.

Pollen, D., and Feldman, S. (1979): Spatial periodicities of periodic complex cells in the visual cortex cluster at one-half octave intervals. *Investigative Ophthalmology and Visual Science* **18** pp. 429-433.

Pollen, D., and Ronner, S. (1981): Phase relationships between adjacent simple cells in the visual cortex, *Science* **212**, pp. 1409-1411.

Pratt, W.K. (1978): *Digital Image Processing*. New York: John Wiley & Sons.

Rosenfeld, A. (Ed.) (1984): *Multiresolution Image Processing and Analysis*. Berlin: Springer-Verlag.

Shannon, C., and Weaver, W. (1949): *Mathematical Theory of Communication*. Urbana: University of Illinois Press.

Srinivasan, M., Laughlin, S., and Dubs, A. (1982): Predictive coding: a fresh view of inhibition in the retina, *Proc. R. Soc. Lond. B* **216**, pp. 427-459.

Therrien, C., Quatieri, T., and Dudgeon, D. (1986): Statistical model-based algorithms for image analysis. *Proc. I.E.E.E.* **74**(4), pp. 532-551.

Issues of representation in neural networks

Elie Bienenstock and René Doursat

ESPCI, 10 rue Vauquelin
75005 Paris, France

1. INTRODUCTION

1.1 Invariant Representations in Perception

This paper addresses the issue of the neural code: What is the format used by the brain to represent as patterns of neural activity those things that need to be represented in order for us to perceive, act, and more generally behave the way we do? One could make a fairly extensive list of these things to be represented, yet we shall be chiefly concerned with visual perception, and in particular with so–called high–level perceptual representations. The existence of such representations is the cornerstone of the cognitive approach to perception. The term "high–level" generally implies a certain amount of invariance with respect to the parameters of the physical *presentation* of the object, person or scene that is perceived. Take for instance the face of a person known to us: what we mean by saying that there is an invariant neural representation of this face in our brain is that the pattern of activity which is elicited by the perception of this person's face is reproduced from one occurrence to another. If I see this person's face now at a certain distance, in a given environment, under a given angle, under given lighting conditions etc, the pattern of activity embodying this percept will be, one should think, closely related to the pattern that would be triggered in my brain tomorrow by the view of the same person's face. Yet tomorrow all these physical parameters, and many others as well, including the context, physical and psychological, are likely to be different. We therefore say that the representation is "invariant" along all those dimensions. Of course the existence of literally invariant representations is unrealistic, on both psychological and neurophysiological grounds. One may nevertheless view as a reasonable working hypothesis that some *aspect* or *property* of the activity pattern is conserved from one occurrence to another. It is this property or aspect that we call a high–level, or invariant, representation, and this paper addresses the question of what this property could actually be.

Let us first remark that in order for a representation to make sense, certainly in order for it to allow what we would like to call a conscious experience, this representation has to span a certain duration in time: the *psychological* time unit may be in the range 0.1–1 sec. The *physiological* time–unit on the other hand is, by all standards, much shorter, say 1–3 ms: this is the range of the duration of the action potential, of the length of the absolute refractory period, of synaptic delays, of axonal conduction times etc. One exception is the conduction time along the somato–dendritic membrane: in a large pyramidal cell, it may perhaps take a dozen of milliseconds, or more, for something which happened on one of the distant apical dendrites to be felt at the axon hillock, the spike–generating site. However, locally on a dendrite, spatio–temporal integration takes place faster, again on the scale of the millisecond: the key figure is the duration of the EPSP (excitatory post–synaptic potential), which typically lasts from 3 to 6 ms. (inhibitory PSP's may last longer.)

There are thus at least two orders of magnitude between the physiological time–scale and the psychological time–scale. How should this gap be filled? It is commonly accepted that the answer lies in time–averaging, which one may denote $E_T[...]$. The symbol $E[...]$ stands for the "expected value", or average, of a random variable, and the subscript T indicates averaging over time. We are thus taking the average of the instantaneous spiking process, perhaps a moving average of some kind, over a suitably defined time–window T, say $T = 100$ ms. T is roughly the psychological time–unit. With this averaging over time, we can reasonably hope to solve two problems at once: bridge the gap between the two scales, and eliminate the variability, or stochasticity, inherent to the instantaneous firing process. As said above, the actual firing patterns will never be identical from one occurrence of a percept to another. It seems then reasonable to postulate that a suitably defined time–average is the invariant aspect, or property, that is used by the brain to represent objects, people, scenes, or actions at a high level.

1.2 First and Higher Order Neural Coding Schemes

We are now facing an interesting and non trivial question: What should be put inside the brackets of the expectation symbol? Consider a collection of neurons labelled $1,...,N$, with their instantaneous firing processes $x_i(t)$, $i=1,...,N$, where t ranges over time in discrete units of, say, 3–5 ms. These variables may be assumed to be binary–valued: $x_i(t) = 1$ indicates that neuron i fired at time t, i.e., emitted at least one spike in the 3 or 5 millisecond interval between t and $t+1$, while $x_i(t) = 0$ indicates that neuron i was silent at time t. The simplest way to make use of the operator E_T is then to define a family of variables y_i as follows: $y_i = E_T[x_i]$, for $i=1,...,N$. These variables, defined on a coarser time–scale, are the time–averaged versions of the individual x_i's. Averages are taken componentwise, i.e., separately for each neuron in this "cell–assembly". We shall call this a *first–order coding scheme*: $(y_1,...,y_N)$ is essentially a vector of first–order statistical moments.

However, nothing should prevent one from envisaging more complicated uses of the operator E_T. Consider for instance the quantity:

$$E_T[x_i(t)x_j(t)],$$

the average over time of the product $x_i x_j$ for a given pair i,j, or, more generally, the expression:

$$E_T[x_i(t)x_j(t-\tau)],$$

where τ is any given time delay, defined on the 3–5 ms time–scale, say in the range 0–100 *ms*. This is a second–order moment, which may constitute the basic building block in a second–order coding scheme. Several authors have raised the possibility that the brain may use second–order statistical moments of this kind to convey and process information. An interesting discussion of this hypothesis can be found for instance in Sejnowski (1981).

Note that the combinatorics of coding schemes in the hierarchy we just defined increases very fast with their order. For instance, a *third–order* scheme would contain expressions of the type:

$$E_T[x_i(t)x_j(t-\tau)x_k(t-\rho)].$$

In such a scheme, every choice of a 5–tuple (i,j,k,τ,ρ) where i, j and k range over the family of N neurons and τ and ρ are given time delays as above, is, *a priori*, a legitimate one. A given collection or vector of such choices —possibly a very large one— may be viewed as the third–order generalization of the first–order scheme $(E_T[x_1],...,E_T[x_N])$ discussed above. Clearly then, higher–order schemes offer immense resources for knowledge representation.

To see how such high–order statistical moments could be exploited by the brain, assume that the value of one of these moments, defined by a careful choice of all indices and time–delays, sometimes becomes overwhelmingly different from the corresponding expected value "by chance", i.e., the value it assumes under the hypothesis of mutual statistical independence of the individual instantaneous firing processes $x_i(t)$ appearing in the expression. This moment is then much larger than expected, which means that a certain spatio–temporal pattern, accurately defined in both space and time, reproduces much more often than it should. Such an event —an overwhelming departure from statistical independence— could be used as a basic information carrying signal in the brain. Note that as the order of the code increases, the need for a long averaging period T decreases: it is easy to tell an unexpected event of high order from an event likely to occur by chance. For a large enough order of the code —which may be as small as 10— there is no need for averaging at all: the probability that the same event reproduces by chance just twice is vanishingly small. On the other hand, such events should leave in the network a somewhat lasting yet immediately readable *trace* of their occurrence. Without such a trace —a form of very short–term memory— high–order temporal patterns would be too short–lived to mutually interact on the *psychological* time scale. As we shall see below, this trace could be laid down in the synapses most directly concerned by these high–order events.

1.3 High–Order Codes and Fast Synaptic Traces as a General Format for Knowledge Representation in the Brain

To summarize, we have considered essentially two types of codes: first–order codes, and higher–order codes. From the mere hardware point of view, that is, if one just considers the time constants involved in the propagation and integration of neuronal activity in cortex, high–order codes are quite plausible. Mutual interactions of post–synaptic potentials impinging on nearby locations on a neuron's membrane are highly dependent on the relative timing of the inputs, measured with an accuracy of a few milliseconds. Thus, the biological machinery is there to perform accurate discrimination between patterns differing only in some of their high–order statistical moments.

We remarked that high–order codes offer virtually unlimited space for representing knowledge. However, much more important than the *amount* of information that can be handled is the *format* of knowledge representation. It has been argued by several authors, in particular by von der Malsburg (1981, 1986; see also von der Malsburg and Bienenstock 1986) that (*a*) high–order statistical moments in neural activity provide a natural format for representing structured, or *relational* knowledge of the kind involved in higher cognitive functions, and that (*b*) *fast and reversible synaptic plasticity* is a biologically likely mechanism for stabilizing, for a short time, these high–order statistical patterns. As already noted, the existence of such a *trace* —which is today a mere hypothesis— is required for high–order codes to be actually useful as a representational device.

The reason why first–order codes appear inadequate as the general data structure in the brain is their lack of *inner structure*. In effect, for some authors (Fodor and Pylyshyn 1988), connectionist models (see below) can by no means provide acceptable architectures for cognition. Their fundamental flaw is that the states or "symbols" they manipulate lack inner structure hence *compositional combinatorics* (but see Smolensky 1988). Evidently, this deficiency is linked to the data format used by connectionist models. As we shall see in Section 3, this format is almost always a first–order code. However, as soon as codes of order higher than 1 are introduced, brain states acquire inner structure and can be composed, i.e., linked with each other in many non–trivial ways.

We shall not enter into a general discussion of this issue. Suffice it to say that, in principle, high–order statistical moments and fast synaptic weights are a system of dual variables offering a coherent framework for explaining much of cognition in neurobiological terms (von der Malsburg 1986; see also Shastri and Ajjanagadde 1989). The main focus of this paper is shape recognition, for which this framework is particularly attractive. Indeed, invariant perceptual representations are conveniently described as relational structures, i.e., labelled graphs. Perception is then conceived of as an operation of matching of the sensory input with high–level representations, all of which are couched in the same labelled–graph format. High–order codes and fast synaptic weights are a straightforward neural implementation both of labelled graphs and of the matching operators acting on them. These ideas, which were first discussed by von der Malsburg (1981, 1986; see also von der Malsburg and Bienenstock 1986, 1987; Bienenstock and Doursat 1989; Buhmann *et al.* 1989), are summarized in the central section of the present paper (Section 4), where we also propose a mathematical formulation. Yet before coming to the description of the model, we first present a very succinct survey of currently popular artificial neural networks (Section 2), and discuss their relevance as brain models (Section 3). Finally, as an illustration of the model proposed in Section 4, Section 5 reports on numerical experiments on the task of handwritten digit recognition.

2. A BRIEF REVIEW OF ARTIFICIAL NEURAL NETWORKS

The field of neural networks, or "connectionist" models, has seen a remarkable increase of activity in recent years. Although the chief motivation was originally the modeling of natural brains, much of today's literature on "artificial neural networks" is actually somewhat remote from biology. A review of this field is out of scope of the present paper, and we shall only briefly mention two neural computing paradigms which have been most studied in recent years. In the next section, we shall put our remarks about neural networks in the perspective of our discussion of coding schemes.

2.1 Neural Networks for Associative Memory

The first paradigm is associative memory. An associative memory network, or content–addressable memory, is generally a feedback network, i.e., a network with recurrent connections (Hopfield 1982). It works as follows. A number of given activity patterns, or memories, are stored in the weights of the network by a synaptic modification rule, most often of Hebbian type. The retrieval of a pattern is done by starting the network in a configuration which is a distorted or incomplete version of one of these memories. Retrieval means that the network's state flows into —is attracted by— this particular pattern. Such networks are sometimes called dynamic, because their state undergoes a sequence of transitions —obeying simple deterministic or stochastic rules— from an initial configuration which is the input, i.e., the incomplete or degraded pattern, to a final stable state which is the retrieved memory, i.e., the complete or error–free stored item.

2.2 Backpropagation Networks

The second neural computing paradigm we wish to mention here is the *learning* of a mapping between an input space and an output space, from a set of training examples which are representative input–output pairs. Layered feedforward networks, also called multilayer perceptrons, are most often used in this context. By contrast with feedback associative networks, these pure feed–forward networks are static: a single feedforward propagation of the activity from the input layer to the output layer is enough to determine the entire network's state. The task of learning generally consists in adjusting the connection weights of the network so that the desired input–output mapping be realized, or best approximated by the network. Sometimes, the architecture of the network is adjusted too.

The most extensively used technique for achieving this goal uses the training examples to construct a function of the weights which measures the "error", that is, the departure from the desired behavior (Rumelhart *et al.* 1986). Learning is the minimization of this function over all possible synaptic weight configurations, generally achieved by gradient descent. This is done in two steps: forward propagation of the "stimulus" from input to output, followed by backward propagation of an error signal, used to compute the partial derivatives necessary to alter all synaptic weights in the network according to the gradient descent prescription. Due to this second step, this learning scheme is called the backpropagation algorithm, and the layered network is called a "backpropagation network". Note that backpropagation is essentially an *algorithm*, that is, a computational recipe. It has no straightforward biological interpretation, since retrograde propagation of impulses along axons of cortical cells does not occur in physiological conditions.

2.3 Backpropagation Networks and Statistical Inference

Backpropagation networks have been the most popular neural computing paradigm for five years or so, as they can be used in many practical applications involving classification learning. However, it is worth pointing out that the problems solved by these networks had actually been studied for some years under the name "statistical inference problems" in mathematical statistics, and that a variety of so-called nonparametric, or model–free methods for this class of problems had been developed in the statistical literature. These nonparametric methods for classification and regression, which include nearest–neighbor or kernel estimation (Duda and Hart 1973, and see below, Section 5.2), tree–structured methods, projection pursuit etc., are very similar in spirit to neurally inspired methods. However, until recently, the similarity between neural networks and nonparametric schemes for inference was somewhat overlooked. The emphasis, in the neural network literature, was put on the novelty of the learning–from–examples scheme combined with the distributed mode of knowledge representation (a single item may be stored in many synapses) *vis–à–vis* the conventional rule–based methods of Artificial Intelligence. The hope was expressed that neural computing systems would succeed where A.I. had failed: neural networks would soon exhibit processing capabilities comparable to those of biological systems. In particular, many authors were confident that feed–forward networks equipped with the back–propagation learning algorithm would be able to automatically discover rules, or regularities, hidden in the deep structure of a given sensory environment, in much the same way as living beings seem to do. At the very least, this should have opened a new avenue in the field of computer vision.

In recent years, the enthusiasm for artificial neural networks has been somewhat mitigated by the progressive acknowledgement that, as a result of the deep analogy between adaptive neural networks and nonparametric statistical inference methods, neural networks suffer from the same, often drastic, limitations as other nonparametric inference schemes (see next section). It remains true that artificial neural networks, being essentially (simulated) parallel–processing machines, have interesting properties of noise–tolerance and fault resistance, and that dedicated hardware implementations of neural networks —which are just now coming into existence— can achieve significant speed improvements over other methods.

3. NEURAL NETWORKS AS MODELS OF BRAIN FUNCTION

3.1 The First–Order Coding Scheme Applied to Shape Recognition

Our chief concern however in the present paper is the modelling of the processes at work in real brains. In spite of the limitations of artificial neural networks mentioned in the previous section, it is one of the most challenging claims of neural network supporters that these will help us to understand real brains. As was already mentioned, the relevance of artificial neural networks to biology has been strongly debated in recent years. Of interest to us is the fact that connectionist systems, when envisaged as models of brain function, fall exclusively —with a few exceptions, e.g. Shastri and Ajjanagadde (1989)— into the first category described in Section 1: they posit first–order coding schemes. For instance, in an associative memory network, a memory trace is encoded as an activity pattern, that is, a vector of firing rates of the form $Y = (y_1,...,y_N)$, where the y_i's are explicitly interpreted as *time–averages* of the instantaneous firing processes: $y_i = E_T[x_i]$. In view of the gap between the physiological and the psychological time scales (see Section 1), there is virtually no alternative to this interpretation. Similarly, in the case of feedforward layered networks, inputs, outputs and hidden–layer states are all *activation* states, that is, in the biological perspective, mean firing rates. As we have seen in the previous section, the paradigm of computing in such networks is the *spreading of activation* across interconnected units.

We proposed in Section 1 to define an invariant perceptual representation as the ensemble of those properties or aspects of brain activity patterns that remain unchanged across all occurrences of a given percept. Thus for instance, the representation of person A's face in person B's brain should be fairly insensitive to the distance and orientation of person A when viewed by person B. How are invariant representations construed in the framework of first–order coding schemes? In general, the solution goes along the following lines. What distinguishes person A's face from the faces of persons C, D etc, is a collection of features (shape, size, colour, of eyes, nose, hair etc...). Each feature can for instance assume a given finite number of values, and person A's brain performs classification in this feature space in order to decide that the face viewed is that of person B, C or D. All these faces are of course stored as feature vectors in A's brain. Reliable classification is made possible by the fact that, in this feature space, the vectors corresponding to all possible views of person B's face are –hopefully– sufficiently well separated from the feature vectors corresponding to all possible views of person C, person D etc. Being committed to first–order schemes, we posit that each feature is coded in the activity rate of a number of specialized cells. The invariant representation is then essentially a distributed pattern of activation of a carefully defined cell assembly. This solution is consistent with the scheme of neural networks for associative memory (Hebb 1949; Hopfield 1982) with the scheme of neural networks for categorical perception (Anderson *et al.* 1977), and it is also very close in spirit to the scheme of feedforward neural networks (Rumelhart *et al.* 1986).

The major difficulty with this first–order code solution is that it is not clear how features are actually extracted from the sensory input in an invariant way. In other words, we may have merely pushed down the problem of invariant representation from the level of full objects or shapes to the level of features, yet the problem remains essentially the same at this lower level.

3.2 Backpropagation Networks Applied to Shape Recognition

The algorithm of backpropagation deserves a further comment at this point, as it is often viewed as a potential solution to the problem of invariant perception: it is argued that a backpropagation network will automatically extract an optimally discriminating set of invariant features, i.e., learn these features from examples.

In reality, this may be true *in theory*, but not realizable *in practice*. Indeed, as was already mentioned in Section 2, backpropagation is only one among many other nonparametric methods for estimation, regression and classification, and these methods, or "machines", are well–known to be severely limited in practice, in spite of the *theorems* that can be proven about their asymptotic convergence to optimal performance (a property called consistency in mathematical statistics). The practical problem is that the only way to ensure good performance with one of these machines is to train it on a set of examples so large that it densely covers the space of all possible inputs. If such a covering is done during learning, the machine, when confronted with a new input, will *interpolate* the correct answer from learned data points. On the other hand, if the covering is not dense enough, the machine will have to *extrapolate*. Yet non–parametric schemes yield essentially unpredictable results when asked to extrapolate. Now if the data space is of high dimension, as is the case with raw images, covering the space is *impossible* in practice (this is known as the "curse of dimensionality"). The number of raw images that would be required to correctly train a non–parametric machine —such as a backpropagation network— to recognize the faces of say 100 individuals at any position, orientation etc, is well beyond what can be handled in a lifetime.

For such classification tasks, training a backpropagation network —or for that matter any other non-parametric scheme— on *raw unprocessed* images is hopeless: appropriate features have to be extracted first, and we are back to the original problem. In brief, it appears that although backpropagation is, in principle, much more powerful than its ancestor the perceptron (Rosenblatt 1962), it suffers from the very same limitations in practice. Of course these remarks apply to *difficult* problems such as the recognition of faces. On easier problems, such as the recognition of normalized segmented handwritten digits (see below, Section 5), nonparametric methods may achieve excellent performances (see e.g. LeCun *et al.* 1989). The reader interested in a more mathematical treatment of this issue, and in its relationship to the so–called bias/variance dilemma, is referred to Geman, Bienenstock, and Doursat (in preparation).

3.3 First and Higher Order Codes in View of Experiment

The following statement, made ten years ago (Sejnowski 1981), is still quite relevant today:

> *The possibility that synaptic events are correlated and synchronized is at present beyond the limits of experimental verification, but its consequences are worth exploring.*

Why is it then that neural modeling in these ten years has focused almost exclusively on first–order codes, which are a tiny, one could almost say "zero–measure subset", of *plausible* brain codes?

The answer lies no doubt in the fact that considerable support to first–order schemes as the universal brain code has always come, and continues to come, from electrophysiology. In a typical electrophysiological experiment, one would record from a single cell in an animal's brain while stimulating the animal in various ways, or having it perform a given task. One would show that the total number of spikes fired in a period of say $250ms$ after stimulation onset or during task performance is a good indicator of the presence of the stimulus, or of a certain attribute of the stimulus such as its orientation, speed of motion etc, or that it correlates well with a certain aspect of the action performed. Whenever a cell is found which exhibits such properties, it seems legitimate to say that this cell "codes for" a particular stimulus, action, or memory. Concerning the visual system, it should be stressed that most of these single–cell data deal with low–level vision. Some examples dealing with higher–level processing are available as well (Perrett *et al.* 1982) yet their interpretation is not as unequivocal: it is not quite clear that the increased activity of these cells should truly be regarded as a specific "representation". Even if one is willing to consider that these cells, *code,* in an invariant way, for faces, for hands, for particular types of motion or action etc, there remains a significant gap between the local bar detectors or motion detectors that we are familiar with in V1 for instance, and these high–level rate–coded representations, or "grand–mother cells" in temporal cortex: the mechanism by which the brain constructs these representations remains mysterious.

In brief then, first–order coding is the default assumption, the simplest and easiest to assess experimentally. It is supported by a good deal of experimental evidence, particularly from peripheral sensory and motor parts of the brain.

On the other hand, in spite of the fact that second– and higher–order codes are much more difficult to study experimentally than first–order ones, many authors have started investigating moments of order higher than 1 in neural activity, and their possible relevance to brain function. For instance, Abeles (1982) has made extensive recordings from the auditory and frontal cortices of awake cats and monkeys, and found third–order patterns of the kind described in Section 1, repeating themselves with a frequency higher than expected had the three spike trains been statistically independent processes. This is an intriguing finding, but unfortunately it has not yet been possible to correlate these patterns with a particular stimulation or task that the animal would be engaged in, i.e., to show that these patterns actually *code* for something.

More recently, in the domain of visual perception, the work of Gray *et al.* (1989; see also Gray's contribution to this volume) and Eckhorn *et al.* (1988) has raised the possibility that accurate temporal correlation could be used to bind together elementary visual stimuli —such as elongated intensity gradients— whenever these elementary stimuli have some attributes in common, e.g. speed. If confirmed, this would be a second–order code based solution to the outstanding problem of figure/ground separation, as proposed ten years ago by von der Malsburg (1981).

As is evident from this short review, experimental support to high–order codes is still scarce. However, the rapidly increasing interest for such codes, and the development of new experimental techniques and of adapted conceptual tools (see e.g. Aertsen *et al.* 1989) could possibly lead to something like a "paradigm shift" in brain study in the not too distant future (see also Ullman, this volume). In this perspective, the theoretical investigation of the possibilities and limitations of high–order codes appears more timely now than ever.

4. A MODEL OF SHAPE RECOGNITION BASED ON HIGH–ORDER CODES

In this section, we describe a model of shape recognition suggested by our approach to brain function based on high–order codes and fast synaptic traces (von der Malsburg 1981, 1986; von der Malsburg and Bienenstock 1986). Related versions are discussed in Buhmann *et al.* (1989) and in Bienenstock and Doursat (1989).

4.1 The Model

The model is based on the following idea. Instead of trying to isolate a collection of *features* that would allow us to characterize a face for instance, we consider the face as a whole, and characterize it by a coherently structured set of *relationships* between appropriately selected subparts. We thus propose to abandon the idea that invariant representations are vectors of invariant features: rather, they should be viewed as relational models. Formally, this means that we should be prepared to work with relational variables and relational structures, i.e., labelled graphs. The biological implementation is as follows. Instead of propagating and combining *activation levels* which embody features, we manipulate —according to rules yet to be specified— *dynamical links*, i.e., statistical moments of order higher than 1 and fast synaptic weights, which embody relationships between parts of objects. Our goal within this framework is to achieve intrinsically invariant representations. Ideally, we are interested in relational properties of an object or shape invariant under changes of size, position, perspective etc. What we want to achieve is thus a "deformable template" which will accommodate a specified family of deformations. However, the model we propose is just a first step in this direction. We shall for instance not concern ourselves at all with size invariance.

What are the relationships between subparts of the object that enter in our description, and what are actually these subparts? The answer is of course problem–dependent, and one may envisage relational representations at all levels, from very low to very high. A low level relational representation typically contains simple geometrical, topological, or metric relationships between "primitives". The simplest representation of this kind is a labelled graph (Fig. 1b) derived from a discretized image (Fig. 1a): each node of the graph corresponds to a pixel in the image, and carries a label which is the grey–level of that pixel, in this case a binary variable. Two nodes are connected if and only if they are neighbors in the square lattice. In Fig. 1, the shape represented is a handwritten digit. Note that the nodes of the graph correspond to a subset of the pixel array, defined, somewhat arbitrarily, as the union of the pixels colored black with all first nearest neighbors of these pixels. The computer experiments we shall report on use this nearly "minimal" labelled–graph representation.

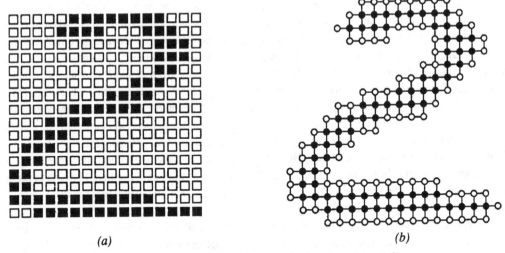

(a)   *(b)*

FIGURE 1. (*a*) **A discretized image of a handwritten digit;**
(*b*) **A labelled–graph representation derived from this image.**

Higher–level relational representations make use of a variety of other features: local features such as edge descriptors, or spatially extended features related for instance to the planar faces of a 3D–object etc. Correspondingly, the relationships embodied in the edges of the graph may be quite sophisticated and may relate specifically to the type of features used. Most relational or structural models discussed in the computer vision literature fall into this category. We will content ourselves with a very low–level description, but we emphasize again that this is meant to illustrate a general principle, which carries over to higer levels.

How do we operate on labelled graphs? At least two different paradigms may be envisaged. The first is the associative recall, or retrieval, of stored memories, where each memory is a labelled graph. The second is labelled graph matching. In this paper we shall focus exclusively on the graph matching paradigm. A model of associative memory for graphs can be found in von der Malsburg and Bienenstock (1987).

Graph matching is the key in our model of shape recognition, and we shall describe it now in detail. We are given two binary–valued images X and X' as in Fig. 1a, and we derive the corresponding graphs G and G' as in Fig. 1b. Nodes in G will be denoted s,t etc., and nodes in G' will be denoted s',t' etc. Each node s in G carries a label $x(s)$, and each node s' in G' carries a label $x'(s')$. In the example shown in Fig. 1, labels assume one of two values, "*Black*" and "*White*". The two labelled graphs will be denoted (G,X) and (G',X'), but we shall often write simply G instead of (G,X) and G' instead of (G',X'). The problem of labelled graph matching may be formulated as follows: find a map φ from the nodes of G to the nodes of G' which preserves both the graph structure and the labels.

Such a map φ has thus to satisfy two types of constraints. The first constraint concerns connectivity: for any s and t in G, if $\varphi(s) = s'$ and $\varphi(t) = t'$, then it should be the case that s and t are connected by an arc in G if and only if s' and t' are connected by an arc in G'. Fig. 2a is an example of the constraint being satisfied for two nodes s and t connected in G (remember that in all our graphs nodes are connected if and only if they are neighbors in the underlying lattice). In Fig. 2b, the constraint is violated: s and t are connected in G, yet s' and t' are not connected by an arc in G'.

The second constraint concerns labels: for any s in G, φ should be such that $x'(\varphi(s)) = x(s)$.

If a map φ exists which satisfies the first constraint for all pairs of nodes s and t in G and the second constraint for all nodes s in G, then G and G' are isomorphic as labelled graphs (more accurately, G and $\varphi(G)$, the labelled subgraph of G' which is the image of G under φ, are isomorphic). In this case, the task of labelled–graph matching amounts to finding the isomorphism from G to G', or one of the isomorphisms if there are several. In many cases of interest though, there is no isomorphism between (G,X) and (G',X'), yet these graphs are related to each other by an "acceptable" *deformation*: there exists a map from G to G' which is, in an appropriate sense, close to a labelled–graph isomorphism. It is of course up to us to define what we mean by "close", in order to capture precisely the right family of deformations. Further, even if an isomorphic map exists, this map may be hard to find numerically; a near–isomorphism may be just as good for practical purposes, yet much easier to find. For both of these reasons, it is desirable to enlarge the notion of matching, to include non–isomorphic maps.

There are many ways of doing this, some of which have actually been proposed for shape recognition. A very convenient definition uses a map functional $H(\varphi)$, called a "cost function". $H(\varphi)$ is an integrated measure of the "badness" of the map φ: the higher $H(\varphi)$, the worse the deformation. Such a function is sometimes called an "energy function" by analogy with some statistical mechanics systems whose energy is minimized under certain conditions. The cost function we shall use takes advantage of the fact that our graphs are actually built on two–dimensional arrays in the plane, equipped with its usual euclidean metric. Thus, the penalty for violating structure will be formulated in terms of distances between nodes in the 2D–array, rather than in terms of arcs in the graph. (There is an equivalent formulation strictly in terms of graphs with labelled arcs, where each arc (s,t) carries a label indicating the relative positions of s and t in the array.) For any pair of nodes s and t in G, the penalty is the squared norm of $\overline{d}$, the difference between two vectors in the plane: the vector $d_{s,t}$ from s to t in the G–plane, and the vector $d_{\varphi(s),\varphi(t)}$ from $\varphi(s)$ to $\varphi(t)$ in the G'–plane (see Fig. 2b).

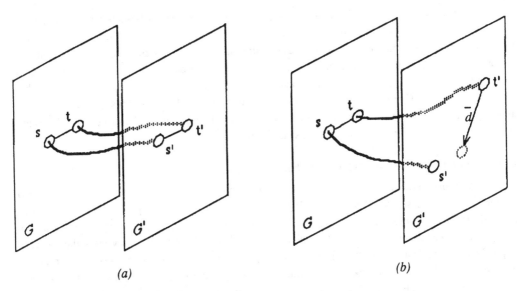

(a)   (b)

FIGURE 2. Mapping neighbors in G to neighbors in G' (a), or to non–neighbors in G' (b).

The cost function will include a term of this kind for each pair of nodes (s,t) in a given family of pairs. In the simulations we shall describe in Section 5, we have chosen to take the sum over all pairs (s,t) such that s and t are neighbors in G. This is a strictly local cost. Other families of pairs could be used: one may for instance include a term for all pairs of nodes s and t in G, possibly weighted by some factor depending on the distance from s to t.

In our definition, the penalty is 0 if and only if the *geometrical* relationship between s and t is identical to the geometrical relationship between $\varphi(s)$ and $\varphi(t)$. Thus for instance, we do penalize rotations: the only zero–cost transformations are translations. We could of course have chosen a different form for the cost function, to allow rotations as well.

Note also that the penalty is quadratic in the amount of local deformation. This is reminiscent of an elastic energy $E_{elast}(d) = k \times d^2$, the energy that is required to stretch a perfect spring of spring constant k by an amount d. Actually, our cost function can be viewed as an approximation of the elastic energy required to deform graph G, viewed as a discrete version of a continuous rubber sheet, into graph G'. The scheme we propose is a form of "elastic matching", a term which has been used in the computer vision literature for related schemes (e.g., Burr 1980).

Of course a cost function for labelled–graph matching should include a second term, relating to the constraint on labels. This term is the sum, over all nodes s in G, of a small cost which is 0 if $x'(\varphi(s)) = x(s)$, and positive otherwise.

To summarize, we formulated the problem of labelled graph matching as a "soft constraint" problem: given the labelled graphs (G,X) and (G',X'), the task is to find a map φ from G to G' which minimizes the following functional:

$$H(\varphi) \;=\; \sum_{s,t \,\in\, G,\, \|\, d_{s,t}\, \| \,=\, 1} \|\, d_{s,t} - d_{\varphi(s),\varphi(t)}\, \|^2 \;+\; \beta \sum_{s \,\in\, G} [\, 1 - \delta\,(x(s)\,,\,x'(\varphi(s)))\,],$$

where $\delta(x,x')$ is 1 if $x = x'$, and 0 otherwise. In this expression, the parameter β is a positive weight used to adjust the size of the label term relative to the structure term. Indeed, unless (G,X) and (G',X') are isomorphic, it is impossible to satisfy all constraints simultaneouly. On the other hand, it is easy to see that one can always satisfy all relational constraints if one doesn't pay attention to labels, or all label constraints if one doesn't pay attention to structure. Thus, the best match is always a compromise between the two terms. One may then want to adjust the relative contribution of these two terms. In our simulations though, β was always set to infinity, which means that we imposed a strict label constraint. Indeed, we found that for this application a strict label constraint didn't hurt.

4.2 Graph–Matching Using Neurons and Synapses

We now return to the biology for a moment: How does the operation of graph–matching relate to temporal correlations, statistical moments of high order, and fast variable synaptic weights? The answer is best illustrated by looking again at Figure 2, which represents a piece of mapping from graph G to graph G' in two situations: a "favorable" one and an "unfavorable" one. We now view the nodes of graph G as neurons in a visual cortical area. Similarly, the nodes of G' are neurons in another area of visual cortex. A dynamical link between a node s in G and a node s' in G', whose existence we have formalized by the condition $\varphi(s) = s'$, is implemented by a high correlation (second–order moment) between the activity of neuron s and the activity of neuron s', and, in the same time, by the fast variable synaptic link between s and s' being at its maximal positive value (we shall come back to this dual implementation below).

To explain the *mechanism* by which graph–matching is performed, which in this case means that neighbors in G should be connected to neighbors in G', we posit that the activities of neighboring neurons in G tend to be mutually correlated, while the activities of distant neurons are uncorrelated. This correlation structure is hard–wired in G: it is enforced by intrinsic connectivity. We assume that a similar local correlation structure is enforced in G'. This assumption is supported by multiple–electrode recordings performed in visual cortex (as well as in the retina) athough observed correlations are, in most cases, rather weak.

In the favorable situation (Fig. 2a), s and t are neighbors in G and s' and t' are neighbors in G'. The mechanism for creating such a situation is essentially *hebbian plasticity*. Indeed, $\varphi(s) = s'$ means that the synaptic link from s to s' is active. This connection will therefore contribute to correlating the activities of s and s'. Since we have assumed that s and t are mutually correlated and that s' and t' are correlated as well, we end up, by transitivity, with t and t' being correlated. Hebb's principle of synaptic modification (Hebb 1949) then tells us that the connection from t to t' should be reinforced, that is, in our notations, $\varphi(t)$ should be t'. In short, provided G and G' have local correlation structure as assumed above, the net effect of hebbian plasticity on dynamic links from G to G' is that any two links which connect neighbors in G to neighbors in G' "cooperate", i.e., reinforce each other, whereas links which connect for instance neighbors in G to non–neighbors in G' do not cooperate. These are respectively the favorable and the unfavorable situations of Figure 2.

The reader who is familiar with the problem of *development of retinotopic maps*, i.e., topology–preserving maps, e.g. from retina to tectum (or cortex, via LGN), will have immediately recognized the similarity between the mechanism proposed here and the basic principles underlying correlation–based models of retinotopy (e.g. Willshaw and von der Malsburg 1976). Of course there is one important difference, and this has to do with the respective time–scales of these processes. Whereas the development of a retinotopic map is a slow process taking place on the time–scale of ontogeny, the topology-preserving mechanism postulated here is a very quick one, which, we assume, may be carried out in a matter of a second or so, the time it takes to recognize a shape. Thus, the basic principle is the same, but it relies on the assumption that there is a form of *very fast* hebbian synaptic plasticity, acting essentially on the psychological time–scale. Under appropriate conditions, as we shall see below, a very weak form of such fast synaptic plasticity should suffice.

As in models of retinotopy development, a form of competition between dynamical links is needed, to complement the cooperative interactions. Otherwise the system would immediately converge to a trivial uninteresting state where all plastic synapses assume their maximum allowed value. In retinotopy models, competition is implemented among all fibres efferent from one and the same cell in the source layer (retina) and/or among all fibres afferent to one and the same cell in the target layer (tectum). In our case, competition among connections efferent from any node s in G is implicit in the very notation $\varphi(s)$. This enforces a hard constraint: only one link is allowed from any given cell s in G. One may either content oneself with this "efferent competition", or introduce an additional third term in the cost function, which will implement "afferent competition". In our simulations, we did include such a term, of the form:

$$\gamma \sum_{s' \in G'} \left[\operatorname{pos} \left(\operatorname{card}\{s \in G \mid \varphi(s) = s'\} - 1 \right) \right]^2,$$

where pos(z), the positive part of number z, is defined as z if $z > 0$, and 0 otherwise. In our simulations, the coefficient γ was equal to 2. Biologically, it is more realistic to use soft constraints, hence to allow multiple incoming and outgoing fibres even in optimal connectivity states. This may be beneficial for increasing the effect of cooperative interactions (see below).

We discussed the biological implementation of graph matching in terms of pairwise correlations. In reality, the build–up of favorable connectivity states implies, and is facilitated by, the build–up of specific statistical moments of order much larger than 2. To see why this is so, note that the mechanism we proposed relies on a simple positive feedback loop: more correlation between two neurons implies enhanced synaptic efficacy (Hebb's principle), which in turn increases even more the correlation, by propagation of activity through the enhanced synapse. Whereas this positive feedback loop is already at work for two neurons with a single plastic synapse between them, it is probably of negligible effect unless it acts in the form of a *mode*, where many growing or reinforced synapses cooperate with many other.

The basic principle of such a cooperation was illustrated in Figure 2a: it involves four neurons and two plastic synapses. It is however quite clear that such a cooperative effect will be much more powerful when taking into account correlations of higher order. For instance, a pyramidal cell in cortex may receive about 10,000 afferents, and it is commonly accepted that these fibres essentially originate from distinct cells. Therefore, the event most likely to trigger the cell's activity is a synchronized discharge of many different presynaptic cells, that is, a very high–order event in the terminology of Section 1. Note that this spatio–temporal pattern has to be very accurate in time but also in space: the presynaptic cells have to be carefully chosen to allow proper summation of local depolarizations on the post–synaptic cell's membrane. The physical location of synapses on the cell body and dendrites plays a crucial role in the conditions required for summation.

In our biological interpretation of the graph–matching paradigm, such high–order interactions are at play: when a perfectly isomorphic one–to–one mapping is established from a rectangular lattice G to a rectangular lattice G', each link (s,s') in this map cooperates directly with 4 neighboring links of type (t,t') as shown in Fig. 2a. It also cooperates somewhat less directly with other links, since correlations do propagate a bit across synaptic connections within each array. Further, if one pays attention to the directionality of synaptic links —formally, one then matches directed graphs rather than undirected ones— large loops including multisynaptic pathways in each layer may be seen to cooperate with each other. Finally, if one allows fan–in and fan–out higher than 1 (see above), this increases even further the chances of links to cooperate, hence of very strong modes to get formed.

To summarize, the mechanism we propose, although illustrated on pairwise interactions, really makes use in a fundamental way of "departures from statistical independence" (see Section 1) of order much higher than 2. Presumably then, the actual modification of each synaptic weight need not be of large amplitude for the system to reach a stable organized state. Indeed, due to cooperation, i.e., specific increase of many mutually reinforcing neighbor–to–neighbor connections, a moderate enhancement of each one of these synaptic links could be enough to reliably stabilize the desired pattern of correlations between neurons in G and neurons in G'. By the same token, due to the positive feedback between correlations and fast synaptic weights, this pattern of cooperating synaptic increases will stabilize itself.

Finally, the second type of constraint in the cost function, relating to labels, has a straightforward biological interpretation: the dynamical links from G to G' are implemented by feature–specific connections (remember that various types of features could be used as labels).

Before we turn to the simulations proper, let us briefly comment on the relationship between the two different accounts of the model given here, the formal mathematical one (Section 4.1) and the neurobiological one. Actually, nothing in the biology should lead one to prefer the particular "elastic" form of graph–matching we have chosen, over many possible other forms. This choice was made for its computational convenience and its adequacy to the practical problem of handwritten digit recognition. As said already, the chief purpose of presenting this detailed formal model is to illustrate the working of a "neurally" inspired approach which differs in a fundamental way from other such approaches, in that it uses dynamical links instead of dynamical activity states.

5. NUMERICAL EXPERIMENTS

In spite of the fact that our elastic matching model was not motivated by computer vision, we have carried out simulations on several data bases of handwritten digits such as the one shown in Fig. 1, to assess the performances of this model used as a shape recognition method relative to other techniques.

5.1 Elastic Matching

In this paper, we report on experiments performed on a data base of discretized handwritten digits kindly provided to us by I. Guyon (AT&T Bell Laboratories). This data–base consists of 1,200 images, 120 for each digit from 0 to 9, written by 12 different people (Guyon 1988). The images are binary–valued, of size 16×16, and each digit has been normalized in size to fit exactly in the 16×16 square. This is a relatively easy data–base, as all scriptors were asked to follow a given template. As a further preprocessing, we have "thinned" the digits, using a standard thinning algorithm. Fig. 3 shows a representative sample of thinned digits. The 1,200 thinned digits were used to derive 1,200 labelled graphs of the type shown in Fig. 1b.

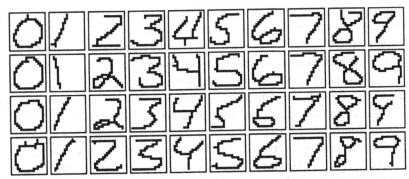

FIGURE 3. A sample of the thinned handwritten digits used in
the simulations.

Given two labelled graphs (G,X) and (G',X') derived from two such images X and X', we need to find the function φ which minimizes the cost $H(\varphi)$ defined in the previous section, relative to these two labelled graphs. Ideally, one should search the space of all possible maps from G to G', to find the least–cost one. However, since the images we use are of size 256 and the number of nodes in the graph is about 100, exhaustive search is unfeasible. We have therefore devised an iterative method to find an acceptable "local minimum" of H. This method is described in detail elsewhere (Bienenstock and Dour-sat 1989). Suffice it to say here that this is a quick, efficient, suboptimal method: while not guaranteed to yield the optimal map, i.e., the φ which minimizes H, it achieves a reasonable approximation to this minimal cost.

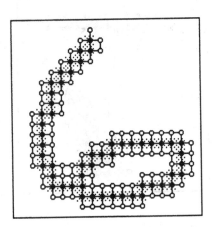
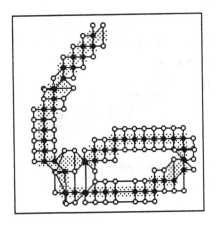

FIGURE 4. Mapping a "6" on a slightly different "6".

Fig. 4 illustrates the operation of elastic matching between two graphs G and G' which are different examples of the same digit "6". In this case the two images, hence the two labelled graphs, are only slightly different from each other. Therefore, there exists a low–cost deformation φ from one to the other. This low–H map φ is represented as follows: the left panel shows one of the two binary–valued images (the shaded region corresponds to black pixels) and, superimposed, the labelled graph G derived from it. The right panel shows the second binary–valued image (again using shading for black pixels) but does *not* show the labelled graph G' derived from it. Rather, it shows the *image of graph* G *under the mapping* φ. Due to the similarity between the two shapes, it is easy to guess which pixel in the left image maps to which pixel in the right, and thus to "recognize" graph G in the G'–plane (right panel).

Note that the map φ is strictly label–preserving: black nodes of G map only to black pixels in the G'–plane, and white only to white. This is due to the fact that the coefficient β, the weight of the second term in the energy, is infinite. Note also that in some cases two distinct nodes s and t in G map onto the same pixel s' in the G'–plane. This is best diagnosed by looking at the number of edges out of a given node in the right panel: if this number exceeds 4, it has to be the case that this "node" is actually an overlap of two or more nodes of G (whenever this happens, the two or more nodes are actually displayed with a very small offset from each other). Such overlaps do contribute a finite penalty to the global cost, through the γ–term, the third term in the functional H. Most of the cost comes from the first term, which penalizes the stretching, the compressing, or the rotation of edges in G. The total cost of the near–optimal φ shown in Fig. 4 is: $H(\varphi) = 73$.

Fig. 5 shows again a near–optimal map φ, but this time from a "5" to a "7". The topology of the "5" can still be identified in the image of the graph G under φ, but only very roughly so: a substantial deformation was needed to accommodate the digit "7". Note the many overlaps in the leg of the "7", and the few very elongated edges. The latter cost a lot, as the price to pay for stretching is quadratic in the amount of stretch. In this example, the cost of φ is: $H(\varphi) = 685$, nearly 10 times as much as in Fig. 4.

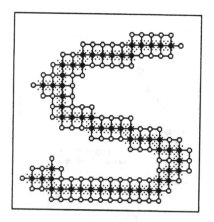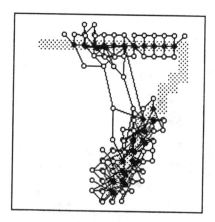

FIGURE 5. Mapping a "5" on a "7".

5.2 Classification

Inasmuch as the two examples of the previous section are representative, the numerical outcome of the operation of elastic matching appears to convey a satisfactory notion of dissimilarity between handwritten digits. To make this statement more quantitative, we need to derive a practical method for shape recognition from this scheme, that is, a method which will correctly *classify* an unknown digit. The idea is of course to measure the similarity of this unknown digit to known shapes, or *prototypes*, representative of all 10 classes. Such classification problems are extensively studied in statistics. One is given N_p prototypes, or "training examples", in the form of pairs (X_i, Y_i), $i=1,...,N_p$, where X_i is a shape —in our case a 16×16 binary–valued image— and Y_i is its class —in our case a label from 0 to 9. One is then asked to classify an unknown shape X, that is, to find its correct label Y. A standard classification method is the k–nearest–neighbor (k–NN) scheme, which works as follows. One computes the distance between X and all prototypes X_i, $i=1,...,N_p$, and selects from the N_p prototypes those k which are closest to X. One then lets these k prototypes "vote": the class of X will be that label Y which occurs most often among the k nearest neighbors of X. The k–NN method is known as a *nonparametric* classification scheme, and many related schemes, such as Parzen–window classification, could be used instead (see Section 2.3).

These nonparametric schemes, although technically different from each other, are similar in principle, and yield similar performances on most classification tasks. In reality, more important than the choice of a particular classification scheme is the choice of a *metric*. In the k–NN scheme for instance, which k of the N_p prototypes will be chosen as the nearest to the unknown shape X strongly depends on the metric used. One can use the straightforward Hamming distance in the raw image space (euclidean distance if more than two grey–levels are used). This is akin to rigid template matching, and generally does not yield excellent results. Therefore, statistical pattern recognition methods almost always proceed by first extracting a number of features from the image: each shape (the unknown one as well as the N_p prototypes) is transformed into a feature–vector. In this feature space, one then uses a conventional metric, such as Hamming distance or euclidean distance if appropriate. Most of the effort in shape recognition thus goes into the choice of appropriate features. This process may be partly automated. For instance, LeCun *et al.* (1989) propose a layered network whose hidden units learn, by an algorithm of "constrained backpropagation", an appropriate set of features for the task of handwritten digit recognition.

Feature extraction followed by non–parametric classification is very close in spirit to the neural model of shape recognition based on first–order codes that we outlined in Section 3.1. In contrast, as explained in Section 4.1, what characterizes our approach is the treatment of a shape as a coherent ensemble of relationships, rather than as a list of extracted features. We shall now see that the operation of graph–matching acting on this relational representation allows us to define a *metric*. For the purpose of classification, we shall simply use this new metric *instead of a feature metric*, in a conventional nonparametric classification scheme. In this paper we shall actually use the simple *first*–nearest neighbor scheme ($k=1$).

The graph–matching based metric is derived as follows. Given two images X and X', and the associated labelled graphs (G,X) and (G',X'), let $H^*(X,X')$ be the lowest cost $H(\varphi)$ among all possible maps φ from (G,X) to (G',X'). Let $\mu = \mu(X,X')$ be defined as follows $\mu = \max \{ H^*(X,X'), H^*(X',X) \}$. μ has all the properties of a metric (in particular, it is symmetric) except for the triangle's inequality. This is of no importance for what follows, but we can actually fix this slight imperfection if we wish, by adding a large enough constant C to $\mu(X,X')$ for every pair of shapes (X,X') such that $\mu(X,X') > 0$ (this trick works only on finite spaces, which is the case here).

In brief, we just defined a *bona fide* metric μ, which, for every pair of shapes (X,X'), measures the distance between these shapes in the sense of the elastic matching operation introduced in the previous section. We are now ready to apply the first–nearest–neighbor classification scheme to the data base of 1,200 handwritten digits equipped with this new metric, and to compare the results to other classification methods.

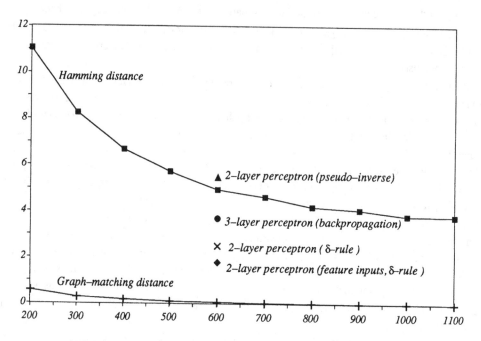

FIGURE 6. Graph–matching classification compared to other classification methods: error rate (percent) *vs* size of training set.

Fig. 6 summarizes such an experiment. The x–axis indicates the number of training examples (prototypes) used, i.e., N_p. Of course, the larger N_p, the better the performance. On the other hand, since we have only 1,200 examples in our data base, N_p has to be strictly less than 1,200. For every value of N_p, the remaining 1,200–N_p shapes in the data base are classified according to the 1–NN rule using either Hamming distance in the raw–image space or graph–matching distance. Fig. 6 shows the percentage of errors (misclassifications) as a function of N_p, for both cases. Note that we are measuring "generalization" performances, as the test sets and the training sets are disjoint. Each data point in these two curves is actually the average of 100 experiments performed with 100 different *random* partitions of the data base into N_p training examples and 1,200–N_p test examples.

Also shown in Fig. 6 are the generalization performances of 4 different layered networks trained and tested on the same data base of 1,200 handwritten digits, for $N_p = 600$. These 4 data points are reproduced from Guyon (1988): they are the best results obtained on this problem using a variety of neural architectures and learning schemes.

It appears from these experiments that the use of the graph–matching based metric μ results in a significant improvement over the use of Hamming distance in raw image space. Further, the graph–matching method appears to yield significantly better results than the best results obtained by layered neural networks, with or without preprocessing (local feature extraction). Preliminary experiments on a larger data base of segmented and normalized handwritten zipcode digits confirm these results.

As mentioned in Section 3.2, the recognition of segmented and normalized handwritten digits is a relatively easy problem which can be addressed by nonparametric methods such as backpropagation networks. The experiments reported in this section demonstrate that even on this relatively easy task, significant improvement can be achieved by adopting a relational representation, and deriving the metric underlying classification from this relational representation.

6. CONCLUDING REMARKS

The graph–matching model presented in Section 4 is certainly not meant to be a faithful account of the way in which the brain recognizes shapes —in particular such familiar shapes as digits. Many details in the model are arbitrary or inacceptable from a biological perspective. On the other hand, its basic working principle is deeply rooted in what we believe is a biologically plausible approach to brain function. This approach may be characterized as the paradigm of computing with dynamical links, i.e., high–order codes and fast synaptic traces (von der Malsburg 1981, 1986). This is an alternative to the usual connectionist paradigm of computing with activation levels, i.e., first–order codes. Whereas in the latter the physical mechanism embodying computation is the *propagation* or *spreading* of activation across carefully adjusted yet frozen synaptic links, computation in the former is embodied as *cooperation* and *competition* between *dynamical links*.

We believe that this mechanism may underlie many kinds of computations performed by our brains. It would seem in particular that a neurobiological account of linguistic functions should require the use of high–order codes. As argued at great length by Fodor and Pylyshyn (1988), connectionist modeling based on first–order coding schemes is inadequate at explaining compositionality, the characteristic property of language. A forthcoming paper will discuss a specific solution to the problem of compositionality in the framework of high–order coding.

Concerning the issue of invariant perceptual representations addressed in the present paper, we have attempted to demonstrate that it might be profitable to construe these neural representations as relational structures. Typically, these are labelled graphs, and a useful operation on them is graph matching. Of course one should envisage more sophisticated representations than image pixels connected neighbor–to–neighbor. Higher–level representations would be at the same time more practical and biologically more plausible. From a practical point of view, it has been suggested to use wavelet transforms as labels (Buhmann *et al.* 1989). In such a representation, each node in the graph carries a label which is the vector of transforms at the corresponding location in the image. Typically, about 50 transforms are used, corresponding to various combinations of the two parameters of the 2D–wavelet: spatial frequency and orientation. Labels are thus 50D–vectors, and the label term in the energy is the euclidean distance between the 50D–vectors carried by the nodes concerned. In all other respects, the operation of labelled–graph matching is the same as discussed in the present paper. In particular, the "structure" term in the energy (first term) has the same quadratic form. This is perhaps an example of a happy marriage between a visual pyramid and a neurally inspired method for shape recognition, as has been called for in this Symposium. Further details on this model, including numerical experiments on a task of face recognition, can be found in Buhmann *et al.* (1989).

In conclusion, we suggest that the best characterization of invariant perceptual representations in terms of brain activity patterns may well be a high–order property, or statistical moment, which, in a precise sense, constitutes a strong departure from statistical independence. Such an event is the repetition of a complex spatio–temporal pattern of action potentials, defined *accurately* both in space —at the level of many individual neurons— and in time —on the millisecond range. Although experimental evidence supporting this hypothesis is still scarce, it should be appreciated that obtaining such evidence is —and will remain for some time— a difficult task. Nevertheless, and contrasting somewhat with the opinion expressed by Shimon Ullman (this volume) in David Marr's tradition of "separation of levels", we believe that this hypothesis may have far–reaching consequences for the future of the computational theory of brain function, in particular computational vision.

Acknowledgment. The research described in this paper was supported by Contract N° ST2J–0416 from the C.E.C. ("B.R.A.I.N." Initiative).

REFERENCES

Abeles, M. (1982) *Local Cortical Circuits: An Electrophysiological Study*. V. Braitenberg ed., Springer–Verlag, Berlin.

Aertsen, A.M.H.J., Gerstein, G.L., Habib, M.K., and Palm, G. (1989) *Dynamics of neuronal firing correlation: Modulation of "effective connectivity"*. J. Neurophysiol. **61**, No. 5, 900–917.

Anderson, J.A., Silverstein, J.W., Ritz, S.A., and Jones, R.S. (1977) *Distinctive Features, Categorical Perception, and Probability Learning: Some Applications of a Neural Model*. Psychological Review, **84**, pp. 413–451.

Bienenstock, E., and von der Malsburg, Ch. (1987) *A Neural Network for Invariant Pattern Recognition*. Europhys. Lett. **4** (1), pp. 121–126.

Bienenstock, E., and Doursat, R. (1989) *Elastic Matching and Pattern Recognition in Neural Networks*. In: *Neural Networks: From Models to Applications*. L. Personnaz and G. Dreyfus eds., IDSET, Paris.

Buhmann, J., Lange, J., von der Malsburg, Ch., Vorbrüggen, J.C., and Würtz, R.P. (1989) Object recognition in the dynamic link architecture — Parallel implementation of a transputer network. In: *Neural Networks: A dynamic systems approach to machine intelligence*. B. Kosko ed., Prentice Hall, New York.

Burr, D.J. (1980) *Elastic Matching of Line Drawings*. IEEE Transactions on Pattern Analysis and Machine Intelligence, Vol. PAMI–3, **6**, pp. 708–713.

Duda, R.O., and Hart, P.E. (1973) *Pattern Classification and Scene Analysis*. John Wiley and Sons, New York.

Eckhorn, R., Bauer, R., Jordan, W., Brosch, M., Kruse, W., Munk, M., and Reitboeck, H.J. (1988) *Coherent oscillations: A mechanism of feature linking in the visual cortex?* Biol. Cybernetics **60**, pp. 121–130.

Fodor, J.A., and Pylyshyn, Z. (1988) *Connectionism and Cognitive Architecture: A Critical Analysis*. Cognition, 28, 1/2, pp. 3–71.

Geman, S., Bienenstock, E., and Doursat, R. (in preparation) *Neural Networks and the Bias/Variance Dilemma.*

Guyon, I. (1988) *Réseaux de neurones pour la reconnaissance des formes : architectures et apprentissage.* Unpublished doctoral dissertation, University of Paris VI, December 1988.

Gray, C.M., König, P., Engel, A.K., and Singer, W. (1989) *Oscillatory responses in cat visual cortex exhibit inter–columnar synchronization which reflects global stimulus properties.* Nature, **338**, pp. 334–337.

Hebb, D.O. (1949) *The Organization of Behavior.* John Wiley and Sons, New York.

Hopfield, J.J. (1982) *Neural Networks and Physical Systems with Emergent Collective Computational Abilities.* Proc. Natl. Acad. Sci. USA, **79**, pp. 2554–2558.

LeCun, Y., Boser, B., Denker, J.S., Henderson, D., Howard, R.E., Hubbard, W., and Jackel, L.D. (1989) *Backpropagation Applied to Handwritten Zip Code Recognition.* Neural Computation, **1**, pp. 541–551.

von der Malsburg, Ch. (1981) *The Correlation Theory of Brain Function.* Internal Report 81–2. Max–Planck Institute for Biophysical Chemistry, Dept. of Neurobiology, Göttingen, W.–Germany.

von der Malsburg, Ch. (1986) *Am I thinking assemblies?* In: *Brain theory.* G. Palm and A. Aertsen eds. Springer–Verlag, Heidelberg.

von der Malsburg, Ch., and Bienenstock, E. (1986) *Statistical Coding and Short–Term Synaptic Plasticity: A Scheme for Knowledge Representation in the Brain.* In: *Disordered Systems and Biological Organization.* E. Bienenstock, F. Fogelman and G. Weisbuch eds., 247–272, Springer–Verlag, Berlin.

von der Malsburg, Ch., and Bienenstock, E. (1987) *A Neural Network for the Retrieval of Superimposed Connection Patterns.* Europhys Lett. 3 (11), pp. 1243–1249.

Perrett, D., Rolls, E.T., and Caan, W. (1982) *Visual Neurons Responsive to Faces in the Monkey Temporal Cortex.* Exp. Brain Res. **47**, pp. 329–342

Rosenblatt, F. (1962) *Principles of Neurodynamics: Perceptrons and the theory of brain mechanisms.* Spartan Books, Washington.

Rumelhart, D.E., Hinton, G.E., and Williams, R.J. (1986) *Learning Internal Representations by Error Propagation.* In: *Parallel Distributed Processing: Explorations in the Microstructure of Cognition.* David E. Rumelhart and James L. McClelland, eds. Volume 1, pp. 318–362. MIT Press, Cambridge.

Sejnowski, T.J. (1981) *Skeleton Filters in the Brain.* In: *Parallel Models of Associative Memory.* G.E. Hinton and J.A. Anderson eds., Erlbaum Associates, Hillsdale, New Jersey.

Shastri, L., and Ajjanagadde, V. (1989) *From simple associations to systematic reasoning: A connectionist representation of rules, variables and dynamic bindings.* Technical Report MS–CIS–90–05. LINC Lab 162. University of Pennsylvania, Philadelphia, PA 19104.

Smolensky, P. (1988) *On the proper treatment of connectionism.* Behavioral and Brain Sciences, **11**, pp. 1–74.

Willshaw, D.J., and von der Malsburg, Ch. (1976) *How patterned neural connections can be set up by self–organization.* Proc. R. Soc. Lond. B **194**, 431–445.

PART 2

Visual cortical processing: from perception to memory

Yves Frégnac and Andrei Gorea

Although visual cortical areas have been a major exploratory field for the past 30 years, using neuroanatomical, electrophysiological and psychological techniques, no consensus has been reached on *how* cortical neuronal activity forms representations of our visual environment. The symposium held at the last ECVP (Paris, September 1990), and the corresponding chapters of the contributors summarize recent trends in the understanding of the neural basis of perceptual and visual memory processes in visual and associative neocortex, and some of their psychophysical correlates.

Different neural codes have been proposed to account for local feature detection and for global form recognition at different stages of integration in the visual pathways, and specially at the cortical level. They can be schematically associated with two major modes of representation of visual information. On one hand, the *neurone doctrine*, which has dominated sensory physiology in the 60-80's, considers neuronal firing as a probability measure of the presence of specific attributes in the receptive field (Barlow, 1972). Perceptual performances of the organism are equated to the detection power of the best neurons. On the other hand, it has also been proposed on both theoretical (see symposium on Visual Pyramids and Neural Networks) and electrophysiological grounds (this symposium) that temporal correlation of nervous activity distributed in different loci of the cortical network could lead to the formation of *dynamic cell assemblies* which participate to the extraction of global visual features of perceived objects.

A provocative defense of the first thesis, entertained by C. Blakemore and collaborators, is that visual cortex can be considered as incorporating a large number of sparsely distributed ensembles of highly specialized units, each corresponding to cells of a specific functional type. Sparse coding implies that information at each point in the visual field is represented without significant summation between independent detectors which belong to the same functional network. This view is supported at least in the primate by the fact that single cells in the primate visual cortex can detect and discriminate elementary aspects of visual stimuli as efficiently as observers (Parker and Hawken, 1985).

In spite of the fact that one or a few photons at the retina might already result in a significant modulation of the firing pattern of individual cortical neurons, local analysis based on signal/noise extraction in the discharge area of the receptive field seems unlikely to account for all the complexities of visual perception. Recent studies in the prestriate cortex of attentive behaving monkeys demonstrate that many neurons can be activated by anomalous contour figures which are not restricted to the classical receptive field (Allman et al., 1985). The ghost of "grandmother cells", sparsely distributed and hierarchically organized in our visual brain, may be reincarnated in these single-units which respond as the human observer to illusory contours: certain classes of orientation selective cells, normally responding to solid contours, can detect also the collinearity or the orientation formed by discontinuous rows of dots (without being activated by the individual presentation of each of them) extending in the "unresponsive surrounding region" along the preferential axis of the receptive field (Peterhans and Von der Heydt, 1989a, 1989b and this volume; Von der Heydt and Peterhans, 1989).

One step further in the chain linking sensory analysis to decision processes in recognition might be achieved using experimental approaches which attempt to relate the variability in visual responses of single neurons to that of the behavioral performance during a discrimination task (Newsome et al., 1989). Microstimulation of a restricted ensemble of specialized units can force in a predictable way the output result computed by the cortical network: for instance, externally imposed changes in the activity of a small cluster of directional selective cells in primate extrastriate areas (MT or V5) are sufficient to bias the perceptual judgement of the organism toward the direction encoded by the stimulated neurons (Newsome, this volume; Salzman et al., 1990). Such causal relationships, demonstrated between cellular activity and perceptual performance, do not necessarily imply a fixed connectivity of the cortical network, but could be acquired through associative processes operating even at the adult age during selective phases of learning. Repeated training to detect fractal contours - which do not belong to the initial visual repertoire - leads to the emergence of newly specialized detectors in associative temporal areas, which are internalized during the formation of a long-term pictorial memory (Miyashita, 1988 and this volume; Miyashita and Chang, 1988).

Studies in the developing brain can also give hints for putative associative mechanisms which link features (orientation, direction, motion, color..) that we know to be detected by separate specialized ensembles of neurones at an adult age. Temporal correlation of nervous activity at different loci of the cortical network is a likely candidate in changing neuronal connections, and it has been shown to play a role in visual cortical epigenesis (Bear and Cooper, 1989; Bear, this volume), and possibly in the formation of pictorial memory in more remote cortical areas (Miyashita, this volume). In addition to classical associative rules which predict potentiation of functional or synaptic links between coactive elements (James, 1890; Hebb, 1949; Brown, 1990; Frégnac and Shulz, in press), recent data - based mostly on *in vitro* studies - suggest the existence of concurrent mechanisms of synaptic depression. The balance between synaptic potentiation and depression could depend critically on the postsynaptic membrane potential at the time of arrival of presynaptic impulses (Artola et al., 1990; Frégnac et al., 1990) and of the voltage-dependent activation characteristics of different classes of postsynaptic receptors. It remains to be understood how attentional processes and extraretinal signals could modulate the expression of synaptic changes, and what type of weight change kinetics could be expected to occur in the attentive behaving organism.

Developmental approaches have comforted supporters of dynamic cell assemblies, who proposed that similar rules of synaptic plasticity hold in the adult as well, but operate on a much faster time-scale, compatible with that of recognition processes (Von der Malsburg, 1981). New hypotheses on cortical mechanisms of figure/ground separation have been proposed, according to which the coding of global features of the visual scene is not realized at the single-unit level, but is expressed in the oscillatory behavior and the temporal coherence of the firing pattern of distributed cell assemblies (Gray and Singer, 1989; Gray et al., 1989; Gray, this volume; see also symposium on Visual Pyramids and Neural Networks). Although the relevance of oscillations in the formation of cell assemblies is still a matter of debate in visual pathways (Frégnac, this volume), transient oscillations in the range of 40-60 Hz found in primary visual areas have been proposed as a plausible physiological substratum mediating rapid coordinated interactions between separate neuronal populations (Gray and Singer, 1989). However the sudden focus that this last finding has raised on rhythm studies in neuronal firing should not make one forget that earlier scenarios in the induction of "synfire chains" (Von der Malsburg, 1981; Abeles, 1982; see also Bienenstock and Doursat, this volume) did not require that cortical cells should oscillate in order to be transiently synchronized. New lines of research, both *in vivo* and *in vitro*, should concentrate on defining more clearly the activity requirements which lead to the fast waxing and waning of functional links in the cortical network, and thus could possibly play a compositional role in the representation of the Whole from that of the Parts.

Abeles, M. (1982) Local cortical circuits. An electrophysiological study. In *Studies of Brain Function*, New-York: Springer-Verlag.

Allman, J., Miezin, F. & McGuiness, E. (1985) Stimulus specific responses from beyond the classical receptive field: neurophysiological mechanisms for local-global comparisons in visual neurons. *Ann. Rev. Neurosci.* **8**, 407-430.

Artola, A., Bröcher, S. & Singer,W. (1990) Different voltage-dependent thresholds for inducing long-term depression and long-term potentiation in slices of the rat visual cortex. *Nature* **347**, 69-72.

Barlow, H.B. (1972) Single units and sensation: a neurone doctrine for perceptual psychology? *Perception* **1**, 371-394.

Bear M.F. & Cooper, L.N. (1989) Molecular mechanisms for synaptic modification in the visual cortex: interaction between theory and experiment. In *Neuroscience and Connectionist Theory*, Eds. M. A. Gluck and D. E. Rumelhart. pp. 65-93. Hillsdale, N.J.: L. Erlbaum Associates.

Brown, T.H., Ganong,A.H., Kairiss, E.W. & Keenan, C.L. (1990) Hebbian synapses: biophysical mechanisms and algorithms. *Ann. Rev. Neurosci.* **1**, 475-511.

Frégnac, Y. & Shulz, D. (in press) Models of synaptic plasticity and cellular learning in mammalian visual cortex. In *Advances in neural and behavioral development.*, Eds. V. Casagrande and P. Shinkman: Neural Ablex Publ.

Frégnac, Y., Smith, D. & Friedlander, M.J. (1990) Postsynaptic membrane potential regulates synaptic potentiation and depression in visual cortical neurons. *Soc. for Neurosci. Abst.* **16**, 798.

Gray, C.M., P. König, A.K. Engel, & W. Singer. (1989) Oscillatory responses in cat visual cortex exhibit inter-columnar synchronization which reflects global stimulus properties. *Nature* **338**, 334-337.

Gray, C.M. & Singer, W. (1989) Stimulus-specific neuronal oscillations in orientation columns of cat visual cortex. *Proc. Natl. Acad. Sci. USA* **86**, 1698-1702.

Hebb, D.O. (1949) *The Organization of Behavior*, New-York: J. Wiley and Sons.

James, W. (1890) *Psychology: briefer course,* Cambridge: Harvard University Press.

Miyashita, Y. (1988) Neuronal correlate of visual associative long-term memory in the primate temporal cortex. *Nature* **335**, 817-820.

Miyashita, Y. & Chang, H.S. (1988) Neuronal correlate of pictorial short-term memory in the primate temporal cortex. *Nature* **331**, 68-70.

Newsome, W.T., Britten , K.H. & Movshon, J.A. (1989) Neuronal correlates of the perceptual decision. *Nature* **341**, 52-54.

Parker, A. & Hawken, M. (1985) Capabilities of monkey cortical cells in spatial-resolution tasks. *J. Opt. Soc. Am. A ssoc.* **2**, 1101-1114.

Peterhans, E. & Von der Heydt, R. (1989a) Mechanisms of contour perception in monkey visual cortex. I. Lines of pattern discontinuity. *J. Neurosci.* **9**, 1731-1748.

Peterhans, E. & Von der Heydt, R. (1989b) Mechanisms of contour perception in monkey visual cortex. II. Contours bridging gaps. *J. Neurosci.* **9**, 1749-1763.

Salzman, C.D., Britten, K.H. & Newsome, W.T. (1990) Cortical microstimulation influences perceptual judgements on motion direction. *Nature* **346**, 174-177.

Von der Heydt, R., & Peterhans, E. (1989) Mechanisms of contour perception in monkey visual cortex. I. Lines in pattern discontinuity. *J. Neurosci.* **9** , 1731-1748.

Von der Malsburg, C. (1981) *The correlation theory of brain function.* Max-Planck Institute for Biophysical Chemistry Goettingen RFA. NTIS, 81-2.

Recent progress toward an understanding of experience-dependent visual cortical plasticity at the molecular level

Mark F. Bear

Center for Neural Science, Box 1953, Brown University, Providence, RI 02912

INTRODUCTION

Our remarkable ability to extract and analyze features of the visual environment depends crucially on the architecture and connectivity of the visual cortex. Research on animals, especially cats and monkeys, over the last 3 decades has shown that the cortical properties on which visual perception depends are not specified at birth, but rather develop postnatally under the influence of visual experience. This was first clearly demonstrated by Wiesel and Hubel; in 1965 they reported that binocular connections in kitten striate cortex could be readily modified by manipulations of the sensory environment. They later determined that this form of visual cortical plasticity was restricted to a "critical period" which, in the kitten, occurs during the second and third postnatal months (Hubel and Wiesel, 1970). This is a period of rapid head and body growth in which plasticity of binocular connections evidently is normally necessary to maintain proper retinal correspondence. It is now generally appreciated that the experience-dependent plasticity of cortical connections during infancy and early childhood is the basis of certain forms of human amblyopia.

The discoveries by Hubel and Wiesel were followed in the 1970's and early '80's by numerous studies of experience-dependent plasticity in the visual system (see reviews by Sherman and Spear, 1982; and Frégnac and Imbert, 1984). This work led to the establishment of various rules that appear to govern visual cortical modifications. For example, it was established that the afferents from the two eyes compete for influence in the visual cortex during the critical period, that the site of this competition is striate cortex, and that this competition requires the postsynaptic activation of cortical neurons. Additional evidence suggested that visual cortical plasticity was also influenced by signals of non-retinal origin that were linked to behavioral state. However, for the most part, these studies were unable to specify the precise mechanisms that underlie synaptic modification in the visual cortex. Fortunately, methodological advances of the last few years have led to a new understanding of the biochemistry of synaptic transmission in visual cortex, and promise to reveal the molecular basis of visual cortical plasticity. Some of our contributions to this rapidly expanding field have been the subject of several recent reviews (Bear and Cooper, 1990; Bear and Dudek, 1990; Bear, 1990, Dykes et al., 1990). Here I shall briefly reiterate the major findings.

MODULATION BY ACETYLCHOLINE AND NOREPINEPHRINE

The classic example of experience-dependent modification of functional circuitry in the striate cortex is the change induced by the temporary closure of one eye (monocular deprivation, MD). Normally, most neurons in area 17 are binocularly activated, and this binocularity can be measured with the physiological assay of ocular dominance (OD) introduced by Hubel and Wiesel (1962). After brief periods of MD, the OD of cortical neurons is shifted dramatically to the eye that remained open - that is, most neurons lose their functional connections to the deprived eye and are activated solely by stimulation through the open eye (Wiesel and Hubel, 1965).

This ocular dominance shift appears to require more than simply an imbalance of afferent activity from the two eyes. Stimulus conditions that are sufficient to activate the sensory (lateral geniculate) afferents but do not activate cortical neurons (diffuse light, for example) will not shift OD. This has led to the idea that postsynaptic activation of cortical neurons is a necessary condition for synaptic modification. However, there is evidence that OD plasticity will not always occur even under conditions where striate cortical neurons are activated robustly by visual stimulation. For example, no OD shift occurs with MD in anesthetized or paralyzed kittens (Freeman and Bonds, 1979; Singer and Rauschecker, 1982). This has led to the concept of an extra-retinal gating signal that depends on the state of the animal and is required for synaptic modifications to occur. Among the candidates examined so far are the projections to striate cortex that arise from the noradrenergic locus coeruleus (Kasamatsu and Pettigrew, 1979), the thalamic intralaminar nuclei (Singer, 1982; Bear et al., 1988) and the cholinergic basal forebrain (Bear and Singer, 1986).

The pioneering work of Kasamatsu and Pettigrew first directly tested the idea that the extrathalamic inputs to the neocortex were important for the modification of circuitry by use. They initially focused their attention on the noradrenergic input that arises from the locus coeruleus in the pons. Their strategy was to destroy this projection and then test for a deficit in the ocular dominance shift that normally results from brief periods of monocular deprivation. They reported that cortical NE depletion prevented ocular dominance plasticity (Kasamatsu and Pettigrew, 1976; 1979). Based on these results, they formulated the "catecholamine hypothesis of visual cortical plasticity" which states that the noradrenergic projection is essential for synapse modification to occur in the striate cortex.

The methods that Kasamatsu and Pettigrew used to deplete NE involved either repeated intraventricular or continuous intracortical administration of the neurotoxin 6-hydroxydopamine (6-OHDA). While these methods have the advantage of restricting the NE depletion to cortex, they both involve the continuous exposure of cortex to 6-OHDA during the period of monocular experience. Bear and Daniels (1983) used another method of catecholamine depletion that had proved effective in rodents, the systemic administration of 6-OHDA to neonates. High pressure liquid chromatography confirmed that the two injections of 6-OHDA given shortly after birth would indeed cause a drastic and long lasting depletion of NE in kitten striate cortex. However, subsequent neurophysiological recording revealed that this method of NE depletion had no detectable effect on the cortical response to MD. Paradiso et al. (1983), in the same laboratory, were able to repeat Kasamatsu and Pettigrew's original experiment by directly infusing the 6-OHDA from an osmotic minipump into visual cortex as the animal was monocularly deprived. Under these conditions, a 50-80% depletion of NE was accompanied by a loss of ocular dominance plasticity (Paradiso et al., 1983).

There are at least two possible explanations for these conflicting results. On the one hand, it is possible that mechanisms exist to compensate for the early destruction of the noradrenergic system. On the other hand, 6-OHDA might directly disrupt plasticity in a way that is independent of the NE depletion. Evidence in favor of the latter explanation has come from the work of Daw and colleagues in St. Louis and of Imbert and colleagues in Paris. Both groups report that methods ranging from locus coeruleus lesions (Adrien et al., 1985) to interruption of the dorsal noradrenergic bundle (Daw et al., 1984) leave visual cortical plasticity intact despite a significant depletion of cortical NE (60-90%). However, this explanation fails to account for the finding that infusion of exogenous NE into the visual cortex apparently will restore susceptibility to lid closure in adult cats (Kasamatsu et al., 1979). Nonetheless, it seems likely that other factors besides NE contribute importantly to critical period plasticity.

The cholinergic projection seemed like a good candidate for 3 main reasons. First, there is biochemical evidence that the striate cortex is heavily innervated by cholinergic fibers during the critical period (Potempska, Skangiel-Kramska and Kossut, 1979; Shaw, Needler and Cynader, 1984). Second, the facilitatory action of ACh on excitatory transmission in striate cortex (Sillito and Kemp, 1983) increases the probability of postsynaptic activation, a condition thought to be required for OD plasticity. Third, the cholinergic projection is thought to be an important component of the ascending reticular activating system, and reticular activation of striate cortex appears to be necessary for OD plasticity (reviewed by Singer, 1982).

The first step in this inquiry was the characterization of the anatomical organization and development of the cholinergic projection to striate cortex (Bear et al., 1985a; b). This work confirmed that striate cortex is heavily innervated by cholinergic fibers during the critical period, and that these fibers arise largely from cells in the basal telencephalon. The next step was to destroy these cholinergic neurons with the excitotoxin N-methyl aspartate, and then test for a deficit in the ocular dominance shift that normally results from MD. Bear and Singer (1986) found that a substantial depletion of cortical ACh had no effect on ocular dominance plasticity. However, when the NE and ACh projections were both destroyed in the same animal, the normal ocular dominance shift was reliably prevented. In these experiments, the NE projection was destroyed with injections of 6-OHDA into the dorsal noradrenergic bundle (DNAB) which normally have no effect on OD plasticity (Daw et al., 1984). Thus, while basal forebrain and DNAB lesions are ineffective alone, the combined destruction of these structures results in a significant loss of OD plasticity in striate cortex.

A second way to deplete both ACh and NE in the striate cortex is to transect the cingulum bundle at the level of the genu of the corpus callosum. Bear and Singer (1986) found that this procedure also prevented the normal OD shift after MD. This result was recently replicated by Gordon et al. (1989) for brief periods of MD.

Bear and Singer also described iontophoretic experiments in the visual cortex which indicated that directly applied 6-OHDA can interfere with cholinergic transmission. Thus, three different methods that retard OD plasticity - *combined lesions of the basal forebrain and DNAB, transection of the cingulum bundle, and intracortical infusion of 6-OHDA* - all apparently disrupt both cholinergic and noradrenergic neurotransmission in the striate cortex. These data suggest that NE and ACh can substitute for one another in the modulation of cortical plasticity, and raise the possibility that they facilitate synaptic modifications via a common molecular mechanism. The precise nature of this mechanism is unknown at present, but it appears to involve the intracellular second messenger signals generated by activation of beta noradrenergic receptors and muscarinic cholinergic receptors.

INVOLVEMENT OF NMDA RECEPTORS

The work on modulators of plasticity leaves unaddressed the question of how specific patterns of retinal activity lead to modifications of synaptic strength in visual cortex. Consider a single cortical neuron receiving input (via the lateral geniculate nucleus) from homotypic points on the two retinae. When patterns of retinal activity in the two eyes occur in register, then the binocular connections in cortex consolidate and strengthen during the critical period. However, when the patterns of retinal activity are brought out of register, either by MD or strabismus, then the binocular connections are disrupted: effective inputs are retained and ineffective inputs are lost, where "effectiveness" is determined by whether an input drives the cortical neuron beyond some threshold. These theoretical considerations have led naturally to the hypothesis that excitatory synapses in visual cortex are consolidated during development when their activity consistently correlates with target depolarization beyond a critical level (Hebb, 1949; Stent, 1973; Bienenstock et al, 1982). What mechanism can support this form of "Hebbian" modification?

One strong candidate is the NMDA receptor mechanism. NMDA receptors are thought to coexist postsynaptically with other types of excitatory amino acid (EAA) receptors (Foster and Fagg, 1985; Stevens and Bekker, 1989). Together, both NMDA and non-NMDA type EAA receptors mediate excitatory synaptic transmission in the kitten striate cortex (Tsumoto et al, 1986). The ionic conductances activated by non-NMDA receptors at any instant depend only on input activity, and are independent of the postsynaptic membrane potential. However, the ionic channels linked to NMDA receptors are blocked with Mg^{2+} at the resting potential, and become effective only upon membrane depolarization (Nowak et al, 1984; Mayer and Westbrook, 1987). Another distinctive feature of the NMDA receptor channel is that it will conduct calcium ions (Dingledine, 1983; McDermont et al, 1986). Hence, the passage of Ca^{2+} through the NMDA channel could specifically signal when pre- and postsynaptic elements are concurrently active. This property led naturally to the hypothesis that synaptic consolidation in the visual cortex occurs when the NMDA receptor-mediated Ca^{2+} flux exceeds some critical level (Bear et al., 1987).

Pioneering work in hippocampus by Collingridge et al. (1983) and others has lent strong support to the idea that NMDA receptor activation is centrally involved in synaptic plasticity. Still, direct tests of the hypothesis that this mechanism contributes to visual cortical plasticity required that NMDA receptors be blocked *in vivo* at the same time that sensory experience is manipulated. Our initial strategy was to infuse the NMDA receptor antagonist D,L-AP5 directly into visual cortex through a chronically implanted cannula for 7 days coincident with MD (Kleinschmidt et al., 1987; Bear et al., 1990). At locations ~6 mm from the infusion site, the extracellular AP5 concentration is estimated to be approximately 130 μM and a normal percentage of neurons are responsive to visual stimulation. At these same sites we found, using single unit recording, that most neurons had retained binocular receptive fields despite the week of MD. This is in sharp contrast to the ocular dominance shift in favor of the non-deprived eye that was observed in the Ringer-treated, control hemispheres.

It should be noted that there is some controversy about the interpretation of the effects of intracortical AP5 infusion. Recently, Miller *et al.* (1989) have shown that infusion of 50 mM AP5 into adult visual cortex can profoundly suppress visual responses at sites close to the infusion cannula. This has led to the question of whether AP5 treatment yields its effects on OD plasticity merely by blocking postsynaptic responses. I would like to stress, however, that the important issue here is not whether AP5 at high concentrations blocks visual responses. Rather, the crucial question is whether the visual responses are blocked at all sites where OD plasticity is disrupted by AP5. The answer to this question is unequivocally negative. At cortical sites 6 mm distant from the infusion cannula, where AP5 clearly disrupts the OD shift after MD, we and Miller et al. (1989) agree that there exists a normal percentage (~90%) of visually responsive neurons. Of course, because NMDA receptors presumably are at least partially blocked at these sites, it would be surprising indeed if the absolute level of visual response was not reduced in many neurons (see, for example, Fox *et al.*, 1989). This fact does not detract from the significance of the results; the data clearly demonstrate that visual responses generated in the absence of sufficient NMDA receptor activation are not sufficient to induce changes in cortical ocular dominance.

We have recently extended the analysis of the effects of intracortical AP5 treatment to include the anatomical consequences of MD (Bear and Colman, 1990). One correlate of the physiological ocular dominance shift is a shrinkage of the neurons in the lateral geniculate nucleus (LGN) that receive input from the deprived eye. It is believed that this change reflects the outcome of binocular competition, probably at the level of layer IV of striate cortex. Using HRP, we labelled the segment of LGN that projects to cortex 6 mm from the site of AP5 infusion. Here we found that there was no change in the size of LGN neurons after MD. However, in the posterior LGN, which projects to a region of visual cortex ~15 mm from the infusion site, we found that deprived LGN neurons were as shrunken as those in control MD kittens. These data provide the first direct evidence that postsynaptic activation of cortical neurons is required for competitive changes in LGN cell size, and suggest a role for NMDA receptors in anatomical as well as physiological plasticity in the mammalian visual system.

Together, these studies demonstrate that blockade of NMDA receptors in visual cortex disrupts the mechanisms of binocular competition. However, from this work alone we cannot specify the precise contribution of NMDA receptors to synaptic plasticity in the visual cortex. Therefore, we and several other labs have turned to a slice preparation of visual cortex with the specific aim of determining the lasting consequences, if any, of NMDA receptor activation (Artola and Singer, 1987; 1990; Kimura et al., 1989; Connors and Bear, 1988; Press and Bear, 1990). The combined results indicate that in visual cortex, like hippocampus, a specific consequence of NMDA receptor activation beyond some threshold level is an enhancement of synaptic efficacy.

The apparent involvement of NMDA receptors in visual cortical plasticity, and the fact that the capacity for plastic changes in area 17 is normally limited to a "critical period" (Hubel and Wiesel, 1970; Olson and Freeman, 1980) suggested that substantial developmental changes might occur in NMDA receptor density in visual cortex. Indeed, Bode-Greuel and Singer (1988) report that NMDA-sensitive [³H]-glutamate binding to visual cortical slices is maximal at 4-6 weeks of age, considered to be the time of maximum sensitivity to visual deprivation. Moreover, Feldman et al. (1990) found a significant decline in NMDA-stimulated ⁴⁵Ca uptake by visual cortical slices prepared from animals older than 5 weeks of age.

Using [³H]MK801 binding, Reynolds and Bear (1989) found that there is an abrupt increase in the density of NMDA receptors between postnatal day 7 and 35 that corresponds to the onset of the critical period. However, while the capacity for visual cortical plasticity is greatly diminished by 12 weeks of age, there is not a corresponding decrease in the density of visual cortical NMDA receptors at this age.

In fact, NMDA receptor density in the adult cortex is only slightly, and not significantly, reduced from the levels found at 5 weeks of age.

These observations lead to the important conclusion that regional and/or developmental changes in the density of NMDA receptors cannot necessarily be taken as evidence for changes in the capabilities for synaptic plasticity, and vice versa. On the other hand, the fact that NMDA receptor density does not decline with increasing age cannot be taken as evidence that they are not involved in visual cortical plasticity. The NMDA receptor, because of its voltage-dependent ionic conductance, can be viewed as a detector of coincident pre- and postsynaptic activity (Mayer and Westbrook, 1987). As mentioned above, there are compelling theoretical arguments for why such a coincidence detector is important for aspects of experience-dependent cortical development (reviewed by Miller, 1990; Bear and Cooper, 1990), and there is direct experimental support for the hypothesis that NMDA receptors play such a role in developmental plasticity in visual cortex and elsewhere (e.g. Bear et al., 1990; Cline et al., 1990; Schmidt, 1990). In adult visual cortex, the same non-linear properties of the NMDA receptor might be utilized for other purposes (i.e. processing of visual information). There is no reason to expect that the same ionic conductance cannot be used both for the mechanisms of visual processing and synaptic plasticity. Viewed in this way, what might distinguish NMDA receptor mechanisms in visual cortex at different ages are the postsynaptic substrates that are activated by Ca^{2+}.

INVOLVEMENT OF EAA-STIMULATED PHOSPHOINOSITIDE TURNOVER

Theoretically, input activity that *fails* to correlate with postsynaptic depolarization should lead to a weakening of synaptic efficacy. Indeed, there is direct evidence that retinal activity promotes synaptic weakening in striate cortex under conditions where cortical neurons are unable to respond normally (Reiter and Stryker, 1988; Frégnac et al., 1988; Bear, 1990). We hypothesize that mechanisms linked specifically to *non*-NMDA receptors support this form of modification.

We have become particularly interested in a type of non-NMDA receptor which is coupled to a phospholipase that hydrolyzes membrane phosphoinositides to form the second messengers diacylglycerol and inositol triphosphate (Sladeczek et al, 1985; Nicoletti et al, 1986; Sugiyama et al, 1987; Dudek et al, 1989). We recently reviewed the pharmacology and development of this receptor in the neocortex (Bear and Dudek, 1990). Briefly, available evidence suggests that PI hydrolysis is stimulated by EAA's primarily at a single receptor site (we call the Q_2 receptor), and that this site is distinct from both the traditional quisqualate (Q_1) receptor and the NMDA receptor. NMDA does, however, inhibit EAA-stimulated PI turnover in visual cortex, confirming that the Q_2 receptor is on cortical neurons. We find that Q_2 receptors in the neocortex are expressed transiently during postnatal development (Dudek et al., 1989). It is particularly noteworthy that the developmental time-course of EAA-stimulated PI turnover correlates precisely with the "critical period" when synaptic modifications are most readily elicited in visual cortex by changes in sensory experience (Dudek and Bear, 1989).

Thus, at the same ages that retinal activity drives synaptic modification in kitten striate cortex, excitatory synaptic activity also stimulates a specific second-messenger cascade in visual cortical neurons. These data are therefore consistent with the hypothesis that Q_2 receptors play a central role in the activity-dependent modification of striate cortex during the critical period. However, direct tests of this hypothesis have been hampered by the lack of a potent and selective antagonist.

One compound that has shown some promise as an inhibitor of EAA-stimulated PI turnover is 2-amino-3-phosphonopropionate (AP3). However, the mechanism of AP3 action appears to be rather complex. For example, Schoepp and Johnson (Schoepp and Johnson, 1989a; Schoepp and Johnson, 1989b) found that the effectiveness of AP3 as an inhibitor of EAA-stimulated PI turnover increases with increasing postnatal age. Also, AP3 appears to exert its inhibitory effects by acting as a partial agonist at the Q_2 receptor (Desai and Conn, 1990).

Despite these limitations, at the present time AP3 remains the most selective pharmacological tool available to test the hypothesis that EAA-stimulated PI turnover plays a specific role in visual cortical plasticity. Therefore, we performed a series of experiments to see if the chronic infusion of AP3 into the visual cortex interferes with the synaptic modifications that are elicited by brief MD during the critical period. We discovered that cortical AP3 treatment in 2-month-old kittens can selectively block the effects of MD in the hemisphere contralateral to the deprived eye (Bear and Dudek, 1990; Dudek et al 1990)

However, we have tried to repeat these experiments under conditions where the effects of MD are more severe, for example in the hemispheres ipsilateral to the deprived eye in younger (4-5 week-old) kittens. So far, the results have been disappointing. Although more work needs to be done, our impression is that the effectiveness of AP3 in blocking the ocular dominance shift after MD varies as a function of age. This perhaps is not surprising considering the age-dependency of AP3's inhibition of EAA-stimulated PI turnover (Schoepp and Johnson, 1989a; Schoepp and Johnson, 1989b).

The interpretation of our results using AP3 is obviously complicated by all the uncertainties concerning the action of AP3 on the Q_2 receptor. There is clearly a need for a pure antagonist of EAA-stimulated PI turnover. Nonetheless, our results are consistent with the hypothesis that the Q_2 receptor is centrally involved in the experience-dependent modification of visual cortex during the critical period. Although it is still too early to determine if the Q_2 receptor contributes to the specific form of modification that we have proposed (Bear, 1988; Bear and Cooper, 1990), preliminary findings suggest that AP3 will block associative long-term synaptic depression in the hippocampus (Chattarji et al., 1990).

CONCLUSION

I divide the problem of visual cortical plasticity into three parts. The first problem concerns the factors that control the onset and duration of the critical period. Although I have not explicitly considered this question in this review, some interesting possibilities include specific patterns of gene expression (e.g. Neve and Bear, 1988) which may be under hormonal control (Daw et al., 1988). The second problem concerns the question of what factors modulate synaptic modification during the critical period. This question was prompted by the observation that many experience-dependent modifications of visual cortex seem to require that animals attend to visual stimuli and use vision to guide behavior (Singer, 1979). And, as reviewed above, the best candidates for "enabling factors" are the neuromodulators acetylcholine and norepinephrine that are released in visual cortex by fibers arising from neurons in the basal forebrain and brain stem. The third problem concerns the question of what controls the direction and magnitude of synaptic modification during the critical period. This is where I believe that the interactions of EAA receptor mechanisms are of crucial importance.

We previously presented a model of synaptic modification that is based on some of the properties of EAA receptors (Bear 1988; Bear and Cooper, 1990). Briefly, we have proposed that whether or not a given pattern of input activity yields an increase or a decrease in synaptic strength depends on the balance between NMDA and non-NMDA (possibly Q_2) receptor activation in the target neuron. Because of the voltage dependency of the NMDA conductance, this balance varies as a function of the postsynaptic response. NMDA receptors are favored when input activity coincides with strong postsynaptic depolarization; under these conditions synaptic efficacy increases. The balance favors non-NMDA receptors when input activity is coincident with relative target inactivity; under these conditions synaptic efficacy decreases.

This simple model closely resembles a formal theory of synaptic modification presented by Bienenstock, Cooper and Munro (1982). According to this theory, input activity yields either increases or decreases in synaptic strength depending on whether it is correlated with postsynaptic activation above or below a critical value termed the "modification threshold" or θ_M. According to the argument presented above, θ_M might represent the critical value of postsynaptic membrane depolarization at which the NMDA and non-NMDA second messenger signals cancel. The similarity of our molecular model with the BCM theory is significant because Clothiaux et al. (1990) have recently demonstrated that this theoretical form of modification, with a few subtle changes, is able to account for the outcome and kinetics of a wide variety of visual deprivation experiments.

One very important feature of the BCM theory is that the value of the modification threshold is not fixed, but rather slides as a non-linear function of the recent history of postsynaptic activity. This gives rise to an obvious question: Is the effectiveness of NMDA receptor activation in triggering a synaptic modification a function of average cortical activity? We are now working on this problem (e.g. Feldman et al., 1990). Our current work is focused on whether synaptic modifications in visual cortex in vitro are influenced by the prior rearing history of the animal, and if so, what molecular mechanisms underlie this change in the "modification threshold".

Acknowledgements: I would like to thank the following individuals for their contributions to the work described in this paper: Serena Dudek, Howard Colman, Amy Bohner, William Press, Dan Feldman, Jon Sherin, Andreas Kleinschmidt, Qiang Gu, Barry Connors, Ford Ebner, Wolf Singer and Leon Cooper. This work was funded by grants from the National Eye Institute and Office of Naval Research.

REFERENCES

Adrien, J., Blanc, G., Buisseret, P., Fregnac, Y. & Gary-Bobo, E. (1985): Noradrenaline and functional plasticity in kitten visual cortex: a reexamination. *J. Physiol.* 367, 73-98.

Artola, A. & Singer, W. (1987): Long-term potentiation and NMDA receptors in rat visual cortex. *Nature* 330, 649-652.

Artola, A. & Singer, W. (1990): The involvement of N-methyl-D-aspartate receptors in induction and maintenance of long-term potentiation in rat visual cortex. *Eur. J. Neurosci.* 2, 254-269.

Bear, M.F. & Daniels, J.D. (1983): The plastic response to monocular deprivation persists in kitten striate cortex after chronic depletion of norepinephrine. *J. Neurosci.* 3, 407-416.

Bear, M.F., Carnes, K.M. & Ebner, F.F. (1985): An investigation of cholinergic circuitry in cat striate cortex using acetylcholinesterase histochemistry. *J. Comp. Neurol.* 234, 411-430.

Bear, M.F., Carnes, K.M. & Ebner, F.F. (1985): Postnatal changes in the distribution of acetylcholinesterase in kitten striate cortex. *J. Comp. Neurol.* 237, 519-534.

Bear, M.F. & Singer, W. (1986): Modulation of visual cortical plasticity by acetylcholine and noradrenaline. *Nature* 320, 172-176.

Bear, M.F., Kleinschmidt, A. & Singer, W. (1988): Experience-dependent modifications of kitten striate cortex are not prevented by thalamic lesions that include the intralaminar nuclei. *Experimental Brain Research* 70, 627-631.

Bear, M.F., Cooper, L.N & Ebner, F.F. (1987): A physiological basis for a theory of synapse modification. *Science* 237, 42-48.

Bear, M.F. (1988): Involvement of excitatory amino acid receptors in the experience-dependent development of visual cortex, In: *Frontiers in Excitatory Amino Acid Research*, ed. E. Cavalheiro, J. Lehman & L. Turski, pp.393-401. New York: Alan R. Liss, Inc.

Bear, M.F. (1990): Involvement of excitatory amino acid receptor mechanisms in visual cortical plasticity. In: *Fidia Research Foundation Symposium Series*, Volume 6, Raven Press, in press.

Bear, M.F. & Colman, H. (1990): Binocular competition in the control of geniculate cell size depends upon visual cortical NMDA receptor activation. *Proc. Nat. Acad Sci. USA*, in press.

Bear, M.F. & Cooper, L.N. (1990): Molecular mechanism for synaptic modification in the visual cortex: Interaction between theory and experiment., In: *Neuroscience and Connectionist Theory*, ed. Gluck M, Rumelhart, pp. 65-93. Livermore, NJ: Lawrence Erlbaum Associates.

Bear, M.F., Kleinschmidt, A., Gu, Q. & Singer, W. (1990): Disruption of experience-dependent synaptic modifications in striate cortex by infusion of an NMDA receptor antagonist. *J. Neurosci.* 10, 909-925.

Bear, M.F. & Dudek, S.M. (1990): Excitatory amino acid stimulated phosphoinositide turnover: Pharmacology, development and role in visual cortical plasticity. *Ann. NY Acad. Sci.*, in press.

Bienenstock, E.L., Cooper, L.N. & Munro, P.W. (1982): Theory for the development of neuron selectivity: Orientation specificity and binocular interaction in visual cortex. *J. Neurosci.* 2, 32-48.

Bode-Greuel, K.M & Singer, W. (1989): The development of N-methyl-D-aspartate receptors in cat visual cortex. *Dev. Brain Res.* 46, 197-204.

Chattarji, S., Stanton, P.K. & Sejnowski, T.J. (1990): 2-amino-3-phosphonopropionate (AP3) blocks induction of associative long-term depression (LTD) in hippocampal field CA1. *Soc. Neurosci. Abs.* 16, 276.20.

Cline, H. T. & Constantine-Paton, M. (1990): NMDA receptor agonist and antagonists alter retinal ganglion cell arbor structure in the developing frog retinotectal projection. *J. Neurosci.* 10, 1197-1216.

Clothiaux, E.E., M.F. Bear & L.N Cooper (1990): A model of visual cortical plasticity with a sliding modification threshold: Comparison of theory and experiment. *Soc. Neurosci. Abs.* 16, 331.12.

Collingridge, G.L., Kehl, S.L. & McLennan, H. (1983): Excitatory amino acids in synaptic transmission in the Schaffer collateral-commisural pathway of the rat hippocampus. *J. Physiol.* 334, 33-46.

Connors, B.W. & Bear, M.F. (1988): Pharmacological modulation of long term potentiation in slices of visual cortex. *Neurosci. Abst.* 14, 298.8.

Daw, N.W., Robertson, T.W., Rader, R.K., Videen, T.O. & Coscia, C.J. (1984): Substantial reduction of cortical noradrenaline by lesions of adrenergic pathways does not prevent effects of monocular deprivation. *J. Neurosci.* 4, 1354-1360.

Daw, N.W., Sato, H. & Fox, K. (1988): Effect of cortisol on plasticity in the cat visual cortex. *Soc. Neurosci. Abs.* 14, 81.11.

Desai, A.J. & Conn, P.J. (1990): Selective activation of phosphoinositide hydrolysis by a rigid analogue of glutamate. *Neurosci. Lett.* 109, 157-162.

Dingledine R. (1983): N-methylaspartate activates voltage-dependent calcium conductance in rat hippocampal pyramidal cells. *J. Physiol.* 343, 385-405.

Dudek, S.M. & Bear, M.F. (1989): A biochemical correlate of the critical period for synaptic modification in the visual cortex. *Science* 246, 673-675.

Dudek, S.M., Bohner, A.P. & Bear, M.F. (1990): Effects of AP3 on EAA-stimulated PI turnover and ocular dominance plasticity in the kitten visual cortex. *Soc. Neurosci. Abs.* 16, 331.6.

Dudek, S.M., Bowen, W.D. & Bear, M.F. (1989): Postnatal changes in glutamate stimulated phosphoinositide turnover in rat neocortical synaptoneurosomes. *Dev. Brain Res.* 47, 123-128.

Dykes, R.W., Tremblay, N. & Bear, M.F. (1990): Cholinergic modulation of synaptic plasticity in sensory neocortex. In: *Activation to Acquisition: Functional Aspects of the Basal Forebrain Cholinergic System*, R.T. Richardson, ed., Birkhäuser, Boston, in press.

Feldman, D., Sherin, J.E., Press, W.A. &. Bear, M.F. (1990): NMDA stimulated calcium uptake by kitten visual cortex *in vitro*. *Exp. Brain Res.* 80, 252-259.

Foster, A. C. & Fagg, G.E. (1985): Amino acid binding sites in mammalian neuronal membranes: Their characteristics and relationship to synaptic receptors. *Brain Res. Rev.* 7, 103-164.

Fox, K., Sato, H. & Daw, N. (1989): The location and function of NMDA receptors in cat and kitten visual cortex. *J. Neurosci.* 9, 2443-2454.

Freeman, R.D. & Bonds, A.B. (1979): Cortical plasticity in monocularly deprived immobilized kitten depends on eye movement. *Science* 206, 1093-1095.

Frégnac, Y. & Imbert, M. (1984): Development of neuronal selectivity in the primary visual cortex of the cat. *Physiol. Rev.* 64, 325-434.

Frégnac, Y., Shultz, D., Thorpe, S. & Bienenstock, E. (1988): A cellular analogue of visual cortical plasticity. *Nature* 333, 367-370.

Gordon, B., Mitchell, B., Mohtadi, E. Roth, E., Tseng, Y. & Turk, F. (1989): Effect of norepinephrine and acetylcholine depletion on plasticity in kitten visual cortex. *Soc. Neurosci. Abs.* 15, 316.16.

Hebb, D.O. (1949) The Organization of Behavior, *John Wiley and Sons*.

Hubel, D.H. & Wiesel, T.N. (1962): Receptive fields, binocular interactions and functional architecture in the cat's visual cortex. *J. Physiol.* 160, 106-154.

Hubel, D.H. & Wiesel, T.N. (1970): The period of susceptibility to the physiological effects of unilateral eye closure in kittens. *J. Physiol.* 206, 419-436.

Kasamatsu, T. & Pettigrew, J.D. (1976): Depletion of brain catecholamines: Failure of ocular dominance shift after monocular occlusion in kittens. *Science* 194, 206-209.

Kasamatsu, T. & Pettigrew, J.D. (1979): Preservation of binocularity after monocular deprivation in the striate cortex of kittens treated with 6-hydroxydopamine. *J. Comp. Neurol.* 185, 139-162.

Kimura, F., Tsumoto, T., Nishigori, A. & Shirokawa, T. (1989): Long-term potentiation and NMDA receptors in the visual cortex of young rats. *J. Physiol.* 414, 125-144.

Kleinschmidt, A., Bear, M.F. & Singer, W. (1987): Blockade of "NMDA" receptors disrupts experience-dependent modifications of kitten striate cortex. *Science* 238, 355-358.

MacDermott, A.B., Mayer, M.L., Westbrook, G.L., Smith, S.J. & Barker, J.L. (1986): NMDA receptor activation increases cytoplasmic calcium concentration in cultured spinal cord neurones. *Nature* 321, 519-522.

Mayer, M.L. & Westbrook, G.L. (1987): The physiology of excitatory amino acids in the vertebrate central nervous system. *Prog. Neurobiol.* 28, 197-276.

Miller, K.D., Chapman, B. & Stryker, M.P. (1989) Visual responses in adult visual cortex depend on N-methyl-D-aspartate receptors. *Proc. Natl. Acad. Sci. USA* 86, 5183-5187.

Miller, K.D. (1990): Correlation-based models of neural development. In: *Neuroscience and Connectionist Theory*. ed. M.S. Gluck & D.E. Rumelhardt, Lawrence Erlbaum, Hillsboro, N.J.

Neve, R.L. & Bear, M.F. (1988): Visual experience regulates gene expression in the developing striate cortex. *Proc. Natl. Acad. Sci. USA.* 16, 331.6

Nicoletti, F., Meek, J.L., Iadarola, M.J., Chuang, D.M. & Roth, B.L. (1986): Coupling of inositol phospholipid metabolism with excitatory amino acid recognition sites in rat hippocampus. *J. Neurochem.* 46, 40-46.

Nowak, L., Bregostovski, Ascher. P., Herbert, A. & Prochiantz, A., (1984): Magnesium gates glutamate-activated channels in mouse central neurones. *Nature* 307, 462-465.

Olson, C.R. & Freeman, R.D. (1980): Profile of the sensitive period for monocular deprivation in kittens. *Exp. Brain Res.* 39, 17-21.

Paradiso, M.A., Bear, M.F. & Daniels, J.D. (1983): Effects of intracortical infusion of 6-hydroxydopamine on the response of kitten visual cortex to monocular deprivation. *Exp. Brain Res.* 51, 413-422.

Potempska, A., Skangiel-Kramska, J. & Kossut, M. (1979): Development of cholinergic enzymes and adenosinetriphosphatase activity of optic system of cats in normal and restricted visual input conditions. *Dev. Neurosci.* 2, 38-45.

Press, W.A. & Bear, M.F. (1990): Effects of disinhibition on LTP induction in slices of visual cortex. *Soc. Neurosci. Abs.* 16, 348.9.

Reiter, H.O. & Stryker, M.P. (1988): Neural plasticity without action potentials: Less active inputs become dominant when kitten visual cortical cells are pharmacologically inhibited. *Proc. Natl. Acad. Sci.* USA 85, 3623-3627.

Reynolds, I.J. & Bear, M.F. (1989): NMDA receptor development in the visual cortex of cats. *Soc. Neurosci. Abs.* 15: 4.7.

Schoepp, D.D. & Johnson, B.G. (1989): Comparison of excitatory amino acid-stimulated phosphoinositide hydrolysis and N-[3H] acytylaspartylglutamate binding in rat brain: Selective inhibition of phosphoinositide hydrolysis by 2-amino-3-phosphonopropionate. *J. Neurochem.* 53, 273-278.

Schoepp, D.D. & Johnson, B.G. (1989): Inhibition of excitatory amino acid-stimulated phosphoinositide hydrolysis in the neonatal rat hippocampus by 2-amino-3-phosphonopropionate. *J. Neurochem.* 53, 1865-1870.

Schmidt, J. T. (1990): Long-term potentiation and activity-dependent retinotopic sharpening in the regenerating retinotectal projection of goldfish: common sensitive period and sensitivity to NMDA blockers. *J. Neurosci.* 10, 233-246.

Shaw C., Needler, M.C. & Cynader, M. (1984): Ontogenesis of muscarinic acetylcholine binding sites in cat visual cortex: Reversal of specific laminar distribution during the critical period. *Dev. Brain Res.* 14, 295-299.

Sherman, S.M. & Spear, P.D. (1982): Organization in visual pathways in normal and visually deprived cats. *Physiol. Rev.* 62, 738-855.

Sillito, A.M. & Kemp, J.A. (1983): Cholinergic modulation of the functional organization of the cat visual cortex. *Brain Res.* 289, 143-155.

Singer, W. (1977): Effects of monocular deprivation on excitatory and inhibitory pathways in cat striate cortex. *Exp. Brain Res.* 134, 508-518.

Singer, W. (1979): Central core control of visual cortex functions., In: Schmitt FO, Worden FG ed. The Neurosciences Fourth Study Program, *MIT Press*, Cambridge, MA, 1093-1109.

Singer, W. (1982): Central core control of developmental plasticity in the kitten visual cortex: I. Diencephalic lesions. *Exp. Brain Res.* 47, 209-222.

Singer, W. & Rauschecker, J. (1982): Central core control of developmental plasticity in the kitten visual cortex: II. Electrical activation of mesencephalic and diencephalic projections. *Exp. Brain Res.* 47, 223-233.

Sladeczek, F., Pin, J.-P., Recasens, M., Bockaert, J. & Weiss, S. (1985): Glutamate stimulates inositol phosphate formation in striatal neurones. *Nature* 317, 717-719.

Stent, G.S. (1973): A physiological mechanism for Hebb's postulate of learning. *Proc. Natl. Acad. Sci.* USA 70, 997-1001.

Stevens, C.F. & Bekker, J.M. (1989): NMDA and non-NMDA receptors are co-localized at individual excitatory synapses in cultured rat hippocampus. *Nature* 341, 230-233.

Sugiyama, H., Ito, I., Hirono, C. (1987): A new type of glutamate receptor linked to inositol phospholipid metabolism. *Nature* 325, 531-533.

Tsumoto, T., Masui, H. & Sato, H. (1986): Excitatory amino acid neurotransmitters in neuronal circuits of the cat visual cortex. *J. Neurophysiol.* 55, 469-483.

Wiesel ,T.N. & Hubel, D.H. (1965): Comparison of the effects of unilateral and bilateral eye closure on cortical unit responses in kittens. *J. Neurophysiol.* 28, 1029-1040.

Synchronous neuronal oscillations in cat visual cortex: functional implications

Charles M. Gray, Andreas K. Engel, Peter Konig and Wolf Singer

Max-Planck-Institute for Brain Research,
Deutschordenstrasse 46, 6000 Frankfurt/M, 71, F.R.G.

Introduction

The elucidation of the neuronal mechanisms underlying the recognition of visual objects is a fundamental problem facing visual neurobiology. Our ignorance of this process is profound and stands in marked contrast to our extensive knowlege of the anatomy and physiology of the mammalian visual system. Three decades of intensive research have revealed a high degree of functional specialization of the visual pathway begining in the retina and continuing throughout the striate and extrastriate cortical areas (Livingstone and Hubel, 1988; DeYoe and Van Essen, 1988, Zeki and Shipp, 1988). This evidence has led to the view that particular categories of visual features are detected and analyzed by anatomically separate and parallel pathways.

Such findings have also, however, led to the notion of the "binding problem". How can the component features of objects be bound together when they are detected and analyzed in spatially separate regions of cortex? This problem is encountered at two different levels of analysis. First, similar features detected at different sites in the visual field must be bound together. Linking the contours of an object is one example of this. Second, binding must also occur among disparate features, for instance, the form, color and movement of an object. Moreover, false conjunctions must be avoided (i.e. the binding of the form of object A with the color of object B). Physiologically, these considerations imply a requirement for specific interactions among feature detecting neurons both within as well as between separate cortical areas.

It is conceivable that the binding of features could depend on the activity of individual neurons receiving convergent input from feature detecting cells at lower levels in the pathway. However, it is difficult to imagine how every possible constellation of features, defining objects and scenes in the real world, could be mapped onto the activity of individual neurons. Such a mapping would lead to a combinatorial explosion in the number of cells needed (Sejnowski, 1986; Malsburg and Singer, 1988) and would require a multiplicative expansion in the number of cellular connections within and between cortical areas.

A well-known alternative proposal suggests that objects may be represented by assemblies of neurons rather than by the activity of individual cells (Hebb, 1949; Edelman, 1978; Malsburg, 1981). In this scheme neuronal assemblies are formed by the transient and selective interaction among a population of cells (Palm, 1982; Gerstein et al., 1989). As changes in the visual image occur neuronal assemblies are thought to form, collapse and reform continuously. Single neurons can, in principle, participate in a multitude of assemblies at different moments in time by dynamically changing the distribution of cells with which they interact. Such a scheme largely avoids the problems associated with high level feature detection and takes advantage of the combinatorial power inherent in the cortical anatomical connectivity (Zeki and Shipp, 1988). Moreover, by enabling such interactions to occur among cells located both within and between different visual areas such a mechanism could potentially solve both forms of feature binding.

The mechanisms for selective assembly formation, however, are not without problems. Foremost among these is the problem associated with the superposition of objects in an image (Malsburg, 1981, 1986; Malsburg and Schneider, 1986; Singer, 1990). If assemblies are formed solely on the basis of enhanced neuronal activity ambiguities arise when a scene contains more than one figure. In such cases a distributed population of neurons become active in response to the features present in the image. However, it becomes impossible to identify which of the many neurons with enhanced responses are coding for the various figures. Separate objects would no longer be distinguishable. These considerations suggest that some mechanism must be present to enable the selective formation of an assembly representing an object and not part of another object or the background. Malsburg (1981) proposed that temporal synchrony among neuronal populations could provide a plausible solution to this problem. According to his model, binding of coherent features within a scene is accomplished by synchronization of feature detecting neurons. In this view segmentation of a visual scene could be accomplished by correlated activity within and asynchrony between different populations of neurons responding to different figures (Malsburg and Schneider, 1986). This temporal coding mechanism, it was suggested, would allow multiple independent assemblies to be formed simultaneously within the same cortical network and thereby avoid false feature conjunctions.

Synchronous Neuronal Responses in Visual Cortex

The first evidence for the prediction that selective temporal correlations among neuronal responses might play a role in cortical information processing came from work on the olfactory system. Freeman and colleagues recorded spatially correlated oscillatory field potential responses with frequencies in the range of 35-90Hz from the olfactory bulb and olfactory cortex of rats, rabbits and cats (for review, see Freeman 1975, 1981, 1985). These findings led to the prediction that similar synchronous oscillatory responses should also exist in neocortical structures. Prompted by such predictions we found that neurons in area 17 of the cat visual cortex exhibit oscillatory responses to optimal visual stimuli in a frequency range of 40-60Hz (Gray and Singer, 1987, 1989). This temporal pattern can be readily seen in the responses of single neurons as well as in the local field potential (Gray and Singer, 1987, 1989; Eckhorn et al., 1988).

Our studies indicate that oscillatory responses can be observed in a large fraction of neurons. Using multi-unit recordings, we found oscillatory firing patterns in 50-70% of the responses recorded from area 17 in the anesthetized cat (Gray and Singer, 1989; Gray et al., 1989; Engel et al., 1990a). We made similar observations in the alert animal (Raether et al., 1989). A subsequent single unit study revealed that mainly the standard complex cells in supra- and infra- granular layers display this type of response (Gray et al.,1990a). Figure 1 shows an example of an oscillatory complex cell. Typically, these cells fire in bursts of two to four spikes which recur at intervals of 15-30ms (Fig. 1D) (Hubel Wiesel, 1965). This rhythmic variation of the firing probability, which we term an "oscillation", is revealed in the periodic modulation of the auto-correlogram of the response (Fig. 1B).

It seems likely that the oscillatory character of these responses can be easily generated at the cortical level through a combination of network interactions and intrinsic membrane conductances (Freeman, 1975; Llinas and Grace, 1989; Silva et al., 1990; Gray et al., 1990a,b). In our studies we found no evidence for 40-60 Hz oscillatory responses in the lateral geniculate nucleus under stimulus conditions designed to evoke optimal responses in cortical neurons (Gray and Singer,1989). However, others have observed rhythmic spontaneous and stimulus-evoked activity in the lateral geniculate nucleus and retina in a similar frequency range (Arnett, 1975; Ariel et al.,1983; Munemori et al., 1984). These latter data suggest that there

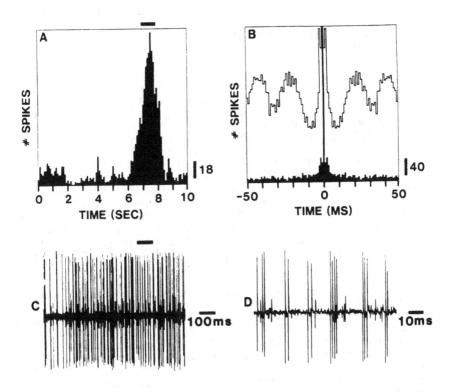

Figure 1. Example of an area 17 standard complex cell which shows an oscillatory firing pattern. (A) Peri-stimulus-time histogram of the response to a light bar of preferred orientation which was moved forward (1-4 sec) and backward (6-9 sec) across the receptive field. The response is selective for the second direction of movement. (B) Auto-correlation histograms computed for the first (filled) and second (open) direction of stimulus movement, respectively. (C) Plot of a spike train recorded during a single stimulus presentation. (D) High resolution plot of the same spike train. Note that the spikes are grouped in bursts which occur at regular intervals. (From Gray et al., 1990a)

may be a significant subcortical contribution to the generation of rhythmic responses in the cortex.

In subsequent studies, we addressed the question of whether synchronization of oscillatory responses across different orientation columns in area 17 could be observed. In order to bind features between different sites in the visual field, such a synchronization should possibly occur over long distances and between cells with nonoverlapping receptive fields. To test this prediction, we recorded with multiple electrodes from area 17 and subjected the data to cross-correlation analysis. We found that in response to visual stimulation oscillatory responses at spatially separate sites would synchronize with little or no phase lag. This response synchronization occurred over distances of up to 7mm within area 17 (Gray et al., 1989; Engel et al., 1990a).

If the recorded cells had overlapping receptive fields, we found response synchronization in a majority of the cases irrespective of the orientation preferences of the recorded cells (Gray et al., 1989; Engel et al., 1990a). If the cells had nonoverlapping receptive fields, synchronization occurred less frequently and was primarily found between cells with similar orientation preferences (Gray et al., 1989; Engel et al., 1990a). Figure 2 illustrates two examples of these long-range interactions. In the first case, (Fig. 2A-C), the two recording sites were separated by 6 mm and the orientation preferences of the cells differed by about 45 degrees (Fig. 2A). As demonstrated by the auto-correlograms, the responses at both sites were oscillatory. However, the cross-correlogram indicates that they were not synchronized (Fig. 2C). In contrast, if the orientation preferences of the recorded cells were similar, we observed a synchronization in about half of the cases tested (Gray et al., 1989; Engel et al., 1990a). An example is given in Figure 2D-F. In this case the recorded cells were separated by 7 mm in the cortex and had almost identical orientation preferences (Fig. 2D). As indicated by the periodic modulation of the cross-correlation histogram (Fig. 2F), the responses were synchronized despite the large inter-electrode distance.

These findings are compatible with the prediction that the synchronization of oscillatory responses may serve to bind similar features in different parts of the visual field. Related studies have demonstrated that oscillatory responses recorded in separate cortical areas (Eckhorn et al., 1988; Engel et al., 1990b) as well as in area 17 of the two hemispheres (Engel et al., 1990b) also exhibit response synchronization. This evidence suggests that such neuronal synchrony may also reflect the binding of different features in an image. Although the mechanisms mediating these synchronous interactions are largely unknown it is likely that tangential intracortical connections serve as the anatomical substrate for response synchronization within a cortical area (Rockland and Lund, 1982; Gilbert and Wiesel, 1983; Martin and Whitteridge, 1984; Luhmann et al., 1986, 1990). However, it is conceivable that intra-areal synchronization is mediated via reciprocal projections from other visual areas (Rosenquist, 1985; Salin et al., 1988; Sporns et al., 1989). These latter connections are also likely to be the substrate underlying the observed inter-areal interactions. Common input from subcortical afferents of the lateral geniculate nucleus can be excluded because the terminal arbors of these fibers do not span sufficiently large distances to account for the long-range interactions observed (Ferster and LeVay, 1978).

If the synchronization of oscillatory responses contributes to the binding of distributed features of objects in the visual field, then it has to be assumed that cross-columnar interactions should not only depend on receptive field properties of the recorded cells, but should change dynamically with variations of the visual stimulus. In the following, we will briefly discuss two experimental paradigms which support the prediction that oscillatory responses should occur and synchronize in a stimulus-specific manner.

In one set of experiments, we investigated whether the occurrence of oscillatory responses within a local group of cells can be affected by changes of the stimulus. In these experiments we compared the responses to optimally oriented light bar stimuli with those recorded using stimuli composed of two light bars having optimal and orthogonal orientations. We then computed auto-correlograms to determine whether the temporal structure of the responses

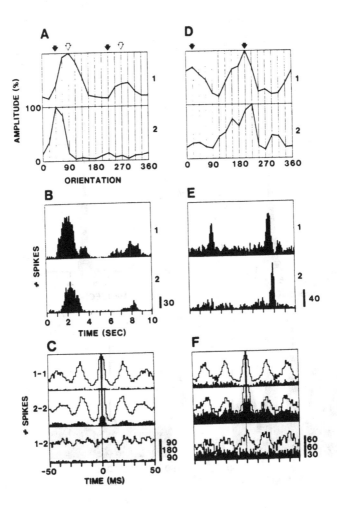

Figure 2. Inter-columnar synchronization of oscillatory responses in area 17. (A) Orientation tuning curves of multi-unit responses recorded at two sites (1,2) separated by 6 mm. The preferred orientation of the two cells differed by about 45 degrees. (B) Peri-stimulus-time histograms of the responses recorded during simultaneous stimulation of each receptive field at its optimal orientation (open and closed arrows in A). (C) Auto- (1-1,2-2) and cross-correlation (1-2) histograms computed for the first (unfilled) and second (filled) directions of stimulus movement. Note that both responses are oscillatory but are not synchronized. (D) Orientation tuning curves of cells at two different recording sites (1,2) separated by 7 mm which had very similar orientation preferences. (E) Peri-stimulus-time histograms of responses recorded with stimuli of optimal orientation (arrows in D). (F) Computation of auto- (1-1,2-2) and cross-correlation (1-2) histograms reveals oscillatory responses at both recording sites which are synchronized with zero phase difference. Unfilled histograms refer to the second direction of stimulus movement.

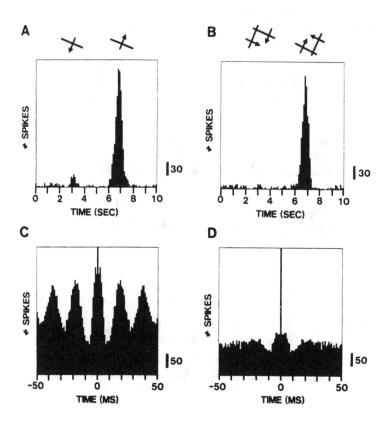

Figure 3. Simultaneous presentation of optimally and orthogonally oriented stimuli reduces the oscillatory modulation of the response. (A) Peri-stimulus-time histogram of the response to a light bar of optimal orientation. (C) Auto-correlation histogram computed for the response to the second direction of stimulus movement. (B) Peri-stimulus-time histogram of the response to the combined stimulus. Note that the overall response amplitude is reduced only slighty. (D) However, the auto-correlation histogram of this response shows only a weak modulation indicating the almost complete absence of oscillatory activity. (From Gray et al.,1990a)

differed between the two stimulus paradigms. The result of one such measurement is illustrated in Figure 3. In this example, the response obtained with the preferred stimulus alone showed a strong oscillatory modulation (Fig. 3C). If a light bar of orthogonal orientation was added to the stimulus configuration, the overall response amplitude was reduced only slightly (Fig. 3B). However, this "conflicting stimulus" altered the temporal structure of the response. As seen in the auto-correlogram, the oscillatory modulation of the response was largely attenuated (Fig. 3D). This occurred even though the orthogonal light bar alone was ineffective in driving the cells. On the average, we observed in our data sample a reduction of the auto-correlogram modulation by about 50% when using the conflicting stimulus paradigm (Gray et al., 1990a). We conclude from this observation that changes of the stimulus can modulate the temporal structure of neuronal responses without affecting the overall response amplitude.

In a second set of experiments, we obtained clear evidence for a stimulus-dependence of cross-columnar synchronization (Gray et al., 1989). The experimental paradigm is illustrated in Figure 4. In the example shown, the cells recorded were separated by 7 mm, preferred vertical stimulus orientations and their receptive fields were aligned co-linearly (Fig. 4B). This enabled us to stimulate the cells either with a single continuous light bar or with two light bars moving in the same or opposite directions (Fig.4B). The differences in the global properties of the stimuli yielded different interactions between the cells. The strongest synchronization was observed with a continuous object (Fig. 4D,III). The interaction was weaker when two independent light bars were used (Fig. 4D,II), and it disappeared when the two stimuli moved in opposite directions (Fig. 4D,I). We conclude from this and related cases that the long-range synchronization within area 17 is sensitive to features of the visual stimulus, in particular to those which reflect the global stimulus coherence, such as continuity of contours and common direction of motion. Thus, the cross-columnar synchronization of oscillatory responses may be particularly suited for the extraction of coherent features of a pattern.

Temporal Properties of Response Synchronization

One of the more impressive aspects of mammalian visual pattern recognition is its rapidity. Psychophysical and behavioral studies have demonstrated that various forms of pattern recognition can be performed within 100-300 ms (Potter, 1975, 1976; Intraub, 1981; Biederman, 1981; Bergen and Julesz, 1983). These findings provide a key temporal constraint on the neuronal mechanisms underlying pattern recognition. They suggest that if oscillatory response synchronization is utilized for the processing of visual information the interactions must be rapid. In fact oscillatory responses of 40-60 Hz should couple within 1-5 cycles of oscillation. Thus, a measure of the temporal properties of the synchronized oscillations becomes an important step in determining their functional significance.

In order to achieve this it became necessary to obtain simultaneous recordings from two clear oscillatory neurons in which the synchronization could be readily observed in the raw data. Figure 5 illustrates an example of such a case. Two neurons of similar orientation preference were recorded from electrodes separated by 1mm in area 17. Their receptive field locations were overlapping but slightly displaced (Fig. 5A). When two light bars of the same orientation and appropriate spatial separation were passed across the receptive fields we could easily evoke responses at both sites which overlapped in time. Moreover, when the bars were initially stationary in the configuration shown in Figure 5A and then moved together, we found that responses could be evoked at roughly the same latency in the two cells after the onset of movement.

This configuration proved useful to address several questions. By examination of the data we could determine 1) the onset latency for oscillations in each cell, 2) the onset latency of the synchrony between the two cells, 3) the time dependent changes in the phase, frequency and duration of the synchrony within individual trials, and 4) the intertrial variation in each of these parameters. As an initial attempt to extract these measures we computed the average auto- and cross-correlation histograms on the two spike trains within two time windows during and after the onset of the responses. The results of these calculations are shown in Figure 5C

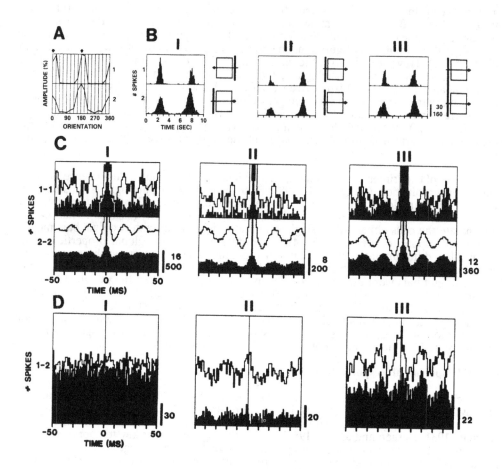

Figure 4. Long-range synchronization in area 17 reflects global stimulus properties. (A) Orientation tuning curves of multi-unit responses recorded at two sites separated by 7mm. At both sites, the responses were tuned to vertical orientations (arrows). (B) Peri-stimulus-time histograms of the responses recorded at each site for three different stimulus conditions: (I) two light bars moved in opposite directions; (II) two light bars moved in the same direction; and (III) one long light bar moved across both receptive fields. A schematic plot of the receptive fields and the stimulus configuration is displayed to the right of each peri-stimulus-time histogram. (C) Auto-correlation (1-1,2-2) histograms computed for the responses recorded at both sites for each of the three stimulus conditions (I-III). (D) Cross-correlation histograms computed for the same responses. Note that the strongest response synchronization is observed with the continuous long light bar. For each pair of correlograms except the two displayed in C (I,1-1) and D (I), the seconf direction of stimulus movement is shown with unfilled bars (From Gray et al.,1989).

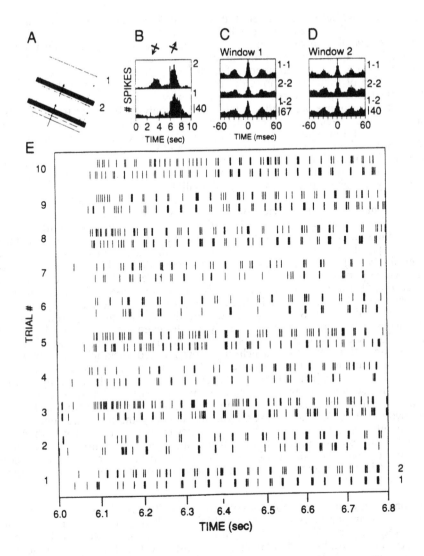

Figure 5. Two cells separated by 1 mm in cortex having the same orientation preference synchronize their oscillatory responses during monocular stimulation. A. Receptive fields of the recorded cells and the spatial placement of visual stimuli. B. Peri-stimulus-time histograms recorded in response to stimulus movment. C,D. Auto- (1-1,2-2) and cross-correlation (1-2) histograms computed from the spike trains recorded during the first (C) and second (D) 500 msec of stimulus movement in the upward direction, respectively. E. Spike trains of the two cells during the second direction of stimulus movement illustrating clear synchronization.

and D. Although the oscillatory character and synchrony of these responses was clearly revealed by this method the time-dependent behavior of the interactions was masked. We found it much more revealing to simply display the spike trains and observe their properties. This revealed a dynamic quality not seen in the average correlograms and one which closely paralleled that observed in our measurements of local field potential correlations (Gray et al., 1991). Close inspection of Figure 5E shows the result. The first 800 msec of the response of each of the two cells to the upward direction of stimulus movement is shown for each of the 10 trials. The onset of the response occurred at a latency of roughly 80-100 msec for each cell. On some trials an additional 50-200 msec was required before the cells began to oscillate clearly (i.e. trials 3,4,5,8 and 10). On other trials, however, the firing pattern of one or both cells was oscillatory at the onset of the response (i.e. trials 1,2,6,7 and 9). In trials 1,2,6 and 7 both cells appear to be oscillating in synchrony from the very start of their responses. The distribution of latencies for the onset of the synchrony shows a marked variability across trials and this is also true of the frequency and phase of the synchronized oscillations.

In order to obtain a more quantitative view of these results we computed the auto- and cross-correlation histograms of the spike trains recorded on each trial. These computations were performed within the same two time windows as shown in Figure 5B. This method revealed the changes in the temporal properties of the correlations within and across separate stimulus trials (Engel et al., 1990a). The results are shown in Figure 6. The histograms displayed in columns 1 and 2 correspond to the data sampled within 6.0-6.5 sec and 6.5-7.0 sec of each trial, respectively (see Fig. 5B for window locations relative to the response). The results displayed within each row show the auto- (A,1-1, B,2-2) and cross-correlation (C,1-2) histograms computed for each trial (1-10). Several properties of the activity are apparent. First, the oscillatory character of the spike trains are clearly visible, even when the correlograms are computed from 500 msec of activity confined to a single trial. Second, the amplitude and frequency of the oscillations and their synchrony show marked variations both within and between trials. For example, within 500 msec the frequency of the oscillations on trial 2 increase from 25 Hz to 35 Hz. Third, there is a marked intertrial variability in the frequency and amplitude of the oscillatory responses. And on some trials (i.e. trial 4) there is little or no indication of an oscillatory response, even though the corresponding spike train in Figure 5E appears to be rhythmic. Fourth, when the variations in frequency and amplitude occur they do so in both cells simultaneously (i.e. compare the frequency change on trial 2 for both channels 1 and 2). Finally, there exist small fluctuations in the phase of the synchronized oscillations. This gives rise to the relatively broad peaks in the correlograms and the changes in phase of the peaks from one trial to the next (i.e. compare the locations of the central peaks in the plots shown in Fig. 6C2).

These results, combined with our previous observations (Engel et al., 1990a), demonstrate that correlated oscillations among pairs of neurons exhibit a high degree of dynamic variability. The amplitude, frequency and phase of the synchronous oscillations fluctuate over time. The onset of the synchrony is variable and bears no fixed relation to the stimulus. Multiple epochs of synchrony can occur on individual trials and the duration of these events also fluctuates from one stimulus presentation to the next. Most importantly, these findings demonstrate that response synchronization can be established on a time scale consistent with behavioral performance on visual discrimination tasks.

Conclusions

In this paper, we have briefly reviewed experimental evidence for a mechanism of feature binding in the visual cortex. Our working hypothesis suggests a solution for the binding problem in the temporal domain. According to this model, coherence in and between feature domains may be encoded by transient synchronization of oscillatory responses. Such interactions, which occur within and between separate cortical areas, may permit a binding of the distributed features of a visual object.

As reviewed in this paper, the available physiological data indicate (i) that oscillatory neuronal responses can be observed in cat visual cortex, (ii) that oscillatory responses can synchronize

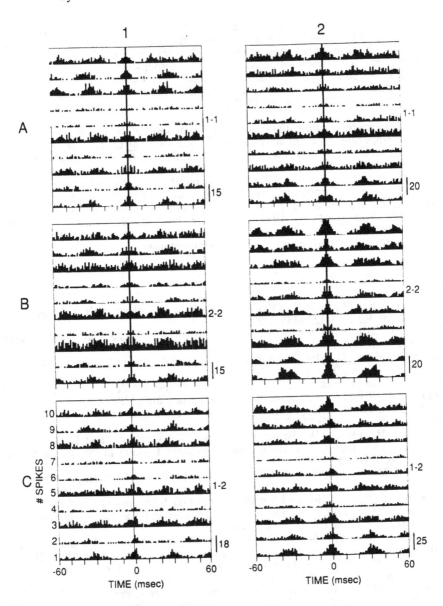

Figure 6. Auto- and cross-correlation histograms computed on individual trials for the spike train data displayed in figure 5. Columns 1 and 2 show the correlograms computed on the data within the time windows 1 and 2 in figure 5, respectively. These times correspond to 6.0-6.5 sec and 6.5-7.0 sec on the peri-stimulus-time histogram displayed in figure 5. Rows A, B and correspond to the auto-correlograms for electrodes 1 (1-1), 2(2-2) and the cross-correlograms between the two signals (1-2), respectively. All 10 trials are displayed within each row. Note the difference in frequency, amplitude and occurrence of rhythmic neuronal activity on different trials as well as at the two different points in time.

across different orientation columns within as well as between separate cortical areas, (iii) that both the occurrence and the cross-columnar synchronization depend on coherent features of the visual stimulus, and (iv) that the oscillatory temporal structure of individual spike trains as well as the synchronization between spatially separate cell groups can be established rapidly and exhibit a high degree of dynamic variability. These results are compatible with the prediction that synchronous oscillations provide a plausible mechanism for feature linking in the visual cortex (Gray et al., 1990b; Singer, 1990).

The data from our cross-correlation studies support the hypothesis that neuronal assemblies are formed by the temporally coherent activation of the participating cells. Since in this model assemblies are not defined by response amplitudes, cells belonging to several different assemblies can be co-activated in the same cortical region and still be distinguished by their temporal relationships (Gerstein et al., 1989). Recent neural network simulations suggest that this mechanism of assembly formation may provide a viable solution for the superposition problem and permit the segmentation of a visual scene (Malsburg and Schneider, 1986; Konig and Schillen, 1990; Sporns et al., 1991). Thus the binding of features by transient synchronization of oscillatory responses can be considered as an attractive proposal to solve the problems of figure-ground segregation and scene segmentation.

References

Ariel,M., Daw,N.W. and Rader,R.K. (1983) Rhythmicity in rabbit retinal ganglion cell responses. Vision Res., 23(12):1485-1493.

Arnett,D.W. (1975) Correlation Analysis of Units Recorded in the Cat Dorsal Lateral Geniculate Nucleus. Exp. Brain Res., 24:111-130.

Bergen, J.R. and Julesz, B. (1983) Parallel versus serial processing in rapid pattern discrimination. Nature 303: 696-698.

Biederman, I. (1981) On the Semantics of a Glance at a Scene. In: Perceptual Organization, M. Kubovy and J.R.Pomerantz (eds.), pp. Erlbaum Associates, Hillsdale.

DeYoe, E.A. and Van Essen, D.C. (1988) Concurrent Processing Streams in Monkey Visual Cortex. Trends Neurosci. 11: 219-226.

Eckhorn, R., Bauer, R., Jordan, W., Brosch, M., Kruse, W., Munk, M. and Reitboeck, H.J. (1988) Coherent Oscillations: A Mechanism of Feature Linking in the Visual Cortex? Biol.Cybern. 60: 121-130.

Edelman,G.M. and Mountcastle,V.B. (1978) The Mindful Brain. Cambridge, Mass., MIT Press.

Engel, A.K., Konig, P., Gray, C.M. and Singer, W. (1990a) Stimulus- Dependent Neuronal Oscillations in Cat Visual Cortex: Inter- Columnar Interaction as Determined by Cross-correlation Analysis. Eur.J.Neurosci. 2: 588-606.

Engel, A.K., Konig, P., Kreiter, A.K. and Singer, W. (1990b) Inter-areal and inter-hemispheric synchronization of oscillatory responses in cat visual cortex. Soc.Neurosci.Abstr. 16:523.1.

Ferster,D. and LeVay,S. (1978) The Axonal Arborizations of Lateral Geniculate Neurons in the Striate Cortex of the Cat. J. Comp. Neurol., 182:923-944.

Freeman, W.J. (1975) Mass Action in the Nervous System. Academic Press, New York.

Freeman,W.J. (1981) A Physiological Hypothesis of Perception. In: Perspectives in Biology and Medicine, University of Chicago, pp. 561-592.

Freeman,W.J. and Skarda,C.A. (1985) Spatial EEG Patterns, Non-linear Dynamics and Perception: the Neo-Sherringtonian View. Brain Res. Reviews, 10:147-175.

Gerstein,G.L., Bedenbaugh,P. and Aertsen,A.M.H.J. (1989) Neuronal Assemblies. IEEE Trans. Biomed. Eng., 36(1):4-14.

Gilbert, C.D. and Wiesel, T.N. (1979) Morphology and Intracortical Projections of Functionally Characterized Neurons in the Cat Visual Cortex. Nature 280: 120-125.

Gray, C.M. and Singer, W. (1987) Stimulus-Specific Neuronal Oscillations in the Cat Visual Cortex: A Cortical Functional Unit. Soc.Neurosci.Abstr. 13: 404.3.

Gray, C.M. and Singer, W. (1989) Stimulus-Specific Neuronal Oscillations in Orientation Columns of Cat Visual Cortex. Proc.Natl.Acad.Sci.U.S.A. 86: 1698-1702.

Gray, C.M., Konig, P., Engel, A.K. and Singer, W. (1989) Oscillatory Responses in Cat Visual Cortex Exhibit Inter-columnar Synchronization Which Reflects Global Stimulus Properties. Nature 338: 334-337.

Gray, C.M., Engel, A.K., Konig, P. and Singer, W. (1990a) Stimulus-Dependent Neuronal Oscillations in Cat Visual Cortex: Receptive Field Properties and Feature Dependence. Eur.J.Neurosci. 2:607-619.

Gray, C.M., Konig, P., Engel, A.K. and Singer, W. (1990b) Synchronization of Oscillatory Responses in Visual Cortex: A Plausible Mechanism for Scene Segmentation. In: Synergetics of Cognition, H. Haken and M. Stadler (eds.), pp. 82-98. Springer, Berlin.

Gray, C.M., Engel, A.K., Konig, P. and Singer, W. (1991) Temporal Properties of Synchronous Oscillatory Neuronal Interactions in Cat Striate Cortex. In: Nonlinear Dynamics and Neural Networks, H.G.Schuster and W. Singer (eds.), in press.

Hebb, D.O. (1949) The Organization of Behavior: A Neuropsychological Theory. Wiley, New York.

Hubel,D.H. and Wiesel,T.N. (1965) Receptive fields and functional architecture in two nonstriate visual areas (18 and 19) of the cat. J. Neurophysiol., 28:229-289.

Intraub, H. (1981) Identification and Naming of Briefly Glimpsed Visual Scenes. In: Eye movements: Cognition and visual Perception, D.F. Fisher, R.A. Monty and J.W. Senders (eds.), pp.. Erlbaum Associates, Hillsdale.

Konig, P. and Schillen, T.B. (1990) Segregation of oscillatory responses by conflicting stimuli - desynchronizing connections in neural oscillator layers. In: Parallel Processing in Neural Systems and Computers, R. Eckmiller, G. Hartmann and G. Hauske (eds.), pp. 117-120. North-Holland, Amsterdam.

Livingstone, M.S. and Hubel, D.H. (1988) Segregation of Form, Color, Movement, and Depth: Anatomy, Physiology, and Perception. Science 240: 740-749.

Llinas, R.R. and Grace, A.A. (1989) Intrinsic 40 Hz Oscillatory Properties of Layer 4 Neurons in Guinea Pig Cerebral Cortex In Vitro. Soc.Neurosci.Abstr. 15: 268.10.

Luhmann, H.J., Martinez-Millan, L. and Singer, W. (1986) Development of horizontal intrinsic connections in cat striate cortex. Exp.Brain Res. 63: 443-448.

Luhmann, H.J., Singer, W. and Martinez-Millan, L. (1990) Horizontal Interactions in Cat Striate Cortex: I. Anatomical Substrate and Postnatal Development. Eur.J.Neurosci. 2: 344-357.

Malsburg, C.v.d. (1981) The Correlation Theory of Brain Function. Internal Report, Max-Planck-Institute for Biophysical Chemistry, Gottingen, West Germany.

Malsburg, C.v.d. (1986) Am I Thinking Assemblies? In: Brain Theory, G. Palm and A. Aertsen (eds.), pp.161-176. Springer, Berlin.

Malsburg, C.v.d. and Schneider, W. (1986) A neural cocktail-party processor. Biol.Cybern. 54: 29-40.

Malsburg, C.v.d. and Singer, W. (1988) Principles of Cortical Network Organization. In: Neurobiology of Neocortex, eds. P.Rakic and W. Singer, pp. 69-99. Wiley, New York.

Martin, K.A.C. and Whitteridge, D. (1984) Form, Function and Intracortical Projections of Spiny Neurons in the Striate Visual Cortex of the Cat. J.Physiol. 353: 463-504.

Munemori, J., Hara, K., Kimura, M. and Sato, R. (1984) Statistical features of impulse trains in cat's lateral geniculate nucleus. Biol.Cybern. 50: 167-172.

Palm,G. (1982) Neural Assemblies - Studies of Brain Functions, Vol.7, New York, Springer.

Potter, M.C. (1975) Meaning in Visual Search, Science 187: 965-966.

Potter, M.C. (1976) Short-Term Conceptual Memory for Pictures. J.Exp.Psychol. 2: 509-522.

Raether, A., Gray, C.M. and Singer, W. (1989) Intercolumnar Interactions of Oscillatory Neuronal Responses in the Visual Cortex of Alert Cats. Eur.J.Neurosci.Suppl.2: 72.5.

Rockland, K.S. and Lund, J.S. (1982) Widespread periodic intrinsic connections of the tree shrew visual cortex. Science 215: 1532-1534.

Rosenquist, A.C. (1985) Connections of visual cortical areas in the cat. In: Cerebral Cortex, Vol. 3, A. Peters and E.G. Jones (eds.), pp. 81-117. Plenum Press, New York.

Salin,P.A., Bullier,J. and Kennedy,H. (1988) Convergence and Divergence in the Afferent Projections to Cat Area 17. J. Comp. Neurol., 283:486-512.

Sejnowski, T.J. (1986) Open Questions About Computation in Cerebral Cortex. In: Parallel Distributed Processing, J.L. McClelland and R.D. Rummelhart (eds.), pp 372-389. MIT Press, Cambridge.

Silva, L.R., Chagnac-Amitai, Y. and Connors, B.W. (1989) Intrinsic Oscillatory Properties of Layer 5 Neocortical Neurons. Soc.Neurosci.Abstr. 15: 268.9.

Singer,W. (1990) Search For Coherence: A Basic Principle of Cortical Self-Organization. Concepts Neurosci. 1: 1-26.

Sporns,O., Gally,J.A., Reeke,G.N. and Edelman,G.M. (1989) Reentrant Signaling Among Simulated Neuronal Groups Leads to Coherency in Their Oscillatory Activity. Proc. Natl. Acad. Sci., 86:7265-7269.

Sporns, O., Tononi, G. and Edelman, G.M. (1990) Modeling perceptual grouping and figure-ground segregation by means of active reentrant connections. Proc.Natl.Acad.Sci.U.S.A., In Press.

Zeki, S. and Shipp, S. (1988) The Functional Logic of Cortical Connections. Nature 335: 311-317.

How many cycles make an oscillation?

Yves Frégnac

Laboratoire de Neurobiologie et Neuropharmacologie du Développement, Bât. 440, University Paris XI, 91 405, Orsay, France

Visual receptive fields have long been considered as spatial windows in the brain, through which a pandemonium of decision making demons (Selfridge and Neisser, 1968) shouted more or less loudly at the appearance of their grandmothers standing out on the horizon . According to the *Neurone Doctrine* (Barlow, 1972), pattern coding was then a more or less complex matter of input / ouput relationships which relate the firing rates, or the sequences of intervals between spikes emitted by single cells, to specific trigger features of the visual stimulus. However most sensory physiologists are still convinced nowdays, at least at the start of their experiments, that by measuring the strength of the visual discharge or the temporal structure of the spike train, they will be able to extract the relevant code which tells them what the neuron "sees". In the visual system of the fly, the *reverse analysis* technique indeed leads to an almost perfect reconstitution of some characteristics of the input signal (such as motion) from the sole knowledge of the time occurence of spikes and "triplets" in the activity of individual central neurons (Ruyter van Steveninck and Bialek, 1988). But in contrast to the large variety of cellular types found in the visual cortex of vertebrates, motion detectors in the blowfly lobula correspond to less than a few tens of identified neurons which span a large part of the visual field, and where most of the integration of the visual message is achieved at the postsynaptic level.

In vertebrate visual cortex, the problem turns into a many-neuron puzzle where most of the integration is carried out by the intracortical network. This network links first- and second-order postsynaptic elements which outnumber by far the actual number of afferent geniculate fibers (the "bottleneck problem" in Barlow, 1981). Laudable attempts have been made to decipher the temporal coding of area 17 single cell activity (Richmond and Optican, 1990, Richmond et al. 1990). Combinatory associative schemes have been proposed in order to construct in more remote cortical areas adaptive grandmothers from selective feature detectors (Rolls, 1989; but see discussion in Sejnowski, 1986). However, although it is admitted that the first step in scene segmentation consists of the extraction of a primal sketch by the parallel processing of various features of the visual input, more and more attention has been given recently to the analysis of the coherence of temporal patterns of activity distributed within and across the many visual cortical areas. Such processes could reflect the dynamics in the formation of functional links relating features between them (review in Singer 1990). Theoreticians have pointed out the importance of phase timing for separating cell assemblies (Hebb, 1949) formed by coactive elements, which otherwise would be indistinguishable and result in a "superposition catastrophy" (Abeles, 1982; Von der Malsburg, 1986). In order to substantiate a reasonable hypothesis for group formation, new functional roles have been attributed to long-distance intracortical connections that violate the retinotopic mapping in the cortical tissue, and which could help in establishing spatial contiguity and coherence binding between cortical feature detectors (Singer, 1990).

In spite of tremendous progress achieved in the time resolution of voltage sensitive dyes, most of the advances made in the understanding of coherent activity in the cortical tissue have been gathered from more conventional techniques, such as EEG, local field potential (Bullock and Basar, 1988) and simultaneous single unit recordings (Gerstein and Perkel, 1969; 1972). Surprisingly, the observations made so far in primary visual areas seem to have stirred out of their dens oscillating dragons that were thought to be forgotten long ago.

Do visual cortical cells sniff?

New trends in the neurophysiological understanding of visual perception are in fact inspired from pilot studies in the olfactory cortex (review in Freeman and Skarda, 1985) where sensory triggered oscillatory activity (or "induced wave" in Adrian (1942)) has been shown to develop during stimulus oriented behavior (sniffing). Similar rhythms have been reported occuring transiently in visual cortex, and have been described mostly in terms of local field potentials (LFP in figure 1 in Gray and Singer, 1989), and to a lesser extent by the firing pattern of single-units: the spectrum of the autocorrelation function - which does not exceed 10 Hz during spontaneous activity - is notably modified during sensory activation, and new energy peaks emerge and are centered around 40-60 Hz (Gray et al., 1989; Eckhorn et al., 1988). However this sudden focus on oscillations in visual cortex might ring bells in the ears of older experimenters, since numerous studies in the 50's and 60's already reported high frequency rhythms arising in response to a flash stimulus (review in Stériade, 1968).

At what frequency do grandmothers oscillate?

The chosen range of frequency, in addition to the similarity of that found in olfactory structures, corresponds to *beta plus* attention rhythms, initially reported in frontal cortex (Bouyer et al., 1981) and more recently described in localized parts of area 18 and (more rarely) 17 in the cat (Rougeul-Buser, personnal communication). It is admitted at least for the frontal cortex that these rhythms originate from thalamic pacemakers (Bouyer et al., 1987). However decisive studies are lacking showing their systematic implication in attentive behavior: EEG studies showed that while *beta* rhythms in frontoparietal cortex appear when the subject focalizes its attention on a prey, lower frequency oscillations develop in the preattentive state when the animal expects some target which is not yet presented (*mû* rhythm: 12 Hz). At the cellular level, beta- or gamma-like rhythmicity was first noted by Singer and Mioche in 1983 during the chronic multi-unit recording of awake cats (Mioche and Singer, 1989). However most data which have been quantitatively reported, were recorded in the anesthetized cat (using halothane). Although it is known that anesthetics such as barbiturates, and possibly synthesis steroids affect the generation of oscillatory behavior, recent voltage sensitive dyes confirm that most of the spectrum of oscillations under pentobarbital anesthesia is restricted below 25 Hz in visual cortex (Arieli et al., 1990). It remains to be clearly established if the 50 Hz oscillations appear only during light anesthesia levels and awake state, and since no significant difference was found between the two latter types of preparations, the direct relevance of oscillatory activity to attention processes and stimulus oriented behavior remains questionable. Nevertheless, let us note that electrical stimulation of the mesencephalic reticular formation, a structure which is thought to play a part in the regulation of arousal states, induces in the anesthetized animal an increase in the amplitude, but also the oscillatory coherence, of both LFP and multiunit responses, when it is paired with visual stimulation (Gray et al., in pressa) .

Another important issue is to understand how specific these oscillations are to intrinsic activity restricted within the cortical tissue. Local field potentials - the origin of which is poorly understood - is the artificial result of the filtering and spatial integration of a largely undefined mixture of intracellular and extracellular signals, which might not be extracted or propagated by the network itself. When restricting the analysis to the single-unit level, 40-60 Hz oscillations have been ascribed mainly to standard complex cells *in vivo* , and are recorded in all layers but to a lesser extent in layer IV (Gray, this volume). Intrinsic oscillatory properties of neocortical neurons reported *in vitro* , which could generate such a range of frequencies, concern layer Vb in somatosensory cortex (Silva et al. , 1989) and layer IV cells in frontal cortex (Llinas and Grace, 1989). In this latter case oscillating cells were identified as GABA-ergic interneurons, with sparsely-spiny stellate cell morphology. The transition to the oscillatory state was triggered once the postsynaptic membrane potential had reached an all-or none plateau potential which generated a

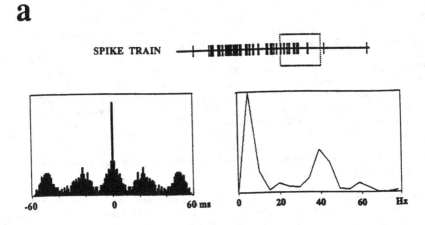

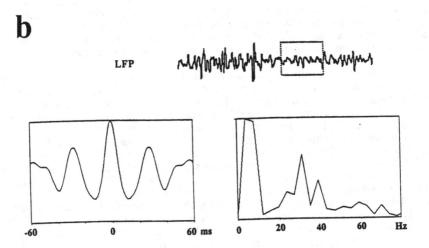

Figure 1: Adapted from Neuenschwander and Varela (1990) with permission.

Autocorrelation functions (left) and corresponding FFT transform (right) of the multiunit activity (spike train, upper row shown in a) and local field potential (LFP, bottom row shown in b) chronically recorded in the optic tectum of an awake adult pigeon. The same temporal window (dotted inset) was used for the analysis of both signals. Oscillatory patterns are seen in both autocorrelograms, but the frequency of the peaks differs in LFP and spike activity, as shown in the corresponding power spectra. Stimulus dependent oscillations could be more easily detected in multi-unit rather than in single unit activity. Occasionally LFP oscillations developed during periods of spontaneous activity.

close-to-fixed frequency oscillation. Fast oscillations could be the result of both network interactions, mostly through recurrent feedback linking antagonist (excitatory/inhibitory) cells (Wilson and Cowan, 1973; Tank, 1990), and intrinsic membrane conductances, although these latter, on their own, tend to produce persistent oscillations in the 6 or 15 Hz range (Stériade et al., 1990; Silva et al., 1990). The cortical tissue would thus possess an interneuronal machinery susceptible to change "bursting" cells into "regular" ones (Mc Cormick et al., 1985). By its action it would transform desynchronized thalamic input into a 40 Hz intracortical modulation. Occasional observations have been made (figure 5 in Gray et al., in pressa) which support simulations of fast-acting loops using negative feedback, based on populations of excitatory and inhibitory neurons oscillating at the same frequency and shifted in phase by 90°. Interestingly, intracellularly imposed oscillations (using a technique developed in thalamus by Yarom and Llinas, 1990), *in vitro* induce a potentiation-like phenomenon sensitizing the conditioned cell to all inputs. Such a mechanism - if put into play by intrinsic oscillations - could offer a way of rapidly and transiently reinforcing previously active connections in the cortical network.

However 50 Hz oscillations are not the apanage of visual cortex: similar rhythms were described in non-visual thalamic nuclei (Bouyer et al., 1981), and also more upstream in the visual pathway: in the LGN (Ghose and Freeman, 1990), and even in the retina (amacrine cells: Dacheux and Raviola, 1986; ganlion cells: Ariel et al., 1983). In a recent comparative study of cellular oscillatory activity between thalamus and cortex, the highest coherence and temporal stability was found in rhythmic generators of the LGN, to a lesser degree in cortical complex cells, and least in some simple cells (Ghose and Freeman, 1990). Neither do oscillations seem to be restricted to the retino-geniculo-cortical pathway, since sensory triggered oscillations have been described in the tectum of awake birds (pigeons) (Neuenschwander and Varela, 1990; see figure 1). They can be observed occasionally independently of visual activation, at least in structures where there exists a tonic level of spontaneous activity (retina, thalamus and to a lesser extent tectum).

Another problem in relating rhythm analysis and perception results from the use of heavy digital filtering which restricts the choice of frequency bandwidth analysis, and possibly masks other potential sources of oscillatory behavior. Lower frequencies of oscillations have been found in the entorhinal cortex in the range of the theta rhythm, and might be also relevant to behavior (Alonso and Llinas, 1989). In the hippocampal formation itt is proposed that the weakening of the inhibition following high frequency trains can lead to the induction of sustained 10 Hz oscillations (Miles and Wong, 1983; Traub et al., 1989). In contrats, most studies in visual cortex eliminate the 10-15 Hz range, which supposedly corresponds to alpha-rythm activity, and very limited information is given on the proportion of the energy spectrum which falls in the 40-60 Hz channel. It should be stressed that in primary visual areas the energy of lower frequency bands (6-25 Hz) is heavily modulated during sensory activity both in anesthetized and awake animals (see figure 2a). These oscillations are not phase-locked to the stimulus since periodicity is lost in trial-shuffled autocorrelograms, and the onset and the duration of the oscillations vary from trial to trial as shown in figure 2. Under Althesin anesthesia (Glaxo, 0.3 ml/kg/hr, i.e. 2.7 mg/kg/hr alfaxalone and 0.9 mg/kg/hr alfadolone acetate), both extracellular and intracellular recordings (figure 3) show sensory triggered oscillations centered on the 8-15 Hz frequency , which confirms previous reports in the literature. These were seen intracellularly both under subthreshold and suprathreshold conditions, and can be explained by a variety of cellular conductances, as demonstrated at the thalamic level (Jahnsen and Llinas, 1984; review in Stériade et al., 1989) or at the cortical level (Walton et al., 1990). More surprisingly, in the course of conditioning experiments to be reported elsewhere, we could observe oscillatory activity in the 8-15 Hz range which developed for specific characteristics of the visual stimulus shown during conditioning, whereas the autocorrelogram for non conditioned characteristics of the visual input was unchanged. Stimulus dependency and trial by trial variability of the 8-15 Hz oscillations were similar to those already documented in the 40-60 Hz range. Figure 4 illustrates differences in the frequency of oscillation in response to opposite contrasts of a bar of a fixed orientation sweeping across the receptive field. Oscillations tended to be more prominent for moving rather than static stimuli.

A plausible network-dependent scheme could be that higher frequency of oscillations needs stronger synaptic activation and higher frequency of postsynaptic firing to be sustained (see

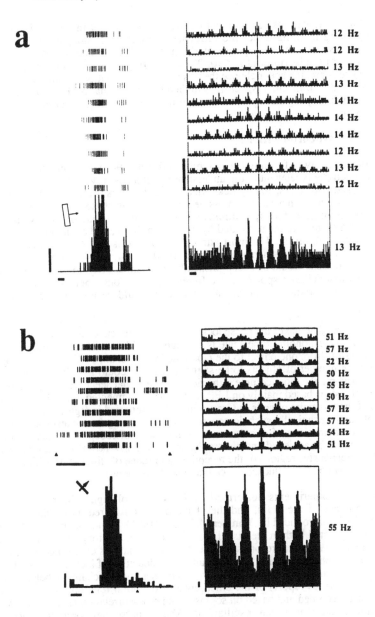

Figure 2: Oscillations and response variability.

Left column: dot display of individual trials and corresponding average PSTHs (time scale 500 ms). Right column: The frequency of oscillation for each trial is indicated on the right side of each individual autocorrelogram computed from the raw data shown on the left (time scale 50 ms).
(a): unpublished data from Debanne, Shulz and Frégnac. This simple cell was recorded in layer VI in a 6 week old normally reared kitten. The mean frequency of oscillation is around 13 Hz.
(b): adapted from figure 10 in Gray et al. (1990). Multiunit activity recorded in the adult cat. The mean frequency of oscillation is around 55 Hz. Oscillatory responses are not time-locked to the stimulus presentation.

also Silva et al., 1989). Simulations indicate that tightly coupled sets of inhibitory interneurons could help phase-locking in firing under the influence of direct excitatory drive (Lytton and Sejnowski, 1990), as has been shown experimentally in the thalamus (Yarom and Llinas, 1990) and the olfactory system (Gelperin et al., 1990; Tank, 1990). The more strongly synaptically activated the cortical cells are, the better chance there is of seeing oscillations of higher frequencies (compare figures 2a and 2b). In this respect, it is difficult to admit that 50 Hz represents some magic number reached during cognitive processes rather than some input-driven frequency limit in the oscillatory behavior.

Oscillation and / or synchronization?

The most obvious way of forming cell assemblies (Hebb, 1949) is to define ensembles of cells whose conjoint activity is relevant to the understanding of a cognitive activity. The rationale developed by Singer and coworkers, and by Eckhorn and collaborators is that oscillatory behavior is used to synchronize groups of neurons. Impressive sets of data, illustrated in the previous paper (Gray, this volume) demonstrate that synchronization in oscillatory behavior can extend across different regions of the cortical tissue separated by more than 7 mm, and even between adjacent visual areas (Engel et al., in press). The most attractive proposal is that synchronization could solve the "binding problem" (Treisman and Gelade, 1980, Crick, 1984, Engel et al., 1990): segmentation of a visual scene would be accomplished by coherent activity within and asynchrony between different cell assemblies responding to different figures. Consequently, synchronization would code global characteristics of the visual input, and would be absent when perceptual coherence is lost.

One of the features that makes the phase locking of 50 Hz attractive "as a mediator of visual awareness" is that the corresponding time scale is compatible with that of recognition processes when attention has to shift from one object to the next (Biederman et al., 1982). Since recognition time increases with the number of objects to be seggregated, one could imagine dynamic phase relationships which become the more complex, the richer is the visual scene. Electrophysiological support for that hypothesis is based mainly on LFP, and is more fragmented at the dual single-unit level (but see one of the two pairs of cells in Gray et al., 1989 (reported in figure 3b) and figure 4 of Gray, in this volume). Technically such demonstration requires very heavy test routines where the complete set of combinations of simultaneous stimulations of receptive fields, or of their surrounding regions has to be achieved and electrophysiologists know how difficult it is to define unambiguously the unresponsive regions of the receptive field (zone of stimulation which does not elicit activity per se but which can influence the central discharge region).

The pionneering experiments performed by Gray and colleagues bear a tremendous attraction for theoreticians of the brain, and they have been considered as validations of the correlation theory initially introduced by Von der Malsburg (1981). A number of neural network models have indeed suggested both in the visual system and in the auditory cortex (e.g. Von der Malsburg and Schneider, 1988) that resonances or coherent oscillations could provide a linking mechanism to construct global perception from simultaneous detection of local features. However it must be noted that no oscillatory behavior is required to form a "synfire chain" (Abeles, 1982) or transient assemblies. Previous reports of the emergence of temporal relations across groups of neurons simultaneously recorded and based on sophisticated crosscorrelation techniques (Aertsen et al., 1989) did not reveal concomitant oscillations (Abeles, personnal communication). The analysis of the sole crosscorrelogram which shows rhythmic activity might be misleading on the interpretation of this oscillation, since, in experiments reported by Gray and colleagues (1989, figure 3b), synchronization appeared in cells whose autocorrelation functions showed already an oscillatory behavior for other stimulus configurations.

A systematic study of simultaneous recordings across different parts of the visual field suggests that synchronization selectively binds cortical loci showing the same functional specificity. This finding agrees with the orientation specificity shown by long-range intracortical horizontal connections which preferentially link isofunctional domains in the adult cortex (Ts'o et al. ; 1986; Luhmann et al., 1986; Gilbert and Wiesel, 1989; Callaway and Katz, 1989). However one should note that most experiments using dual stimulation to test the effects of stimulus

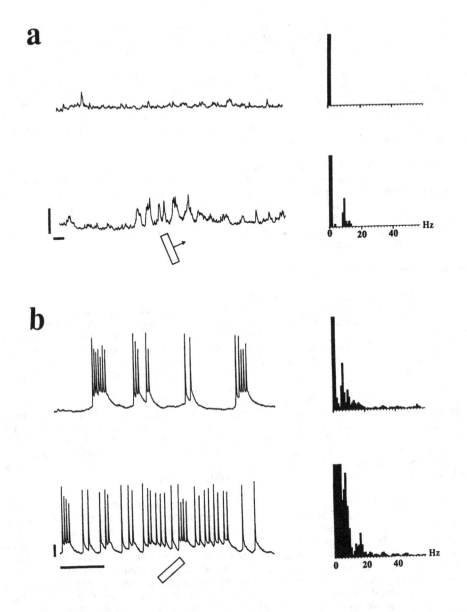

Figure 3: Intracellular subthresholds and suprathresholds oscillations in cat area 17 (unpublished data from Baranyi, Debanne, Shulz and Frégnac).

Left, intracellular recordings (upper line, spontaneous activity; lower line, visual activation). Calibration bars: 100 ms, 10 mV. Right, FFT of the analog signal.
(a): complex cell recorded in the adult cat. A 9 Hz subthreshold oscillation was triggered by the sweep of a oriented stimulus across the receptive field. Note the rapid ascending and descending phases indicative of strong recurrent excitation and inhibition.
(b): complex cell recorded in a 6 week old normally reared kitten. Oscillations shift from 6 Hz to 16 Hz under visual activation (ON) with a static stimulus.

coherence, chose test stimuli pairs consisting of orientation bars which were adjusted to fit with the preference exhibited locally by each recording locus. It would be interesting to simultaneously test different orientation columns with two stimuli of the same orientation (as is the case when the two receptive fields overlap and only one stimulus is used), which has been shown at least in certain cases (figure 11 in Ts'o et al., 1986) to reveal apparent connectivity for a very selective range of orientation preference, intermediate to that encoded by each cortical locus. The question remains as to know whether specific pairs of spatially separated colums get synchronized when presenting a coherent dual-stimulus, the orientation of which is independent of the functional preference of the recorded columns. Interestingly coherent motion does trigger oscillation of some MT neurons (Newsome, personnal communication).

Complexity might add to our understanding when the distance between synchronous loci of oscillatory activity surpasses the normal extent of intracortical connexions. However this contradiction might be only apparent. Simulations have shown that the longer the time delays in intrinsic connectivity, the lower will be the maximum frequency where no phase shifts will occur between spatially separated clusters of cells (Sompolinsky et al. in press; see also Sporns et al., 1989). Dynamic synchrony is probably mostly influenced by activity arising from large number of unseen units ("network plasticity" in Aersten et al., 1989)

How many cycles make an oscillation?

Additional problems arise when looking at the phase relationships across assemblies, since most cells tend to fire on average together (zero phase). How should one deal with superposition of activities arising in different cell assemblies, since the same couple of cells could transiently participate to different functional assemblies during the recognition of one visual stimulus (see Von der Malsburg, 1986)? This problem was solved by a sophisticated but cumbersome estimation of the significance of the correlation by Gray (this volume). Dynamic in phase locking was extracted by assessing how many cycles an oscillation lasts (estimated from LFP after digital filtering between 20 and 100 Hz). Rough calculations suggest that if oscillations are utilized for pattern recognition, only a limited number of cycles (1 to 5) are allowed to reach synchronization (assuming a recognition time within 100-300 ms). Phase locking ranges from + 5 to -5 msec (Gray et al., in pressb). Duration of phase coupling is variable and lasts from 40 to 430 ms. Half of these periods are less than 60 ms (3 cycles), and 90% of them less than 150 ms. Too elusive coupling might become a problem, since one could also expect spurious synchronisation appearing during a minimum number of cycles under the simple assumption that each neuron was triggered by independent jittering oscillators beating at slightly different frequencies, .

In summary, the most plausible scheme that emerges from these recent data is that representation of the visual scene would be a dynamic process, where various properties are combined only transiently, rather than their association being encoded by a grandmother waiting for some demon to be called upon. However more systematic comparative EEG, LFP and spike rate studies, correlated with visual attention (see Livingstone and Hubel, 1981), are needed to clarify the status per se of oscillations. The question stands whether cellular oscillatory behavior is an epiphenomenon, which has been upsetting electrophysiologists for the past 30 years, or should it be considered unambiguously as a modulation carrier for linking cell assemblies. In this latter case, fast oscillations would be ideal as providing a computation cycle basis for the network.

Several issues remain unsolved in justifying the necessity of this modulation carrier during perception. Why do oscillations occur only for moving stimuli rather than static ones (although these are perceived as well)? Why do oscillations do not seem to hold when a physical discontinuity is introduced that does not degrade the perception of the bar as a single object (Peterhans, personal communication)? It could be that oscillations signal only the absence of spatial discontinuities, which for co-oriented detectors is met only when the receptive fields are aligned.

Intrinsic cortical oscillations might be not the only way to form a cell assembly or to put a group of neurons on the alert. Crick (1984) suggested for instance that neurons of the thalamic reticular nucleus could illuminate all visual cortical neurons activated by a given object by a high

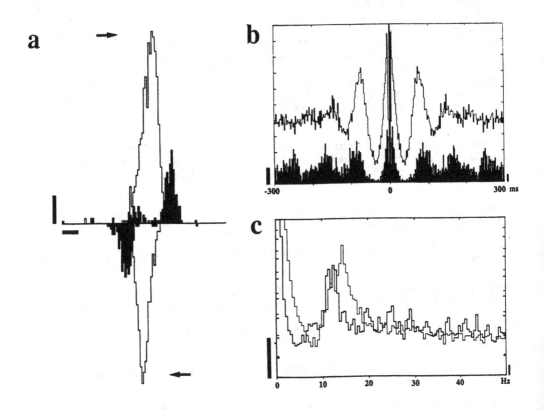

Figure 4: Stimulus dependency of low frequency oscillations (10-15 Hz).
(Unpublished data from Debanne, Shulz and Frégnac)

Simple cell recorded in a 6 week old normally reared kitten in response to a white bar (empty histograms) or to a dark bar (filled histograms). (a) PSTHs for the bar moving in opposite directions (see the respective arrows). (b) Average autocorrelograms for opposite contrasts of the stimulus. (c) FFT of the autocorrelograms shown in (b). A significant difference in the frequency of oscillations is found between the two stimulus conditions.

frequency burst. Other mechanisms could be involved as well, such as slow directed attention potentials which have been described in EEG recordings from the human occipital lobe, and the action of which is unknown at the cellular level (review in Harter and al., in press). Oscillations could also have a much simpler function: it could help cells to be depolarized and increase the probability of the summation of their output signal on common target neurons (Stryker, 1989). Then why not propose that oscillations *per se* are unrelated to the encoding of global versus local features, but are instead the expression of a collective and selective mechanism used to reduce noise level in restricted assemblies of the cortical network (see also Solimpsky et al., in press). It may be ironic to history that the recent boom of oscillatory activity in primary visual areas is exploring again "rhythms of the brain revered by the pioneers of electroencephalography" but which were previously dismissed as irrelevant to neural information processing. With or without oscillations, synchrony remains a key to perception.

Acknowledgement. Research illustrated in figures 2-4 was supported by contract N° ST2J-0416 from the C.E.C and HFSP. I thank Charles Gray, Sergio Neuenschwander and my collaborators Daniel Shulz, Dominique Debanne, Vincent Bringuier for helpful discussions.

Abeles, M. (1982) *Local cortical circuits. An electrophysiological study.* Studies of Brain Function. New-York: Springer-Verlag.

Adrian, E.D. (1942) Olfactory reactions in the brain of the hedgehog. *J. Physiol. (London)* **100**, 459-473.

Aersten, A.M.H.J., Gerstein, G.L., Habib, M.K. and Palm, G. (1989) Dynamics of neuronal firing correlation: modulation of "effective connectivity". *J. Neurophysiol.* **61**, 900-917.

Alonso, A. and Llinas, R.R. (1989) Subthreshold Na+-dependent theta-like rhytmicity in stellate cells of entorhinal cortex layer II. Nature 342, 175-177.

Ariel, M., Daw, N.W. and Rader, R.K. (1983). Rythmicity in rabbit retinal ganglion cell responses. *Vision Res.* **23**, 1485-1493.

Arieli A., Shoham, D., Hildesheim, R. and Grinvald, A. (1990) Oscillations of on-going and evoked activity in neuronal assemblies revealed by real-time optical imaging in cat visual cortex. *Soc. Neurosci. Abst.* **16**, 1220.

Barinaga, M. (1990) The mind revealed? (Research News) *Science* **249**, 856-858.

Barlow, H.B. (1972) Single units and sensation: a neurone doctrine for perceptual psychology? *Perception* **1**, 371-394.

Barlow, H.B. (1981) Critical factors in the design of the eye and visual cortex (The Ferrier Lecture, 1980). *Proc. R. Soc. Lond. B* **212**, 1-34.

Biederman, I., Mezzanotte, R.J. and Rabinowitz, J.C. (1982) Scene perception: detecting and judging objects undergoing relational violations. *Cognitive Psychology* **14**, 143-147.

Bouyer, J.J., Montaron, F.F. and Rougeul, A. (1981) Fast fronto-parietal rhythms during combined focused attentive behavior and immobility in cat: cortical and thalamic localizations. *Electroencephal. Clin. Neurophysiol.* 51, 244-252.

Bouyer, J.J., Montaron, M.F., Vahnee, J.M., Albert, M.P. and Rougeul, A. (1987) Anatomical localization of cortical beta rhythms in cat. *Neuroscience* 22, 863-869.

Bullock, T. and Basar, E. (1988) Comparison of ongoing compound field potentials in the brains of invertebrates and vertebrates. *Brain Res. Rev.* **13**, 57-75.

Callaway, E.M., and L.C. Katz. (1990) Emergence and refinement of clustered horizontal connections in cat striate cortex. *J. Neurosci.* **10**, 1134-1153.

Crick, F. (1984) Function of the thalamic reticular complex: the searchlight hypothesis. *Proc. Natl. Acad. Sci. USA* **81**, 4586-.

Dacheux, R.F. and Raviola, E. (1986). The rod pathway in the rabbit retina: a depolarizing bipolar and amacrine cell. *J. Neurosci.* **6**, 331-345.

Eckhorn, R., Bauer, R., Jordan, W, Brosch, M., Kruse, W., Munk, M. and Reitboeck, H.J. (1988). Coherent oscillations: a mechanism of feature linking in the visual cortex? *Biol. Cybernetics* **60**, 121-130.

Engel, A.K., König, P., Gray, C.M. and Singer, W. (1990). Stimulus-dependent neuronal oscillations in cat visual cortex: inter-columnar interaction as determined by cross-correlation analysis. *Eur. J. of Neurosci.* **2**, 588-606.

Engel, A.K., König, P., Kreiter, A.K., and Singer, W. (in press) Temporal coding by coherent oscillations as a potential solution to the binding problem: physiological evidence. In *Nonlinear dynamics and neural networks*, Eds. H.G. Schuster and W. Singer.

Freeman, W.J. and Skarda, C.A. (1985) Spatial EEG patterns, non-linear dynamics and perception: the neo-Sherringtonian view. *Brain Res. Rev.* **18**, 147-175.

Gelperin, A., Flores, J. and Tank, D.W. (1990) In situ intracellular recordings from procerebral neurons during coherent network oscillations in the olfactory processing system of limax limulus. *Soc. Neurosci. Abst.* **16**, 595.

Gerstein, G.L. and Perkel, D.H. (1969) Simultaneous recorded trains of action potentials: analysis and functional interpretation. *Science* **164**, 828-830.

Gerstein, G.L., and D.H. Perkel (1972) Mutual temporal relationships among neuronal spike trains: statistical techniques for display and analysis. *Biophys. J.* **12**, 453-473.

Ghose, G.M. and Freeman, R.D. (1990) origins of oscillatory activity in the cat's visual cortex. *Soc. Neurosci. Abst.* **16**, 1270.

Gilbert, C.D. and Wiesel, T.N. (1989) Columnar specificity of intrinsic horizontal and cortico-cortical connections in cat visual cortex. *J. Neurosci.* **9**, 2432-2442.

Gray, C.M. and Singer, W. (1989) Stimulus-specific neuronal oscillations in orientation columns of cat visual cortex. *Proc. Natl. Acad. Sci. USA* **86**, 1698-1702.

Gray, C.M., König, P., Engel, A.K. and Singer, W. (1989) Oscillatory responses in cat visual cortex exhibit inter-columnar synchronization which reflects global stimulus properties. *Nature* **338**, 334-337.

Gray, C.M., Engel, A.K., König, P. and Singer, W. (1990). Stimulus-dependent neuronal oscillations in cat visual cortex: receptive field properties and feature dependence. *Eur. J. of Neurosci.* **2**, 607-619.

Gray, C.M., Engel, A.K., Konig, P. and Singer, W. (in pressa) Mechanisms underlying the generation of neuronal oscillations in cat visual cortex. In *Induced Rhytmicities in the Brain.* Eds. T. Bullock and E. Basar.

Gray, C.M., Engel, A.K., Konig, P. and Singer, W. (in pressb) Temporal properties of synchronous oscillatory neuronal interactions in cat striate cortex. In *Nonlinear Dynamics and Neural Networks.* Eds. H.G. Schuster and W. Singer.

Harter, M.R., Miller, S.L., Price, N.J., LaLonde, M.E. and Keyes, A.L. (in press) Neural processes involved in directing attention. *J. Cogn. Neurosci.*

Hebb, D.O. (1949) *The Organization of Behavior*. New-York: J. Wiley and Sons.

Jahnsen, H. and Llinas, R. (1984). Ionic basis for the electroresponsiveness and oscillatory properties of guinea-pig thalamic neurones in vitro. *J. Physiol. (London)* **349**, 227-247.

Livingstone, M.L. and Hubel, D.H. (1981). Effects of sleep and arousal on the processing of visual information in the cat. *Nature* **291**, 554-561.

Llinas, R.R. and Grace, A.A. (1989) Intrinsic 40 Hz oscillatory properties of layer 4 neurons in guinea pig cerebral cortex in vitro. *Soc. Neurosci. Abst.* **15**, 660.

Llytton, W.W. and Sejnowski, T.J. (1990) Inhibitory interneurons may help synchronize firing of postsynaptic cells. *Soc. Neurosci. Abst.* **16**, 468.

Luhmann, H.J., L. Martinez-Millan, and W. Singer. (1986) Development of horizontal intrinsic connections in cat striate cortex. *Exp. Brain Res.* **63**, 443-448.

Mc Cormick, D.A., Connors, B.W., Lightball, J.W. and Prince, D.A. (1985). Comparative electrophysiology of pyramidal and sparsely spiny stellate neurons of the neocortex. *J. Neurophysiol.* **54**, 782-806.

Miles, R. and Wong, R.K.S. (1983) Single neurons can initiate synchronized population discharge in the hippocampus. *Nature* **306**, 371-373.

Mioche, L., and W. Singer. (1989) Chronic recordings from single sites of kitten striate cortex during experience-dependent modifications of receptive-field properties. *J. Neurophysiol.* **62** , 185-197.

Moran, J. and Desimone, R. (1985). Selective attention gates visual processing in the extrastriate cortex. *Science* **229**, 91-93.

Neuenschwander, S. and Varela, F.J. (1990) Sensory triggered oscilaltions in the visual system of birds: possible implications to perception. *Soc. Neurosci. Abst.* **16**, 109.

Richmond, B.J., Optican, L.M. and Spitzer, H. (1990) Temporal encoding of two-dimensional patterns by single units in primate primary visual cortex. I. Stimulus-response relations. *J. Neurophysiol.* **64**, 351-369.

Richmond, B.J., and Optican, L.M. (1990) Temporal encoding of two-dimensional patterns by single units in primate primary visual cortex. II. Information transmission. *J. Neurophysiol.* **64**, 370-380.

Rolls, E. (1989) The representation and storage of information in neuronal networks in the primate cerebral cortex and hippocampus. In *The Computing Neuron,* Eds. R. Durbin, C. Miall and G. Mitchison. pp. 125-159. Wokhingam: Addison-Wesley Publishing Company.

Ruyter van Steveninck, R. de, and Bialek, W. Real-time performance of a movement-sensitive neuron in the blowfly visual system: coding and information transfer in short spike sequences. *Proc. Roy. Soc. London B* **234**, 379-414.

Sejnowski, T.J. (1986) Open questions about computation in cerebral cortex. In *Parallel Distributed Processing*, vol. 2, Eds. JL. Mc Cleland and D.E. Rumelhart. pp. 372-389. Cambridge (MA): MIT Press.

Selfridge, O.G. and Neisser, U. (1968) Pattern recognition by machine *Sci. American* **203**, 60-68.

Silva, L.R., Chagnac-Amitaj, Y. and Connors, B.W. (1989) Intrinsic oscillatory properties of layer 5 neocortical neurons. *Soc. Neurosci. Abst.* **15**: 660.

Singer, W. (1990) Search for coherence: a basic principle of cortical self-organization. *Concepts in Neurosci.* **1**, 1-26.

Sompolinsky, H., Golomb, D. and Kleinfeld, D. (in press) Global processing of visual stimuli in a neural network of coupled oscillators. *Proc. Natl. Acad. Sci. USA.* **87**, 7200-7204.

Sporns, O., Gally, J.A., Reeke, G.N. and Eselman, G.M. (1989) Reentrant signaling among simulated neuronal group leads to coherency in their oscillatory activity. *Proc. Natl. Acad. Sci. USA,* **86**, 7265-7269.

Steriade, M. (1968) The flash-evoked afterdischarge (Review Article). *Brain Res.* **9**, 169-212.

Steriade, M., Jones, E.G. and Llinas, R. (1990) *Thalamic oscillations and signalling.* Neurosciences Institute Publication, New York, John Wiley and Sons.

Stryker, M.P. (1989) Is grandmother an oscillation? *Nature* **338**, 297-298.

Tank, W. (1990). Computations performed with oscillatory dynamics in invertebrate and vertebrate olfactory systems. In *Neural Computation.* Soc. for Neuroscience pp. 53-66.

Treisman, A.M. and Gelade, G. (1980) A feature-integration theory of attention. *Cogn. Psychol.* **12**, 97-136.

Traub, R.D., Miles, R. and Wong, R.K.S. (1989) Model of the origin of rhythmic population oscillations in the hippocampal slice. *Science* **243**, 1319-1325.

Ts'o, D.Y., C.D. Gilbert, and T.N. Wiesel. 1986. Relationships between horizontal interactions and functional architecture in cat striate cortex as revealed by cross-correlation analysis. *J. Neurosci.* **6** , 1160-1170.

Von Der Malsburg, C. 1981. *The correlation theory of brain function.* Max-Planck Institute for Biophysical Chemistry Goettingen RFA. NTIS, 81-2.

Von der Malsburg, Ch. and Schneider, W. (1986) A neural cocktail-party processor. *Biol. Cybernetics.* **54**, 29-40.

Von der Malsburg, ch. (1986) Am I thinking assemblies? In *Brain Theory.*, Eds: G. Palm and A. Aertsen, pp. 161-176.

Yarom, Y. and Llinas, R. (1990). Intracellular autostimulation of in vitro guinea-pig thalamic neurons (Th) utilizing a hardware bio-electric reentry system. *Soc. Neurosci. Abst.* **16**, 955.

Walton, K.D., Yarom, Y. and Llinas, R. (1990). Intrinsic subthreshold 10-50 Hz membrane oscillations in interneurons in the fourth layer of the frontal cortex. *Soc. Neurosc. Abst.* **16**, 1134.

Wilson, H.R. and Cowan, J.D. (1972) Excitation and inhibitory interactions in localized populations of model neurons. *Biophys. J.* **12**, 1-24.

Elements of form perception in monkey prestriate cortex

E. Peterhans and R. von der Heydt

Department of Neurology, University Hospital Zürich, Frauenklinikstr. 22, CH-8091 Zürich, Switzerland

INTRODUCTION

A critical step in form vision is the definition of object borders. In situations of spatial occlusion, for example, borders imply discontinuities in depth, but on the retina they correspond to a variety of image features, typically to discontiuities in color, luminance, texture, motion or binocular disparity. Contrast borders are represented at the first stage of cortical processing, in striate cortex (Hubel & Wiesel, 1968), but relatively little is known about the neural processing of borders defined by color contrast, texture discontinuity, or differences in motion and binocular disparity. In perception it has been shown that orientation discrimination for anomalous contours and contrast lines is similar (Vogels & Orban, 1987), and the two types of contour suffer similar tilt aftereffects (Smith & Over, 1975; Paradiso et al., 1989). Further, contours defined by stereoscopic depth produce geometrical illusions as contrast lines do (cf. Julesz, 1971). With bars defined by anomalous contours parallel visual search is possible (Gurnsey & Humphrey, 1990) as with bars defined by luminance contrast (Treisman, 1988). However, orientation as a dimension of visual coding plays a more general role than contour processing implies - orientation may also represent axes of objects (Johansson, 1973; Marr & Nishihara, 1978). A recent study by Westheimer (1989) supports this view. He found that anomalous contours as well as other oriented elements like ellipses, rows of dots, and even lines defined by a single oscillating dot are subject to tilt illusions, just as contrast lines are. These findings suggest that the visual system generalizes for oriented elements, and that these elements are processed by similar mechanisms.

We have investigated these aspects of form perception in the visual cortex of the alert monkey. In striate cortex neurons generally required continuous contrast borders for response, whereas in prestriate cortex we found neurons that seemed to generalize for oriented elements. They signalled the orientation of contrast borders as well as anomalous contours, dotted lines, or elongated figures defined by the coherent motion of dots.

METHODS

Only a brief description of our methods shall be given here; for details see von der Heydt & Peterhans (1989). We worked with alert rhesus monkeys trained to fixate their gaze on a small visual target that reinforced foveal viewing. During the periods of fixation we presented visual stimuli and studied the receptive fields of cortical neurons. The stimuli were generated by means of analog and digital circuits on a high resolution x-y oscilloscope (Ferranti 7/21) and presented to the animal via a stereoscope. The stimulus luminance was 10 cd/m^2 in early experiments and later 36 cd/m^2, added to a uniform background of the same luminance. Spike trains were recorded on instant film and stored by computer for immediate displays and off-line analysis. The microelectrode tracks were reconstructed histologically from series of sections stained alternatingly for Nissl substance and cytochrome oxidase.

ANOMALOUS CONTOURS

In situations of spatial occlusion objects are perceived even if they have the same color, texture or luminance. They are seen as entities, bounded completely by contours. Such contours have been called "anomalous contours" because they lack contrast, but appear as sharp "subjective" or "illusory" contours in perception (see Kanizsa, 1979, and Petry & Meyer, 1987 for reviews). Examples are shown in Fig. 1.

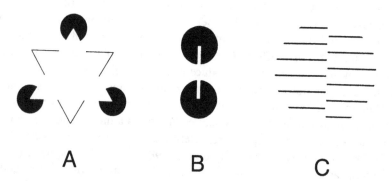

A   B   C

Fig. 1. "Illusory contours". A triangle (A), a vertical bar (B), and a vertical edge (C) are perceived although these shapes are not defined by contrast borders. Such "anomalous contours" are common in every-day vision, especially in situations of spatial occlusion (cf. Petry & Meyer, 1987, for a review). (A) is Fig. 4.2 of Kanizsa (1979).

We studied the responses to anomalous-contour stimuli in areas V1 and V2. Striate neurons generally failed to respond to anomalous contours; they preferred continuous contrast borders for response. By contrast, about one third (55/150) of the neurons of area V2 that were selective for the orientation of contrast borders also responded to anomalous contours (Peterhans & von der Heydt, 1989b; von der Heydt & Peterhans, 1989). We called these "contour neurons". Figure 2 shows an example. The neuron preferred long dark bars with oblique orientation (A), but was also activated by stimulus (B) that perceptually produced an illusory bar (cf. Fig. 1B). Anomalous contours between abutting line-gratings (C) evoked similar responses. In general, both types of anomalous-contour stimuli were effective, or neither of them. The responses of contour neurons were reduced or abolished in stimulus conditions that destroyed the illusion. For example, half of stimulus (B) failed to activate these neurons as did line-gratings without discontinuity (not shown).

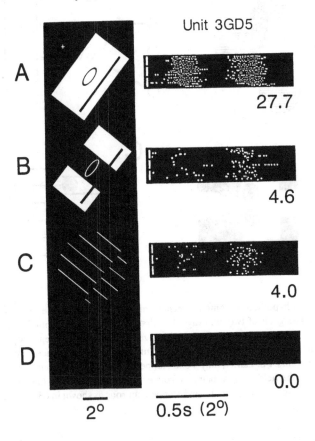

Fig. 2. Responses of a neuron of area V2 that signalled illusory contours. (A) shows the responses to a dark bar of the preferred orientation. An ellipse indicates the response field, a cross marks the fixation point of the monkey. The neuron also signalled the illusory bar (B) and the contour defined by abutting line-gratings (C). The displays show activity during 24 cycles of stimulus movement (forth and back sweep on the left and right halves, respectively). Each dot marks an action potential, the mean numbers per motion cycle are written underneath. The spontaneous activity was zero (D).

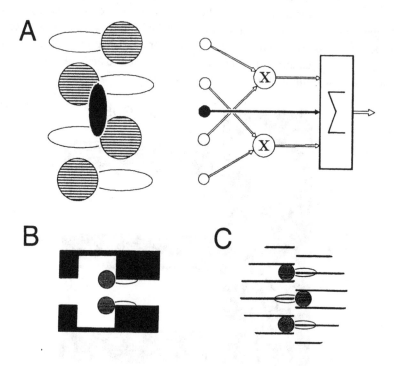

Fig. 3. Model of a hypothetical contour mechanism. (**A**) The contour neurons of V2 (Σ) are thought to sum the signals of two pathways. The first, with a simple or complex receptive field (vertical ellipse), enables detection of lines and edges; the second, with end-stopped receptive fields (horizontal ellipses with hatched areas indicating inhibitory end-zones), enables detection of anomalous contours. For clarity, only 4 end-stopped fields are shown, but they are thought to overlap and cover the field. Distant pairs are connected by gates (x). The line-ends and corners of illusory-contour figures stimulate the end-stopped neurons as shown in (**B**) and (**C**).

Neural mechanism

We have suggested a model that explains the responses of contour neurons of V2 (Peterhans et al., 1986). The circuit is shown in Fig. 3A. We assume that each neuron sums (Σ) the signals of two input pathways, one activated by contrast borders, the other by anomalous contours. The first pathway depends on the activity of neurons with simple or complex receptive fields (vertical ellipse), the second combines signals of neurons with end-stopped receptive fields (horizontal ellipses with hatched disks symbolizing inhibitory end-zones) that are aligned along the length axis of the field of the first input and oriented perpendicular to it. Only four fields are shown, but they are thought to be densely packed. Distant pairs are connected by gates (x) such that one half of the stimulus of Fig. 3B fails to produce a signal at the output. Line-gratings without discontinuity also fail to activate this circuit because they stimulate the inhibitory end-zones.

The role of end-stopped cells: The model raises the question whether end-stopped cells can be activated by occlusion features such as line-ends and corners, and if so, how frequent are asymmetric fields? In their original description Hubel & Wiesel (1965, 1968) noted that end-stopped cells responded to short, but not to long lines or edges, and that some preferred corners. We studied end-stopped cells from this point of view and found that they nearly always responded to line-ends or corners. Often, these responses were equally strong as those to short stimuli (Peterhans & von der Heydt, 1987). About half of the cells studied quantitatively (32/61) had asymmetric fields. They gave strong responses when the terminating line or edge covered one end-zone, and less than half of that response or none when it covered the other, thus indicating asymmetry of strength of end-inhibition. An example is shown in Fig. 4. The neuron preferred edges; responses to different stimulus lengths are shown in (A). The optimum length was 0.8 deg, further lengthening reduced the response. The inset on the right shows the edge of optimal length (dark rectangle) in relation to the receptive field; the dotted line indicates the variation of length up to 12 deg. The open ellipse represents the response field for the optimal edge, the hatched disk the inhibitory end-zone (size arbitrary). (B) shows the responses to a 2 deg long edge in different positions of the field. This stimulus produced the strongest response in the upper half of the field, when its lower corner just covered the response field (b), but virtually no response in the corresponding position of the lower half of the field (a), and in the center. This indicates that the receptive field was asymmetric with a single inhibitory zone at the lower end. Such neurons can detect the direction of terminations.

The results suggest that end-stopped cells do respond to terminations of lines and edges and that some, those with asymmetric fields, signal the direction of terminations. Therefore, in situations of spatial occlusion, end-stopped neurons can detect occlusion cues. In this function they may contribute to the representation of anomalous contours according to the model outlined above, thus enabling a more stable perception of occluding contours.

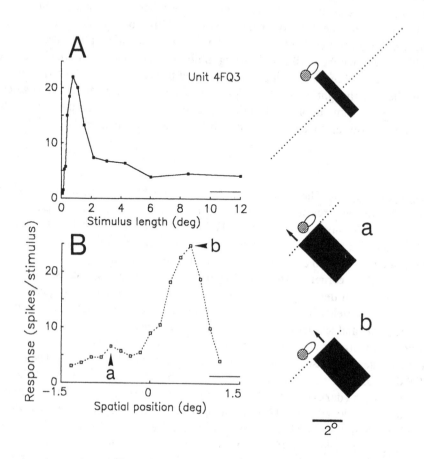

Fig. 4. End-stopped cell of V2. The neuron preferred short oblique edges; the responses for different lengths are shown in (**A**). The level of spontaneous activity is marked by a horizontal line. The inset on the right shows the edge of optimal length (0.8 deg) in relation to the receptive field. The ellipse indicates the excitatory region, the hatched disk the inhibitory end-zone (size arbitrary). The dotted line marks the extent of the length variation. (**B**) shows the responses to a 2 deg long edge in different positions of the field. The stimulus produced the strongest response when its lower corner covered the response field (**b**). Virtually no response was evoked by the upper corner in the corresponding position (**a**), thus indicating that the field was asymmetric. Such neurons can serve the detection of occlusion cues.

SHAPES FROM DOT-STIMULI

Human observers often recognize shapes from dotted lines just as well as from solid lines, as shown for the triangle of Fig. 5A, B. However, with a dotted background, the triangle disappears, it is camouflaged (C). It becomes visible again when the dots forming the outline are moved in synchrony and relative to the background. We have studied these aspects of form perception in areas V1, V2 and V3-V3A.

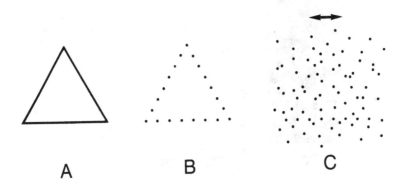

Fig. 5. Form perception tolerates a high degree of abstraction. Line drawings are often sufficient to recognize the form of an object, and solid and dotted lines can be equivalent (**A, B**). With a dotted background the object is camouflaged (**C**), but it can be perceived if the outlining dots move together (< - >), relative to the background.

While continuous contrast borders were effective stimuli in striate neurons, interrupted borders were not (Peterhans & von der Heydt, 1989c). For example, neurons that preferred bright bars often gave weaker responses or none to dashed bars of the same mean luminance. Even a single gap of 2-8 min arc in the center of a solid line could abolish the response. Similar effects have been observed in cat striate cortex (Hammond & McKay 1983, 1985). Thus, gaps seem to be encoded in striate cortex. By contrast, in areas V2 and V3-V3A, we often found neurons that responded equally well or even better to dotted lines than to bars, even when the mean luminance of the dot-stimulus was much lower than that of the bar (Peterhans & von der Heydt, 1989c). They were similarly selective for the orientation of dotted lines as they were for solid lines. Stimulus conditions that vitiated the perception of a rigid, oriented element also reduced or abolished the responses of these neurons. Figure 6 shows an example. An oscillating narrow bar (A) evoked similar responses as 5 dots in a row moving back and forth in synchrony (B). When the dots moved with the same amplitude, but out of phase (C), the responses were strongly reduced. About a third (33/92) of the neurons of V2 and about half (17/31) of those of V3-V3A showed this behavior.

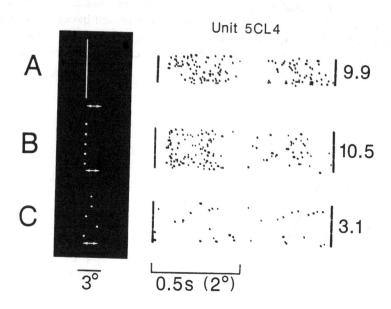

Fig. 6. Equivalence of bar and row of dots in an orientation selective cell of area V3-V3A. The neuron gave similar responses to a moving bar (**A**) and a row of dots moving back and forth in synchrony (**B**). The responses were abolished when the dots moved with the same amplitude but out of phase (**C**). For illustration, the dots (1 min arc) have been enlarged. Conventions as for Fig. 2, except that only 16 motion cycles are shown in (**A**).

Two kinds of mechanism could produce this result, one that relies on the spatial arrangement (collinearity) of dots, and another that depends on the synchronous movement (coherent motion) of dots. We have separated the two mechanisms experimentally and found that many neurons relied on coherent motion, and only few on collinearity.

Lines defined by collinearity

An example of a neuron sensitive to collinearity is shown in Fig. 7. It responded to solid bright lines and to stationary or moving rows of dots. When the dots were still moved in synchrony but misaligned within a range of ± 2 min arc the response strength was reduced to about half, larger displacements abolished the response. This result was not simply due to changing the width and mean contrast of the stimulus. In this and other neurons we tested the effect of stimulus width and contrast separately and found that these parameters were not critical.

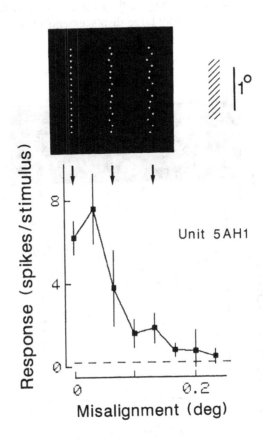

Fig. 7. Sensitivity to collinearity. Responses of an orientation selective neuron of area V2 for which solid and dotted lines were equivalent. When the dots were misaligned (but still moving in synchrony) the responses were reduced. The ranges of misalignment are indicated on the abscissa scale; within each range the dots were distributed evenly. The dots were 1 min arc in size; the hatched rectangle shows the size of the response field as plotted with a solid line. Note that misalignments as small as ± 2 min arc (center stimulus) reduced the response to half the maximum, larger displacements abolished the response.

Shapes from coherent motion

Neurons sensitive to coherent motion, on the other hand, tolerated considerable displacements of the dots, provided the dots moved in synchrony, thus producing a rigid, elongated figure in perception. Figure 8 shows an example. A row of dots produced similar responses as a solid bar (cf. Fig. 5), and when the dots were misaligned the responses remained virtually unchanged up to displacements of ± 30 min arc. Only large displacements of ± 1 deg or more reduced or abolished the responses.

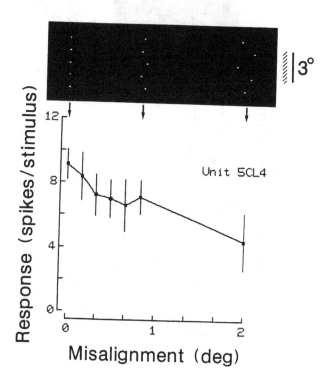

Fig. 8. Neurons sensitive to coherent motion tolerated large misalignments of dots. The Figure shows the responses of cell 5CL4 of Fig. 6. The neuron gave similar responses to solid and dotted lines, and to stimuli with misaligned dots as long as the dots moved in synchrony and the misalignment did not exceed ± 1 deg. The spatial configuration of the dots was changed after each fixation period such that, on the average, the dots were evenly distributed within the ranges indicated on the abscissa. Note the difference in scale when comparing with Fig. 7. These neurons seem to encode the orientation of the elongated shapes perceived by human observers.

These neurons also segregated dotted lines from dotted backgrounds by motion. Figure 9 shows an example. Five dots in a row (size enlarged for ease of discrimination) produced a response when moving relative to the background that was stationary (A), or moving in antiphase (B). The neuron remained silent when both the background and the row of dots were kept stationary (C) or moved together in phase (D). This result parallels perception - in condition (A) and (B) one perceives a rigid line element, but not in (C) and (D). Also, the orientation selectivity for such lines defined by motion was similar as for lines defined by luminance contrast (not shown). This fits in well with the psychophysical findings of Regan (1989) who showed that orientation discrimination is similar for motion-defined bars and for bars defined by luminance contrast.

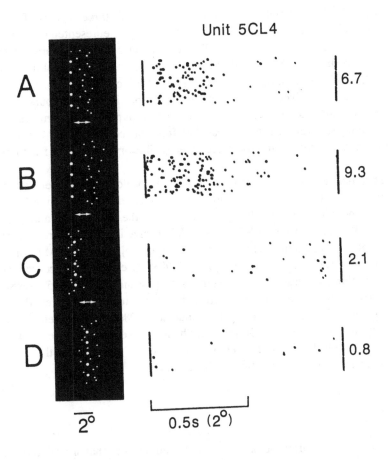

Fig. 9. Sensitivity to coherent motion. The Figure shows again responses of cell 5CL4. Five dots in a row produced a response when they moved in synchrony and relative to a dotted background that was stationary (**A**), or moving in antiphase (**B**). The neuron failed to respond when the row and the background moved in phase (**C**) or were stationary (**D**). This result parallels perception since a vertical row is perceived in conditions (**A**) and (**B**), but not in (**C**) and (**D**). All dots were 1 min arc in size, for ease of discrimination the dots of the row have been enlarged, conventions otherwise as in Fig. 2.

ANATOMICAL SEGREGATION

It has been shown that in area V2 of the monkey different aspects of vision are processed in separate regions that can be identified anatomically (Hubel & Livingstone, 1987; Zeki & Shipp, 1988). When stained for the enzyme cytochrome oxidase area V2 shows a pattern of alternating thin and thick dark stripes separated by pale stripes. The thin dark stripes have been proposed to process color, the pale stripes form, and the thick stripes stereoscopic depth and movement. Our results in the alert monkey revealed a segregation of function in accordance with this scheme (Peterhans & von der Heydt, 1989a).

We found selectivity for orientation of lines and edges in all three regions of V2, though less frequently in the thin stripes. Also line-ends and corners were represented in all parts of V2. On the other hand, we found neurons that signalled anomalous contours or dotted lines only in the pale and in the thick, but not in the thin stripes (thought to process color). This result has analogues in perception (cf. Gregory & Heard, 1989). With continuous borders (line drawings, cartoons) shapes are perceived even at equiluminance, but illusory contours and surfaces are lost, or nearly so, and fractured images that require grouping of distant elements become virtually meaningless. Our results suggest that form processing based on color borders is anatomically separated from form processing based on illusory contours or fractured images.

Neurons that responded to the synchronous movement of dots and weaker or not at all when the dots moved out of phase were found in areas V2 and V3-V3A. In V2, these neurons concentrated in the thick stripes; only few were found in the pale stripes, and virtually none in the thin stripes. The thick stripes of V2 and area V3 receive input from direction selective cells of lamina 4B of V1, are anatomically connected, and project to area MT (or V5), an area involved in motion processing (cf. Zeki & Shipp, 1988). Area MT is concerned with direction and speed of motion (Maunsell & Van Essen, 1983; Movshon et al., 1985; Newsome & Paré, 1988), foreground-background segregation from motion (Allman et al., 1985) and 3-dimensional structure from motion (Siegel & Anderson, 1986), but MT-neurons seem to show little selectivity for stimulus shape. On the other hand, the neurons that we studied in areas V2 and V3-V3A showed little direction selectivity. Rather, our data suggest that these regions of prestriate cortex use motion information for form analysis.

CONCLUSIONS

In prestriate cortex of the alert monkey we have found neurons that signalled certain elements of form perception that require mechanisms of processing we could not identify in striate cortex.

In area V2, we found neurons that were selective for the orientation of contrast borders as well as anomalous contours and lines defined by collinear dots. While neurons of V1 required continuous contrast borders for response, some neurons of V2 seemed to generalize for oriented borders irrespective of the actual stimulus pattern producing these borders. These neurons were found in the pale and in the thick cytochrome-oxidase stripes of V2, but not in the thin stripes thought to process color. This result parallels perception since illusory contours and shapes from fractured images cannot be perceived at equiluminance. Other prestriate neurons were sensitive to coherent motion. They signalled the orientation of rigid, elongated shapes produced by the synchronous movement of groups of dots. They also segregated dotted lines from dotted backgrounds and were orientation selective for such motion-defined lines as for contrast lines. These neurons concentrated in the thick dark stripes of V2 and in V3-V3A suggesting that motion processing in these regions of prestriate cortex involves form analysis.

Thus, some neurons in monkey prestriate cortex seem to generalize for oriented elements that can be defined by a variety of image features such as contrast border, anomalous contours, collinear dots or groups of coherently moving dots. We suggest that these neurons play a basic role in form vision. Some may contribute to the encoding of object borders, others to the encoding of axes of objects, or object elements, as has been proposed by studies of perception and theories of visual object representation.

REFERENCES

Allman, J., Miezin, F., and McGuinness, E. (1985): Direction- and velocity-specific responses from beyond the classical receptive field in the middle temporal visual area (MT). Perception 14: 105-126.

Gregory, R. L., and Heard, P. (1989): Some phenomena and implications of isoluminance. In Seeing Contour and Colour. eds J. J. Kulikowski, C. M. Dickinson, and I. J. Murray, pp. 725-728. Oxford: Pergamon Press.

Gurnsey, R., and Humphrey, G. K. (1990): Preattentive discrimination of subjective contours. Invest. Ophthalmol. Vis. Sci. 31: 107.

Hammond, P., and MacKay, D. M. (1983): Influence of luminance gradient reversal on simple cells in feline striate cortex. J. Physiol. (Lond) 337: 69-87.

Hammond, P., and MacKay, D. M. (1985): Influence of luminance gradient reversal on complex cells in feline striate cortex. J. Physiol. (Lond) 359: 315-329.

Hubel, D. H., and Livingstone, M. S. (1987): Segregation of form, color, and stereopsis in primate area 18. J. Neurosci. 7: 3378- 3415.

Hubel, D. H., and Wiesel, T. N. (1965): Receptive fields and functional architecture in two nonstriate visual areas (18 and 19) of the cat. J. Neurophysiol. 28: 229-289.

Hubel, D. H., and Wiesel, T. N. (1968): Receptive fields and functional architecture of monkey striate cortex. J. Physiol. (Lond) 195: 215-243.

Johansson, G. (1973): Visual perception of biological motion and a model for its analysis. Percept. Psychophys. 14: 201- 211.

Julesz, B. (1971): Foundations of Cyclopean Perception. Chicago: University of Chicago Press.

Kanizsa, G. (1979): Organization in Vision. Essays on Gestalt Perception. New York: Praeger.

Marr, D., and Nishihara, H. K. (1978): Representation and recognition of the spatial organization of three-dimensional shapes. Proc. R. Soc. Lond. B 200: 269-294.

Maunsell, J. H. R., and Van Essen, D. C. (1983): Functional properties of neurons in middle temporal visual area of the macaque monkey. I. Selectivity for stimulus direction, speed, and orientation. J. Neurophysiol. 49: 1127-1147.

Movshon, J. A., Adelson, E. H., Gizzi, M. S., and Newsome, W. T. (1985): The analysis of moving visual patterns. In Pattern Recognition Mechanisms. eds C. Chagas, R. Gattass, and C. Gross, pp. 117-151. Vatican City: Pontifical Academy of Science.

Newsome, W. T., and Paré, E. B. (1988): A selective impairment of motion perception following lesions of the middle temporal visual area (MT). J. Neurosci. 8: 2201-2211.

Paradiso, M. A., Shimojo, S., and Nakayama, K. (1989): Subjective contours, tilt aftereffects, and visual cortical organization. Vision Res. 29: 1205-1213.

Peterhans, E., and von der Heydt, R. (1987): The role of end-stopped receptive fields in contour perception. In New Frontiers in Brain Research: Proceedings of the 15th Göttingen Neurobiology Conference. eds N. Elsner, and O. Creutzfeldt, p. 29, Stuttgart: Thieme.

Peterhans, E., and von der Heydt, R. (1989a): Elements of form perception in monkey V2 - A correlation with the cytochrome oxidase pattern. Soc. Neurosci. Abstr. 15: 161.

Peterhans, E., and von der Heydt, R. (1989b): Mechanisms of contour perception in monkey visual cortex. II. Contours bridging gaps. J. Neurosci. 9: 1749-1763.

Peterhans, E., and von der Heydt, R. (1989c): The whole and the pieces - cortical neuron responses to bars and rows of moving dots. In Seeing Contour and Colour. eds J. J. Kulikowski, C. M. Dickinson, and I. J. Murray, pp. 125-130. Oxford: Pergamon Press.

Peterhans, E., von der Heydt, R., and Baumgartner, G. (1986): Neuronal responses to illusory contour stimuli reveal stages of visual cortical processing. In Visual Neuroscience. eds J. D. Pettigrew, K. J. Sanderson, and W. R. Levick, pp. 343-351. Cambridge: Cambridge University Press.

Petry, S., and Meyer, G. L. (1987): The perception of illusory contours. New York: Springer.

Regan, D. (1989): Orientation discrimination for objects defined by relative motion and objects defined by luminance contrast. Vision Res. 29: 1389-1400.

Siegel, R. M., and Andersen, R. A. (1986): Motion perceptual deficits following ibotenic acid lesions of the middle temporal area (MT) in the behaving rhesus monkey. Soc. Neurosci. Abstr. 12: 1183.

Smith, A., and Over, R. (1975): Tilt aftereffects with subjective contours. Nature (Lond) 257: 581-582.

Treisman, A. (1988): Features and objects: The forteenth Bartlett memorial lecture. Quart. J. Exp. Psychol. A 40: 201-237.

Vogels, R., and Orban, G. A. (1987): Illusory contour orientation discrimination. Vision Res. 27: 453-467.

von der Heydt, R., and Peterhans, E. (1989): Mechanisms of contour perception in monkey visual cortex. I. Lines of pattern discontinuity. J. Neurosci. 9: 1731-1748.

Westheimer, G. (1989): Line orientation considered as a separate psychophysical domain. In Neural Mechanisms of Visual Perception. eds D. M. K. Lam, and C. D. Gilbert, pp. 237-246. The Woodlands: Portfolio Publishing Company.

Zeki, S., and Shipp, S. (1988): The functional logic of cortical connections. Nature (Lond) 335: 311-317.

ACKNOWLEDGEMENTS

We thank Dorothea Weniger for help with the English and Peter O. Bishop for stimulating discussions. This research was supported by SNF grant 3.939.84.

Manipulating perceptual decisions by microstimulation of extrastriate visual cortex

William T. Newsome, C. Daniel Salzman, Chieko M. Murasugi and
Kenneth H. Britten

Department of Neurobiology, Stanford University School of Medicine,
Stanford, California 94305, U.S.A.

Neurons in mammalian visual cortex are typically analyzed and classified with respect to a number of physiological properties such as selectivity for orientation, direction of motion, wavelength and retinal disparity. Implicit in such classification schemes is the notion that physiological properties provide clues to the functional role played by each neuronal class in visual perception. Thus the classes of neurons mentioned above would figure prominently in our perception of form, motion, color and depth, respectively.

While this notion is attractive, it has been notoriously difficult to test in a rigorous manner. One satisfying test would involve quantitative measurement of perceptual effects while the responses of a specific class of neurons are manipulated by an investigator. Unfortunately, such manipulations are not straightforward since neurons with different physiological properties can be intermixed within a single visual area in a highly intricate manner (e.g. Livingstone and Hubel, 1984). However, relative homogeneity of physiological properties can be found at a finer scale - that of the cortical column. For instance, an orientation column in striate cortex consists largely of orientation selective neurons that respond preferentially to a common orientation (e.g. Hubel and Wiesel, 1977). Similarly, a direction column in the middle temporal visual area (MT, or V5) contains direction selective neurons that share a common preferred direction (Zeki, 1974; Albright, *et al.*, 1984).

It is therefore possible, in principle, to activate selectively a physiologically homogeneous population of neurons by exciting a single cortical column with trains of weak electrical pulses delivered with an appropriately located microelectrode. Clearly, though, it is an open question whether activation of one, or even a few, cortical columns from among the hundreds or thousands available in a given area of visual cortex can have effects that are measurable at the perceptual level. This outcome may not be unreasonable, however, since the number of columns devoted to analyzing any specific portion of the visual field is substantially smaller than the total number in a given visual area.

We have now employed this strategy to test directly the role of direction selective cortical neurons in perceptual judgements of motion direction. All experiments were carried out in alert rhesus monkeys trained to report the direction of motion in a near-threshold motion display. While the monkeys performed the direction discrimination task, we stimulated columns of direction selective neurons in the middle temporal visual area (MT), an extrastriate area that plays a prominent role in processing motion information (e.g. Zeki, 1974; Van Essen *et al.*, 1981; Albright, 1984 ; Newsome and Pare, 1988; Newsome *et al.*, 1989). In many cases microstimulation caused substantial changes in the monkey's judgements of motion direction. When such effects occurred, the monkey's judgements were virtually always biased towards the preferred direction of the stimulated neurons. Thus microstimulation appeared to enhance the intracortical signal related to a particular direction of motion, and the monkey responded to that signal in a meaningful way in the context of the direction discrimination task. The results demonstrate that physiological properties measured at the neuronal level can be causally related to a specific aspect of perceptual performance.

METHODS

The experiments were conducted in three alert rhesus monkeys (*Macaca mulatta*). Prior to the start of the experiments, each monkey was surgically implanted with head stabilizing hardware, a scleral search coil for measuring eye movements (Robinson, 1963), and a recording cylinder which permitted microelectrode access to MT. The monkeys were seated in primate chairs for daily experimental sessions which lasted from 3-6 hours each. The monkeys were returned to their home cages following each experimental session.

Electrophysiological recording and microstimulation. Our electrophysiological methods were based on those of Wurtz and his colleagues (e.g. Mikami *et al.*, 1986). Commercially obtained tungsten microelectrodes (Microprobe, Inc.) were used; uninsulated tip exposures varied between 10-30 μm. Electrodes were inserted into MT using a hydraulic microdrive mounted on the recording cylinder, and the multiunit signal was inspected with an oscilloscope and an audio monitor.

Electrical microstimulation was administered using a biphasic pulse generator in series with an optical stimulus isolation unit (Bak Electronics). We employed trains of biphasic stimulating pulses, cathodal phase leading (each phase was .2 ms in duration, and the two phases were separated by a .1ms interval). Unless stated specifically in the Results, current amplitude was 10 μA (20 μA peak-to-peak) and train frequency was 200 Hz. We used 10 μA currents in the initial series of experiments because previous studies in primate motor cortex indicated that such currents stimulate neurons within 85 µm of the microelectrode tip, approximating the width of a cortical column (Stoney *et al.*, 1968). We chose a frequency of 200 Hz because it approximated the best responses to visual stimuli that we have observed in MT neurons.

<u>Selection of microstimulation sites.</u> Each experiment began with selection of a microstimulation site in MT. Our electrodes generally travelled at an oblique angle through the cortex in MT, and we qualitatively identified the receptive field boundaries, preferred direction of motion and preferred speed for multiunit clusters of neurons at regular intervals along each penetration. We selected a site for microstimulation when the receptive field and preferred direction remained constant for at least 150 µm of electrode travel; sites were frequently 200-400 µm wide. We then positioned the electrode tip in the middle of the site and proceded with psychophysical testing.

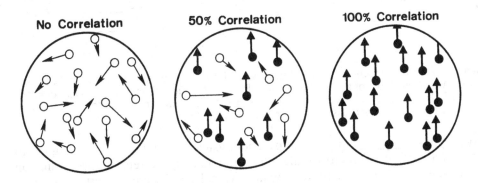

Figure 1. A schematic representation of the motion display employed in the present study. The display consisted of a stream of randomly positioned dots, each of which appeared briefly and was then replaced by a partner dot. In the 0% correlation state (left panel), replacement dots appeared at randomly selected screen locations. In this state, the display consisted of random motion noise with no net motion signal in any direction. In the 100% correlation state (right panel), replacement dots appeared with a constant offset in space and time relative to their partners. Thus each dot provided a unidirectional motion signal that was consistent in direction and speed across the entire display. In the 50% correlation state (center panel), half of the dots provided random motion noise while the remaining half provided a unidirectional motion signal. The proportion of correlated dots could be set to any desired level by the computer software. Reprinted with permission from Newsome and Pare, 1988.

<u>Visual stimuli.</u> The monkeys were trained to report the direction of motion in a dynamic random dot display presented on an oscilloscope. In the display, a specifiable proportion of the dots moved in a single direction while the remaining dots moved in random directions. The unidirectional motion signal could be varied from a strong signal which the monkey discriminated perfectly to a very weak signal for which the monkey failed to perform above chance. In the right hand panel of Fig. 1, for example, each dot in the display moved in the same direction, a state that we refer to as the 100% correlated state. This state provided a robust motion signal which generally elicited perfect performance

from the monkeys. At the other extreme, motion in the display could be completely random as illustrated for the 0% correlation state in the left hand panel of Fig. 1. The display could assume any intermediate state between these two extremes. In the center panel of Fig. 1, for example, half of the dots carried the unidirectional motion signal while the remaining dots comprised random motion noise (50% correlation state).

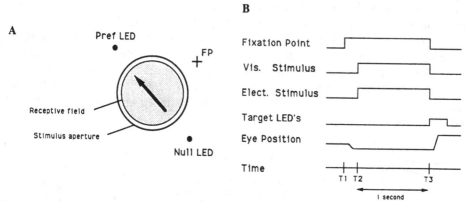

Figure 2. A schematic representation of the psychophysical procedures employed in the present study. **A.** Spatial layout of the fixation point (FP), receptive field, visual stimulus aperture, and target LED's. **B.** Temporal sequence of events in a single microstimulation trial. See text for details. Reprinted with permission from Salzman *et al.*, 1990.

Psychophysical procedures. The psychophysical procedures were designed to measure the effect of microstimulation on the monkey's direction discrimination performance. We measured psychophysical performance with the two-alternative, forced choice method illustrated in Fig. 2 (after Newsome and Pare, 1988). The monkey was required to fixate a point of light (FP) while viewing the eccentrically located motion display. The motion display was confined within a circular aperture (thick circle - Fig. 2A) placed directly over the multi-unit receptive field (thin circle) of the neurons at the stimulation site. The dimensions of the aperture were matched to the dimensions of the receptive field, and the speed of the motion signal was set equal to the preferred speed of the neurons at the stimulation site. On each trial, motion could occur in the neurons' preferred direction (arrow) or in the direction 180 degrees opposite (null direction), and the monkey was required to report which direction of motion occurred. In this manner we attempted to tailor the demands of the psychophysical task to the information likely to be supplied by neurons at the stimulation site.

Figure 2B illustrates the temporal sequence of events in each trial. The fixation point appeared at time T1, and the monkey transferred its gaze to establish fixation ("Eye Position" trace). After a brief delay, the motion display ("Vis. Stimulus") appeared at time T2 and remained on for one second. At the end of the one second viewing period, the fixation point and the motion display disappeared, and two target LED's appeared ("Pref LED" and "Null LED" - Fig. 2A) corresponding to the two possible directions of motion. The monkey indicated its judgement of motion direction by making a saccadic eye movement to the corresponding LED. Correct judgements were rewarded with a drop of water or juice for both stimulated and non-stimulated trials; incorrect judgements were punished with a brief time-out period. A PDP 11/73 computer monitored the monkey's eye movements continuously throughout each trial. If the monkey broke fixation at an inappropriate time, the trial was terminated and the data were discarded. On microstimulation trials, the train of stimulating pulses ("Elect. Stimulus" - Fig. 2B) began and ended simultaneously with onset and offset of the motion display.

Psychophysical data were obtained in blocks that varied in length from 640 to 1200 trials. We presented an equal number of trials at four correlation levels: three near psychophysical threshold and zero percent correlation. For the near-threshold correlation levels, motion was presented in the preferred direction on half of the trials and in the null direction on the remaining half (being completely random, motion direction was undefined at 0% correlation). Similarly, microstimulation was applied on half the trials and withheld on the other half. All trials were presented in random order within a block so that the monkey had no basis for anticipating the correlation level, the direction of motion, or the occurence of microstimulation. The effect of microstimulation was assessed by comparing psychophysical performance on stimulated vs. non-stimulated trials.

RESULTS

By stimulating physiologically characterized columns of neurons in MT, we attempted to enhance the cortical representation of the preferred direction of motion of the stimulated neurons. If the monkey bases his judgements of motion direction on the responses of these MT neurons, microstimulation would cause the monkey to choose the preferred direction more frequently on stimulated trials than on non-stimulated trials.

Figure 3 illustrates two experiments in which this result was obtained. The abscissa shows the strength of the motion signal in percent correlated dots. Weak motion signals, near 0% correlation, are represented near the middle of the abscissa while strong motion signals are represented at either end. Positive correlation values indicate stimulus motion in the preferred direction while negative values indicate motion in the null direction. The ordinate shows the proportion of "preferred decisions" made by the monkey for each stimulus condition; a "preferred decision" is defined as decision in favor of the direction that corresponds to the preferred direction of the neurons. The open symbols and dashed line indicate the monkey's performance on non-stimulated trials; the closed symbols and solid line represent performance on stimulated trials. Each

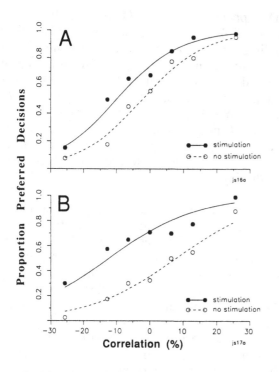

Figure 3. Effect on psychophysical performance of microstimulation at two sites in MT. The proportion of preferred decisions is plotted against the strength of the correlated motion signal. The open symbols and dashed line represent performance on trials with no electrical stimulation. The closed symbols and solid line represent performance on stimulated trials. In both experiments, microstimulation biased the monkey's choice behavior toward the preferred direction of motion of the stimulated neurons, resulting in a leftward shift of the psychometric function. **A.** An experiment which resulted in a leftward shift of 7.7% correlated dots. **B.** An experiment which resulted in a leftward shift of 20.1% correlated dots. See text for details. Reprinted with permission from Salzman *et al.*, 1990.

symbol is based on 40 trials, with the exception of the points at 0% correlation which are based on 80 trials.

In each experiment in Fig. 3, the primary effect of microstimulation was to increase the proportion of preferred decisions. Such effects occurred at every correlation level in both experiments, though the effect was much larger in the experiment of Fig. 3B. In both cases, the effect can be conveniently described as a leftward shift of the psychometric function. This metric has the advantage of expressing the effect in units of the visual stimulus - percent correlated dots.

We fitted each data set with sigmoidal curves of equal slope and employed logistic regression analysis (Cox, 1970) to measure the magnitude and statistical significance of the leftward shift. In Fig. 3A, the curve in the stimulated condition shifted leftward by 7.7% correlated dots relative to the unstimulated condition. In Fig. 3B, the "stimulated" curve shifted leftward by 20.1% correlated dots. Both effects were highly significant ($p \leq .0001$). Intuitively, one can think of the leftward shift as indicating that microstimulation added a "signal" equivalent to a particular amount of the visual stimulus in a given experiment. In other words, the leftward shift indicates how strong a visual stimulus would have been necessary to offset the effect of microstimulation and superimpose the two curves.

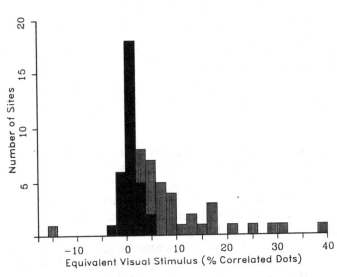

Figure 4. A frequency histogram of the microstimulation results from an initial series of 62 experiments in 3 monkeys. The abscissa indicates the magnitude of the shift of the psychometric function observed in each experiment. The magnitude of the shift can be thought of as an "equivalent visual stimulus", that is, the visual stimulus (in percent correlated dots) that would have offset the effect of microstimulation. Positive values represent leftward shifts of the psychometric function (increased preferred decisions); negative values represent rightward shifts (increased null decisions). Shaded bars indicate experiments in which the shift of the psychometric function was statistically significant (logistic regression analysis, p < .05). Reprinted with permission from Salzman et al., 1990.

In an initial set of experiments, we made such measurements at 62 sites in 3 monkeys (Salzman et al., 1990). We obtained significant effects (p < .05) at 30 of the 62 sites, in similar proportions in each animal. Figure 4 shows the magnitude of the shift of the psychometric function ("Equivalent visual stimulus") for each of these 62 experiments. Positive values indicate leftward shifts of the psychometric function; negative values indicate rightward shifts. The shaded bars represent experiments in which the effect of microstimulation was statistically significant; solid bars represent experiments lacking a significant effect. The most important result in Fig. 4 is that microstimulation caused the monkeys to increase their choices in favor of the preferred direction of the stimulated neurons (leftward shift of the psychometric function) in 29 of the 30 experiments with statistically significant effects. The result shows that activation of direction selective MT neurons can influence perceptual judgements in a way that is predictable from the physiological properties of the activated neurons.

The results in Fig. 4 are consistent with the notion that microstimulation actually enhances the sensory representation within the cortex of a particular direction of motion. By this interpretation, the observed changes in the monkey's behavior occur because the monkey bases his choices accurately on the sensory signals provided by direction selective neurons. However, an alternative interpretation is that microstimulation may directly affect motor signals related to the operant response - a saccadic eye movement - rather than sensory signals related to a particular direction of motion.

For several reasons, we consider this interpretation to be highly unlikely. First, the data in Fig. 4 represent experiments in which receptive field position and preferred direction changed from experiment to experiment. Thus there was a variable relationship between the preferred direction of the stimulated neurons and the direction of the saccade made by the monkey on "preferred decision" trials. If the microstimulation effects resulted from direct action on motor signals, the effects would not be associated consistently with the preferred direction of motion.

Secondly, there is no evidence that motor signals related to saccadic eye movements exist in MT. Physiological recordings during saccades have not revealed such signals (Newsome *et al.*, 1988), and lesions of MT have no effect on saccadic eye movements to stationary targets (Newsome *et al.*, 1985). In addition, we have observed no eye movements, either saccadic or pursuit, associated with microstimulation of MT during the intertrial interval. Finally, the long latency (> 1 sec) between the onset of microstimulation and execution of the saccade argues against a direct effect on motor signals. It therefore appears highly probable that microstimulation in MT affects perceptual decisions by modifying the sensory representation of motion direction.

It is worth noting that microstimulation failed to yield a significant effect in roughly half of the experiments in Fig. 4. Several factors may contribute to such failures, the foremost being our lack of knowledge concerning the local geometry of direction columns around the electrode tip. Although we explored the width of each column along the line of electrode travel, we had no way of determining columnar structure to the right or left of the electrode track. In some experiments, then, the stimulating current may have had an unintended effect on neighboring columns encoding other directions of motion. This seems a likely explanation for the single counterintuitive result as well (far left of Fig. 4) since neighboring columns sometimes have opposite preferred directions (Albright *et al.*, 1984).

Spatial extent of the stimulation effects. A major question of interest in these experiments concerns the effective spread of the microstimulation current within the cortex. We deliberately employed weak stimulating currents in an effort to restrict the region of neuronal excitation to the approximate width of a cortical column. Although we have no direct measurements of current spread in our experiments, several observations suggest that the microstimulation effects were highly local.

First, we occasionally encountered a point within a penetration at which the preferred direction of the neurons shifted abruptly by 180 degrees following a short (100 μm) advance of the microelectrode. In one experiment we carried out microstimulation experiments on *both* sides of such a transition point, successfully biasing the monkey's choice behavior in *opposite* directions at stimulation sites separated by only 250 μm. In another session, we mapped out a stimulation site in MT and then conducted a microstimulation experiment with

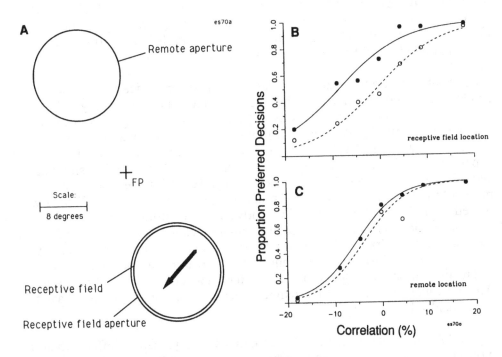

Figure 5. Methods and results of an experiment in which psychophysical measurements were made at two topographic locations in the visual field during microstimulation in MT. One location corresponded to the receptive field at the microstimulation site, but the second location was remote from the receptive field. **A. Methods.** In random order, visual stimuli were presented in either the receptive field location in the lower right quadrant of the visual field (FP = fixation point) or in the remote aperture in the upper left quadrant. The absolute location of the aperture was actually constant during the experiment, and the topographic shift was achieved by moving the fixation point. Half of the trials at each location were accompanied by microstimulation. **B.** Results of psychophysical testing at the receptive field location. Again, open symbols and the dashed line represent behavior on trials with no electrical stimulation. Closed symbols and the solid line represent behavior on stimulated trials. Each point is based on 25 trials with the exception of the 0% correlation points which are based on 50 trials each. Microstimulation shifted the psychometric function leftward by 7.9% correlated dots (logistic regression, p ≤ .0001). **C.** Microstimulation had no significant effect on choice behavior at the remote location (logistic regression, p > .05).

the electrode tip positioned at the edge of the site. After obtaining no effect at the first location, we advanced the electrode 100 μm into the site and obtained a very large effect at the second location.

Both of these observations suggest that the *direct* excitatory effects of microstimulation are quite local and that precise positioning of the electrode tip is critical to the success of the experiment. However, it is possible that the effects of microstimulation may spread trans-synaptically over a wider area of cortex (Jankowska et al, 1975). Trans-synaptic spread does not necessarily imply a loss of functional specificity since recent experiments in striate cortex have shown that orientation columns are preferentially interconnected with other columns having a similar preferred orientation (T'so and Gilbert, 1988; Gilbert and Wiesel, 1989). Thus the directionally specific signal introduced into MT by the stimulating current may be amplified by trans-synaptic activation of additional columns with similar preferred directions.

To test whether trans-synaptic activation was topographically restricted, we conducted several experiments using the procedure illustrated in Fig. 5A. The essential idea of this experiment was to stimulate at one topographic location in MT while obtaining psychophysical measurements both at the stimulated location and at a topographic location remote from the stimulation site. In the experiment of Fig. 5, for example, the receptive field at the microstimulation site (shaded) was located in the lower right quadrant of the visual field with respect to the fixation point (FP). In half of the trials, the motion display was placed over the receptive field ("Receptive field aperture"); in the remaining half, the motion display was placed at a remote location of equal eccentricity in the upper left quadrant ("Remote aperture"). As usual, the motion display was tailored to the preferred direction and speed of neurons at the stimulation site. Microstimulation was applied on half of the trials conducted at each location. All trials, including those at the remote location, were randomly interleaved in one extended block.

Figure 5A shows that microstimulation had a strong perceptual effect when the motion display was presented within the receptive field of the stimulated neurons. The leftward shift of the psychometric function (7.9% correlated dots) was highly significant (logistic regression, $p \leq .0001$). However, Fig. 5B shows that microstimulation had no effect when the motion signal was presented at the remote location (logistic regression, $p > .05$). Thus trans-synaptic spread of microstimulation effects within MT was not sufficiently extensive to influence psychophysical performance at the remote location.

Clearly, the experiment of Fig. 5 represents an extreme case which would have required trans-synaptic spread to the opposite hemisphere in order to affect performance at the remote location. However, we have conducted similar experiments over a broad range of inter-aperture distances. In 15 experiments we obtained a statistically significant microstimulation effect on performance with the receptive field aperture. In 12 of these 15 experiments, including 4 in which the two apertures were actually abutting, there was *no* effect of microstimulation on performance with the remote aperture. In the 3 experiments in which significant effects occurred with both apertures, the effect was much smaller at the remote location.

For comparison, we conducted 13 experiments in which the receptive field and remote apertures overlapped by half their diameters. We obtained significant effects on performance with the receptive field aperture in 9 of 13 experiments; for 7 of the 9 significant cases, microstimulation also had a significant effect on performance with the remote aperture. The effect at the remote aperture was smaller in all 7 of these cases, and the difference in the size of the effect with the receptive field and remote apertures was statistically significant across the group of 13 experiments (paired t-test, p < .01).

These experiments show that the microstimulation effects are spatially selective to a remarkable extent. Microstimulation signals that have robust effects when visual stimuli are presented within the receptive field aperture can be completely ineffective for stimuli presented in an abutting aperture. In other words, the monkey was able to ignore the signals introduced into MT via microstimulation when he attended to a spatial location that was non-overlapping with the receptive field at the stimulation site. In the experiments conducted thus far, microstimulation effects usually transferred to the remote location when there was substantial overlap between the receptive field and the remote aperture. While these results certainly do not eliminate the possiblity of trans-synaptic spread of stimulation effects, they do place rather sharp restrictions on the possible extent of such spread.

Dose-response relationships. To further characterize the effects of microstimulation in MT, we have measured the effect of varying the amplitude and frequency of the stimulation pulses. To accomplish this, we conducted microstimulation experiments in the usual manner except that 5 different microstimulation conditions were randomly interleaved with the no-stimulation condition (1200 trials, total). In the amplitude experiments, we employed current pulses ranging from 5 µA to 80 µA in amplitude while holding stimulation frequency constant at 200 Hz. In the frequency experiments, we used stimulation frequencies ranging from 25 to 500 Hz (25 to 200 Hz in early experiments) while maintaining amplitude at 10 µA.

Figure 6A illustrates the results of a current amplitude experiment. We obtained a separate psychometric function for each current level, and Fig. 6A shows the leftward shift of the psychometric function relative to the unstimulated case ("Equivalent visual stimulus") as a function of amplitude. In this experiment, the effect was tightly tuned with statistically significant shifts (open circles) occurring at three current levels. The largest effect occurred at 40 uA, and the size of the effect fell off steeply on either side of the optimal. In general, we found that optimal effects occurred for currents between 10 µA and 40 uA. 80 uA currents were always less effective than 40 uA, suggesting that current spread from 80 uA pulses may be sufficiently extensive to prevent selective enhancement of a single direction of motion. We rarely observed effects for current levels as low as 5 uA.

The frequency tuning curve illustrated in Fig. 6B is typical of those we obtained. We often observed significant effects for frequencies of 100 Hz or more, but significant effects were rare at lower frequencies. The tuning curves generally rose steeply between 50 and 200 Hz, but saturation of the effect was usually evident above 200 Hz. We never observed a significant reduction in the size of the effect at high frequencies.

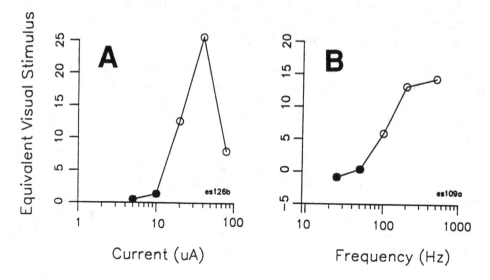

Figure 6. Dose-response relationships measured in two experiments. The stimulation-induced shift of the psychometric function (equivalent visual stimulus) is shown on each ordinate in percent correlated dots **A**. Size of the microstimulation effect as a function of current amplitude. **B**. Size of the microstimulation effect as a function of stimulation frequency.

DISCUSSION

This study demonstrates a direct relationship between the physiological properties of single cortical neurons and a particular aspect of perceptual performance. Specifically, stimulation of directionally selective neurons in MT biases a monkey's judgements of motion direction toward the preferred direction of the stimulated neurons. The results provide compelling evidence that direction selective neurons comprise the physiological substrate for perceptual judgements of motion direction.

During the course of these experiments, we were surprised to find that weak electrical currents applied to single cortical columns could have such robust effects on perceptual performance - the larger effects in Fig. 4 represent equivalent visual stimuli that are well above psychophysical threshold. As suggested above, it is possible that the large size of some effects may result from trans-synaptic activation of a network of columns with similar preferred directions. However, interleaved measurement of psychophysical performance at different topographic locations indicates that trans-synaptic activation does not spread across a large extent of MT (see Results). Knowing the extent of such spread is clearly integral to a precise description of the effective cortical circuits for perception. A reliable assessment of trans-synaptic microstimulation effects may be obtainable in future experiments in which neural activity is measured over a wide range of cortex using optical imaging methods or 2-deoxyglucose methods.

The approach employed in these experiments, combining microstimulation with multi-unit physiological recording and an appropriate perceptual task, provides an opportunity to answer additional questions in the near future. For example, the present experiments show that the neural circuits underlying performance on a direction discrimination task can be accessed by stimulating columns of neurons in MT. However, MT appears to be only one node in an extended motion pathway that begins in layer 4B of striate cortex and extends into the parietal lobe (see Maunsell and Newsome, 1987). Can the circuits responsible for motion perception be accessed in a similar manner by stimulating other visual areas in this pathway, or is MT unique in this respect? In a similar vein, can we identify and manipulate the cortical circuits responsible for other aspects of visual perception such as form, color and depth? Such experiments promise to extend the functional analysis of visual cortical circuits in new and interesting ways.

A most intriguing question that is frequently asked about the present experiments is: "what does the monkey actually *see* during microstimulation?" Unfortunately, answering this question remains problematic since we have no direct information concerning the monkey's subjective visual sensations. It seems possible, however, that microstimulation produces a sensation of motion that is attributed perceptually to the pattern viewed by the monkey. Such a phenomenon might well resemble the sensation of motion that we attribute to stationary objects when experiencing a motion aftereffect (waterfall illusion). However, satisfactory answers to such questions would probably require that similar experiments be carried out in humans. While the opportunities for such experiments are severely limited at present, the advent of non-invasive magnetic stimulation techniques (e.g. Mills et al, 1987) may eventually permit investigation of the perceptual roles of distinct cortical circuits in humans.

ACKNOWLEDGEMENTS. We are grateful to Drs. J.A. Movshon and M. Shadlen for helpful suggestions during the course of these experiments. The work was supported by the National Eye Institute (EY 5603), the Office of Naval Research (K-0161), and by a McKnight Development Award to W.T.N. C.D.S. is supported by a Medical Student Research Training Fellowship from the Howard Hughes Medical Institute.

REFERENCES

1. Albright, T.D. (1984): Direction and orientation selectivity of neurons in visual area MT of the macaque. *J. Neurophysiol.* **52**, 1106-1130.

2. Albright, T.D., Desimone, R. & Gross, C.G. (1984): Columnar organization of directionally selective cells in visual area MT of the macaque. *J. Neurophysiol.* **51**, 16-31.

3. Cox, D.R. (1970): *Analysis of Binary Data* (Methuen, London).

4. Gilbert, C.D. & Wiesel, T.N. (1989): Columnar specificity of intrinsic horizontal and corticocortical connections in cat visual cortex. *J. Neurosci.* **9**, 2432-2442.

5. Hubel, D.H. & Wiesel, T.N. (1977): Functional architecture of macaque monkey visual cortex. *Proc. R. Soc. London Ser. B* **198**, 1-59.

6. Jankowska, E., Padel, Y. & Tanaka, R. (1975): The mode of activation of pyramidal tract cells by intracortical stimuli. *J. Physiol.* **249**, 617-636.

7. Livingstone, M.S. & Hubel, D.H. (1984): Anatomy and physiology of a color system in the primate visual cortex. *J. Neurosci.* **4**, 309-356.

8. Maunsell, J.H.R. & Newsome, W.T. (1987): Visual processing in monkey extrastriate cortex. *Ann. Rev. Neurosci.* **10**, 363-401.

9. Mills, K.R., Murray, N.M.F., & Hess, C.W. (1987): Magnetic and electrical stimulation: physiological mechanisms and clinical applications. *Neurosurgery* **20**, 164-168.

10. Mikami, A., Newsome, W.T. & Wurtz, R.H. (1986): Motion selectivity in macaque visual cortex. I. Mechanisms of direction and speed selectivity in extrastriate area MT. *J. Neurophysiol.* **55**, 1308-1327.

11. Newsome, W.T., Britten, K.H. & Movshon, J.A. (1989): Neuronal correlates of a perceptual decision. *Nature* **341**, 52-54.

12. Newsome, W.T. & Pare, E.B. (1988) A selective impairment of motion perception following lesions of the middle temporal visual area (MT). *J. Neurosci.* **8**, 2201-2211.

13. Newsome, W.T., Wurtz, R.H., Dursteler, M.R. & Mikami, A. (1985): Deficits in visual motion processing following ibotenic acid lesions of the middle temporal visual area of the macaque monkey. *J. Neurosci.* **5**, 825-840.

14. Newsome, W.T., Wurtz, R.H. & Komatsu, H. (1988): Relation of cortical areas MT and MST to pursuit eye movements. II. Differentiation of retinal from extraretinal inputs. *J. Neurophysiol.* **60**, 604-620.

15. Robinson, D.A. (1963): A method of measuring eye movement using a scleral search coil in a magnetic field. *IEEE Trans. Biomed. Eng.* **10**, 137-145.

16. Salzman, C.D, Britten, K.H. & Newsome, W.T. (1990): Cortical microstimulation influences perceptual judgements of motion direction. *Nature* **346**, 174-177.

17. Stoney, S.D., Jr., Thompson, W.D. & Asanuma, H. (1968): Excitation of pyramidal tract cells by intracortical microstimulation: effective extent of stimulating current. *J. Neurophysiol.* **31** 659-669.

18. Ts'o, D.Y. & Gilbert, C.D. (1988): The organization of chromatic and spatial interactions in the primate striate cortex. *J Neurosci.* **8**, 1712-1727.

19. Van Essen, D.C., Maunsell, J.H.R. & Bixby, J.L. (1981): The middle temporal visual area in the macaque: Myeloarchitecture, connections, functional properties and topographic representation. *J. Comp. Neurol.* **199**, 293-326.

20. Zeki, S.M. (1974): Functional organization of a visual area in the superior temporal sulcus of the rhesus monkey. *J. Physiol.* **236**, 549-573.

Primal long-term memory in the primate temporal cortex: linkage between visual perception and memory

Yasushi Miyashita, Naohiko Masui and Sei-ichi Higuchi

Department of Physiology, The University of Tokyo, School of Medicine, Hongo, Tokyo 113, Japan

INTRODUCTION:
LONG-TERM MEMORY HAS STAGES WITH DISTINCT LOCALIZATIONS

Memory has stages. The distinction between short- and long-term memory has been supported not only with respect of their capacity, optimal code and time parameters but also by the double dissociation of neuropsychological deficits (Milner et al., 1968; Warrington, 1982). Long-term memory itself has been claimed to have at least two components; one is the recently acquired, labile memory which can be readily disrupted by head injury, as retrograde amnesia demonstrates clinically (Russell, 1971). The other is the remote, fully consolidated memory (Squire et al., 1975). Drug applications selectively depress or facilitate the labile component (McGaugh and Herz, 1972).

Bilateral damage to the medial temporal region, which includes the hippocampus, amygdala, and adjacent cortex, accompanied a short-span retrograde amnesia (Milner et al., 1968). Bilateral lesions of the hippocampal CA1 field (case R.B.) produced little retrograde amnesia, although the possibility remains that the patients could have suffered some retrograde amnesia for a period of a few years prior to this surgery (Zola-Morgan et al., 1986).

These data suggest the possibility that the labile component of the long-term memory is localized in the hippocampus and/or adjacent cortex. Possible contributions of the hippocampal neural circuits were examined previously (Miyashita et al., 1989a; Cahusac et al., 1990). In this article, we propose a hypothesis that the anterior ventral temporal cortex contains a group of neurons which encode one component of the visual associative long-term memory. The neurons are first of all characterized by 1) the code of recently learned visual-visual association (Miyashita, 1988a), and by 2) limited spatial distribution along the border between the area 35 and area TE (Miyashita, 1990). This area is tightly connected with the hippocampus via the entorhinal cortex (area 28) and via the perirhinal cortex (area 35) (Insausti et al., 1987). I propose to name the mnemonic code on these neurons <u>"primal long-term memory (primal LTM)"</u>. Recording the activities of these neurons <u>(primal LTM neurons)</u> in monkeys which learn and perform visual memory tasks will give vivid concrete shape to the presumed multiple representation of LTM, complementary to that obtained by neuropsychological observations.

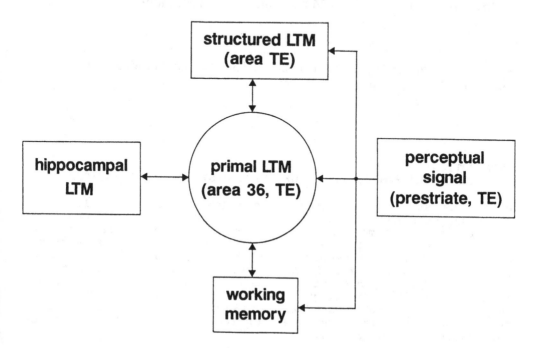

Fig. 1: A hypothesis: the primal LTM and related visual memory systems of the primate.

THE PRIMAL LTM

Figure 1 represents a hypothetical structure of the visual memory system of the primate.

The physical properties of a visual object (such as its size, color, texture and shape) are analyzed in the multiple subdivisions of the prestriate-posterior temporal complex (Mishkin, 1982; Zeki and Shipp, 1988). The anterior part of the inferior temporal cortex (area TE) synthesizes the analyzed attributes into a unique configuration and forms a central image of the object (Gross, 1982).

The primal LTM neurons are presumed to receive a synthesized perceptual images of the object from the area TE, since the neurons' activity is highly selective for coded pictorial information and is independent of the physical attributes such as size, orientation, color or position of the object (Miyashita and Chang, 1988; see below). The region where the primal LTM neurons are located indeed receives afferent projections from more lateral part of the TE (Van Hoesen and Pandya, 1975).

Some of the primal LTM neurons are also presumed to interact with visual working memory since they exhibited shape-selective sustained activity during the delay period of a delayed matching-to-sample task (Miyashita and Chang, 1988). The delay activity diminishes with memory load, as is expected for working memory (Miyashita et al., 1989b; see below).

Since localization of the primal LTM neurons is limited to a region not far from the rhinal sulcus, it is likely that the LTM system has another group of neurons to store the remote, fully consolidated long-term memory. Columnar organization of the primal LTM neurons supported the idea. By systematic electrophysiological mapping, the primal LTM neurons were found to cluster along the direction perpendicular to the cortical surface (Miyashita, 1990 and to be published), forming a column-like organization. However, when we simultaneously recorded activities of two adjacent neurons in the column with a single microelectrode, both of which were selective to learned fractal patterns, their optimal pictures were neither common nor related to each other through associative relations among the pictures (Miyashita, to be published). Thus, the primal LTM neurons were incompletely structured from the viewpoint of the functional columnar organization. By contrast, shape-selective neurons in the lateral surface of the TE (Saito et al., 1987), as well as the face neurons in the superior temporal sulcus (Perrett et al., 1987), seem to form a cluster with common optimal pictures. For this reason, I named the latter stage of LTM "structured LTM" in Fig.1. In the following, I describe evidence supporting the brief sketch given above, and examine some detailed properties of the primal LTM, obtained by unit recording in monkeys which learned and performed visual memory tasks.

BEHAVIORAL TASK AND RECORDING OF NEURAL ACTIVITIES

In a trial of our visual memory task (delayed matching-to-sample task), sample and match stimuli were successively presented on a video monitor, each for 0.2sec at a 16sec delay interval. Three adult monkeys (Macaca fuscata) were trained to memorize the sample stimulus during a delay period and then to decide whether the match stimulus was the same as the sample. A correct response was rewarded with fruit juice. A set of 97 color patterns was generated by a fractal algorithm with a 32-bit seed of random numbers (Miyashita, 1988b); the set was repeatedly used during an overtraining session ("learned stimuli") in a fixed sequence according to an arbitrary attached number (serial position number, SPN). While extracellular discharges of a neuron were recorded, a sample stimulus was selected not only from the 97 learned patterns but also from a new set of 97 patterns ("new stimuli"). Different sets of new stimuli were created for each neuron using the same algorithm but a different seed. The learned stimuli and new stimuli were used at random, independent of the SPNs.

SELECTIVITY OF NEURAL DISCHARGES IS ACQUIRED THROUGH LEARNING

A few of the 97 learned stimuli reproducibly activated a particular neuron in the anterior ventral temporal cortex during the sample period and/or during the delay period of the task. The optimal picture differed from cell to cell, and the entire population of optimal pictures for the tested cells covered a substantial part of the repertory of the pictorial stimuli.

Intuition told us that learning but not genetic determinants had formed the stimulus selectivity of these neurons, since 1) the optimal stimuli were computer-generated artificial patterns

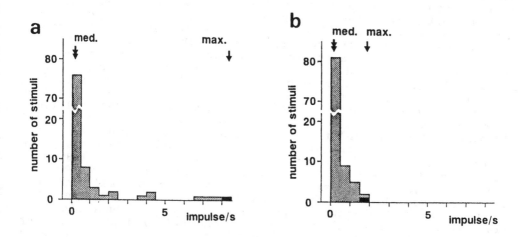

Fig. 2: Selectivity of neural discharges of a cell to the "learned" and "new" stimuli. a, frequency distribution of the average firing rate measured during the delay period following 97 "learned" sample pictures. ↓max. & ↓med., maximum and median value in the distribution. b, similar illustration to a, but "new" pictures are used as sample stimuli in the same cell.

(fractal patterns) which the monkeys would rarely have seen in their life before the experiments, and 2) the responsive neurons were not scattered randomly in the temporal cortex but formed a high density cluster in a limited area related to the hippocampus.

We obtained several lines of experimental evidence supporting our intuition. First we compared the discharge selectivity of a neuron for 2 different sets of the fractal stimuli: the learned stimuli which the monkeys stored in their long-term memory through the training sessions and the new stimuli which they had never seen before the recording session. Figure 2 shows a result. A few of the 97 learned stimuli were effective in activating high-frequency sustained delay discharges in a cell (Fig.2a). By contrast, the 97 new patterns produced only weak delay responses. The distribution of the delay discharge rate to the new stimuli (Fig.2b) lacked the

small population of high-frequency responses (> 2 impulse/s) which
characterized the distribution to the learned stimuli (Fig.2a).

Of the 206 neurons recorded in the anterior ventral temporal cortex
of the overtrained monkeys, 57 neurons exhibited shape-selective
delay discharges, and 17 of the 57 cells were successfully tested
with both learned and new stimuli. In these 17 cells, the maximum
delay discharge for learned stimuli (e.g., the arrow in Fig.2a) was
always larger than that for new stimuli (e.g., the arrow in
Fig.2b), and the difference was highly significant (P<0.001, Wil-
coxon test). When the selectivity of a neuron to a set of 97 patterns was represented by the sharpness of the response distribution, the kurtosis of the distribution (Snedecor and Cochran,
1980), the selectivity to the learned patterns was almost always
(15/17) higher than that to the new patterns (P<0.005). In spite of
the presence of a few effective pictures in the learned stimuli,
there was no difference between the median responses to the learned
and new stimuli (e.g., the double arrows in Figs.2a & b) for the 17
cells (P>0.5). These results suggest that the sharpness of the
response selectivity of these neurons to the learned patterns was
formed throughout the course of training.

ACQUIRED SELECTIVITY REPRESENTS THE ASSOCIATION AMONG MEMORIZED OBJECTS

In human long-term memory, ideas and concepts become associated
through learning. We ask whether the neurons with acquired selectivity can code such associations. To be more specific, is it
possible that training determines not only how sharply the effective learned patterns are represented in each neuron (as shown
above) but also which patterns are conjointly chosen as the few
optimal stimuli? I first examined the geometric similarity of the
optimal stimuli of a cell, and found that the stimuli were often
completely different, as seen in Fig.3 for 6 cells, suggesting a
non-geometric criterion in choosing these patterns as the optimal
stimuli of a cell. I then examined the effect of a fixed-order
presentation of the patterns during the training session according
to an arbitrarily assigned serial position number (SPN). If the
consecutively presented patterns tended to be associated together,
and if the association was fixed in the choice of effective patterns for a cell, we could expect to find the effective patterns
correlated along the SPN, in spite of a random presentation of the
stimuli during the unit-recording session.
Figure 4 shows the results. The effective responses to the learned
stimuli indeed cluster along the SPNs and the clustering was not
due to an artifact in the testing procedure because the responses
simultaneously obtained from the new stimuli were not clustered. In
the 17 cells which were tested with both learned and new stimuli,
the responses to the learned stimuli were significantly correlated
(Fig.4, •) in the nearest-neighbor of the SPNs, as compared with
the responses to the new stimuli (o) (P<0.01, Kolmogorov-Smirnov
test). The 57 cells tested by the learned stimuli exhibited similar
correlated responses along the SPN (▲). The nearest-neighbor correlation for the learned stimuli differed from cell to cell, and was
significant (P<0.05) in 28 of the 57 cells according to Kendall's

correlation test (Snedecor and Cochran, 1980), while that for the
new stimuli was not significant in any of the 17 cells. Thus, we

Fig. 3: Colored fractal patterns which produced the strongest and
second strongest delay discharges in learned stimuli for 6 different cells (a-f). Note that the two optimal stimuli for a neuron
have no similarities in their geometric patterns. See the original
color pictures in Miyashita, Y. (1988a).

conclude that these neurons are a good candidate for one of the
visual associative LTM stores (i.e., the primal LTM).

CATEGORIZED PERCEPT OF A PICTURE IS MEMORIZED

The anterior ventral temporal cortex has been designated as the
last link from the visual system to the hippocampus (Van Hoesen and
Pandya, 1975; Insausti et al., 1987; Yukie and Iwai, 1988). Thus
the primal LTM neurons would receive final-stage information in
serial visual processing along occipito-temporal cortices (Fig.1).

Indeed the primal LTM neurons encode highly abstract pictorial
properties of the stimulus, as demonstrated by the following analy-

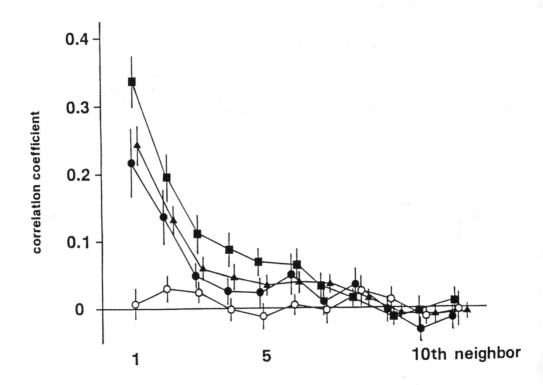

Fig. 4: Stimulus-stimulus association among the learned fractal
patterns. Ordinate, autocorrelations of the delay discharge rate
along the SPN of the stimuli. ● & o, average autocorrelogram for
the learned and new stimuli in the 17 cells. ▲ , that for the
learned stimuli in the 57 cells. ■ , that for the learned stimuli
in the 28 cells for which the nearest-neighbor correlation along
the SPN was significant (P<0.05) according to Kendall's test. Error
bars, standard errors. ***:P<0.001, **:P<0.01, *:P<0.05, according
to the Kolmogorov-Smirnov test in comparison with the value for new
stimuli.

ses of triggering features of the delay responses (Miyashita and
Chang, 1988). Sample pictures were manipulated in 3 different
manners : 1) the stimulus size was reduced by one half, 2) the
stimuli were rotated by 90 degrees in a clockwise direction, 3)
colored stimuli were transformed into monochrome by referring to a
pseudo-color look-up table. Figure 5 illustrates the responses of a
neuron which consistently fired during the delay after one particular picture but not after others, irrespective of stimulus size

(Fig.5b), orientation (Fig.5c), or color (Fig.5d). Similar tolerance of responses was observed in a majority of the tested delay
neurons: to size in 16/19, to orientation in 5/7, to color-
monochrome in 15/20, and to position in 8/13 cells. In other neurons, manipulation of the most effective stimulus reduced or abolished the thereby-evoked delay discharge.

INTERACTION WITH THE WORKING MEMORY

Cooling the anterior ventral temporal cortex of monkeys impairs
performance of the delayed matching-to-sample task even in a trial
when the monkey has already learned the rule of the memory task
(Horel and Pytko, 1982). A primal LTM neuron was activated selec-

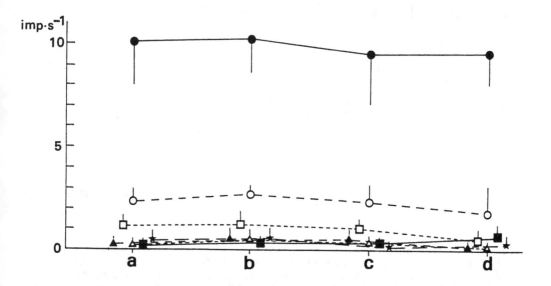

Fig. 5: Response tolerance under stimulus transformation in size,
orientation or color. (a), control responses. (b), (c) and (d) show
the effects of stimulus size reduction by half, stimulus rotation
by 90 degrees in a clockwise direction, and color-to-monochrome
transformation. Ordinate, average delay spike frequencies as a
function of stimulus transformation. Responses to 7 different
sample pictures are plotted with different symbols. Error bars
indicate standard deviations for 4-15 trials.

tively by a few of the 97 learned stimuli with high-frequency
sustained discharges during the delay period of the task, that is,
when the sample stimulus was off and no picture was presented on
the video monitor. To test the possibility that these discharges
are related to working memory, we examined the effects of memory

load. Human short-term memory is limited in its capacity; the number of chunks which can be memorized at a time does not exceed "7±2". We developed a listed delayed matching-to-sample task, in which several sample stimuli are successively presented and a match stimulus is given after a 16sec delay interval. Monkeys were trained to memorize the sample stimuli and choose whether the match stimulus was included in the samples. In all 20 neurons tested, the optimum discharge rate during the delay period was weaker when the monkey memorized several stimuli at a time than when he memorized one optimal stimulus alone (Miyashita et al., 1989b). This close parallelism in the limitation of mnemonic capacity between the delay discharges and working memory suggests an interaction between the primal LTM and working memory.

MEMORY DISORDER AND THE PRIMAL LTM

The primal LTM neurons were most frequently found along the border between the areas TE and 35. Interestingly, this strip-like area largely overlaps with the "PR/PH area" of Zola-Morgan, Squire, Amaral and Suzuki (Zola-Morgan et al., 1989). They found that monkeys with PR/PH lesions were as impaired or more impaired on delayed nonmatching-to-sample task than the comparison group of monkeys with H^+A^+ lesion (which includes the hippocampal formation, amygdala, and adjacent cortex). They argued that "the more severe impairment associated with the PRPH lesion indicates that the perirhinal cortex itself must contribute significantly to memory functions". The memory disorder found in the PR/PH monkeys may result, at least in part, from the loss of the primal LTM neurons, whose unique properties are reviewed in this article.

CONCLUSIONS

In this article, we proposed a hypothesis that the anterior ventral temporal cortex contains a group of neurons which encode one component of the LTM, i.e., the primal LTM (Fig.1). Highly abstract visual information is inputted to the primal LTM through the occipito-temporal visual cortices. Bidirectional communication with the hippocampal system may play a key role in forming the associative organization of the memory code in the primal LTM rather than the organization along geometrical similarity.

Experimentally-controlled associations among fractal patterns were demonstrated in primal LTM neurons. There are indirect clues indicating the occurrence of such association during daily life; in the polysensory areas deep within the superior temporal sulcus (mainly, area TPO) which receive a projection from the IT cortex (Jones and Powell, 1970), some cells were selective for a person's different perspective views (face or body movement) (Perrett et al., 1987), which might be interpreted by an associative experience between the perspective views that are visually distinct but temporally related in many occasions. Interestingly, in area TPO, face cells form a column-like organization (Perrett et al., 1984, 1987). The primal LTM neurons also form a cluster, but closely adjacent neurons do not have common optimal patterns (Miyashita, to be published), which suggests a primitive, incompletely structured organization of the primal LTM. The structured LTM (Fig.1) of faces, for example, may be stored in area TPO. The general pertinence of the distinction between the primal LTM and structured LTM should be examined in future research.

In the present stage of investigation, it is not clear whether the primal LTM described here is identical with the labile component of LTM which can be revealed in retrograde amnesia or with application of drugs after learning. Since a hippocampal neuron can encode a newly formed association (Miyashita et al., 1989; Cahusac et al., 1990), it is urgent to compare it with the primal LTM and to assess their susceptibilities to electric shock and/or convulsant drugs. It will be possible to confirm or reject the present hypothesis on the primal LTM (Fig.1) by an appropriate combination of neuropsychological and electrophysiological experiments.

ACKNOWLEDGMENT

This work was supported by a grant from the Japanese Ministry of Education, Science and Culture (02102008).

REFERENCES

Cahusac, P.M.B., Rolls, E.T., Miyashita, Y. and Niki, H. (1990): Modification of the responses of hippocampal neurons in the monkey during the learning of a conditional spatial response task. (In press).

Gross, C.G.(1972): Visual functions of inferotempoporal cortex. In Handbook of Sensory Physiology, Vol.VIII/3B, ed. R. Jung, pp.451-482. Berlin: Springer-Verlag .

Herzog, A.G. and Van Hoesen, G.W.(1976): Temporal neocortical afferent connections to the amygdala in the rhesus monkey. Brain Res. 115: 57-69.

Horel, J.A. and Pytko, D.E. (1982): Behavioral effects of local cooling in temporal lobe of monkeys. J.Neurophysiol. 47: 11-22.

Insausti, R., Amaral, D.G. and Cowan, W.M. (1987): The entorhinal cortex of the monkey: II. Cortical afferents. J. Comp. Neurol. 264: 356-395.

Jones, E.G. and Powell, T.P.S.(1970): An anatomical study of converging sensory pathways with in the cerebral cortex of the monkey. Brain 93: 793-820.

McGaugh, J.L. and Herz, M.J. (1972): Memory consolidation. San Franscisco: Albion.

Milner, B.(1968): Visual recognition and recall after right temporal-lobe excision in man. Neuropsychologia 6: 191-209.

Milner, B., Corkin, S. and Teuber, H.L.(1968): Further analysis of the hippocampal amnesic syndrome:14-year follow-up study of H.M. Neuropschychologia 6: 215-234.

Mishkin, M.(1982): A memory system in the monkey. Phil. Trans. R. Soc. London, Ser. B 298: 85-95.

Miyashita, Y.(1988a): Neuronal correlate of vistal associative

long-term memory in the primate temporal cortex.
Nature <u>335</u>: 817-820.
Miyashita, Y.(1988b): Neuronal representation of pictorial working
memory in the primate temporal cortex. In <u>Dynamic Interactions
in Neural Networks: Models and Data</u>, ed. M.A. Arbib and S.
Amari, pp.183-192. Berlin: Springer-Verlag.
Miyashita, Y. (1990): Localization of primal long-term memory in
the primate temporal cortex. eds. by L.R. Squire, G. Lynch,
N.M. Weinberger and J.L. McGaugh. New York: Oxford University
Press. (In press).
Miyashita, Y., Cho, K. and Mori, K.(1987): Selective pictorial
information is retained by neurons in the ventral temporal
cortex of the monkey during the delay period of a matching-to-
sample task. <u>Abstr. Soc Neurosci.</u>   <u>13</u>: 608.
Miyashita, Y. and Chang, H.S.(1988): Neuronal correlate of
pictorial short-term memory in the primate temporal cortex.
<u>Nature 331</u>: 68-70 .
Miyashita, Y., Roll, E.T., Cahusac, P.M.B., Niki, H. and
Feigenbaum, J.D.(1989a): Activity of hippocampal neurons in
the monkey related to a stimulus-response association task.
<u>J. Neurophysiol.</u> <u>61</u>: 669-678.
Miyashita, Y., Sakai, K. and Higuchi, S. (1989b): Memorized objects
mutually interfere on delay discharges for pictorial short-
term memory in neurons of primate temporal cortex. <u>Abstr. Soc.
Neurosci.</u> <u>15</u>: 303.
Perrett, D.I., Smith, P.A.J., Potter, D.D., Mistlin, A.J., Head,
A.S., Milner, A.D. and Jeeves, M.A. (1984): Neurones
responsive to faces in the temporal cortex: studies of
functional organization, sensitivity to identity and relation
to perception. <u>Human Neurobiol.</u> <u>3</u>: 197-208.
Perret, D.I., Mistlin, A.J. and Chitty, A.J.(1987): Visual neurones
responsive to faces. <u>Trends in Neuroscience</u> <u>10</u>: 358-364.
Russel, W.R.(1971): The traumatic amnesia. London: Oxford
University Press.
Saito, H., Tanaka, K., Fukumoto, M. and Fukada, Y. (1987): The
inferior temporal cortex of the macaque monkey: II. The level
of complexity in the integration of pattern information.
<u>Abstr. Soc. Neurosci.</u> <u>13</u>: 628.
Seltzer, B. and Pandya, D.N.(1978): Afferent cortical connections
and architectonics of the superior temporal sulcus and
surrounding cortex in the rhesus monkey. <u>Brain Res.</u> <u>149</u>: 1-24.
Snedecor, G.W. and Corchran, W.G. (1980): Statistical Method,
Ames: Iowa University Press .
Squire, L.R., Slater, P.C. and Chace, P.M. (1975): Retrograde
amnesia: Temporal gradient in very long term following
electroconvulsive therapy. <u>Science</u> <u>187</u>: 77-79.
Van Hoesen, G.W. and Pandya, D.N.(1975): Some connections of the
entorhinal (area 28) and perirhinal (area 35) cortices of the
rhesus monkey. I. Temporal lobe afferents. <u>Brain Res.:</u> <u>95</u>,
1-24.
Warrington, E.K. (1982): The double dossociation of short- and
long-term memory deficits. In <u>Human memory and amnesia</u>, ed. by
L.S. Cermak. Hillsdale: Lawrence Erlbaum Associates.
Yukie, M. and Iwai, E (1988): Direct projections from the ventral
TE area of the inferotemporal cortex to hippocampal field CA1
in the monkey. <u>Neurosci. Lett.</u> <u>88</u>: 6-10.
Zeki, S. and Shipp, S.(1988): The functional logic of cortical

connections. <u>Nature</u> <u>335</u>: 311-317.
Zola-Morgan, S., Squire, L.R. and Amaral, D.G.(1986): Human amnesia
    and the medial temporal region: Enduring memory inpairment
    following a bilateral lesion limited to field CA1 of the
    hippocampus. <u>J. Neurosci.</u> <u>6</u>: 2950-2967.
Zola-Morgan, S., Squire, L.R. Amaral, D.G. and Suzuki, W.A. (1989):
    Lesions of perirhinal and parahippocampal cortex that spare
    the amygdala and hippocampal formation produce severe memory
    impairment. <u>J. Neurosci.</u> <u>9</u>: 4355-4370.

PART 3

Eye-movements and vision

John Findlay and Zoi Kapoula

Do workers in visual perception need to know about eye movements? Several texts in the area of vision barely mention the mobility of the eye. Likewise for many workers in vision, eye movements, insofar as they are considered at all, are regarded as something of a nuisance. This is reflected in the use of short stimulus durations to avoid 'artefacts' and in the emphasis given to mechanisms of 'suppression' which supposedly eliminate retinal disturbances accompanying movements of the eye. We believe that this view is very limited. Vision involves much more than simply processing retinal images, and exploring the links between vision and eye movements is both challenging and rewarding.

In one rather obvious sense, eye movements can be said to be vital for vision. If retinal image motion is eliminated by a stabilisation procedure, then vision also is lost. It is thus something of a paradox that one major category of eye movements has as its objective exactly such a stabilisation. The vestibulo-ocular and optokinetic reflexes have long been known to serve the function of keeping the eye fixed in space as the head and body move. However only recently has the full complexity and power of these reflexes been appreciated, thanks in no small measure to the work of Fred Miles. Miles reviews this work in the first article, showing how a set of fast responses based on otolith function overlay the slower canal responses, and how a similar pair of responses appears in optokinetic following. The finding that accurate tracking responses of the eye can be made with a latency as short as 50 ms, quite apart from its intrinsic interest, is of major methodological concern for workers in visual motion.

The absence of stabilisation fade-out can largely be ascribed to the saccadic movements of the eye, which shift the gaze position several times each second under normal viewing. Each time such a movement occurs, there is a total remapping of the visual world on the retina and on the collection of retinotopically mapped areas in the posterior part of the cortex. How does the brain deal with these continual changes in mapping?

The traditional 'solution' suggests a re-mapping from retinal co-ordinates to some hypothetical head centred or body-centred co-ordinates, based on information about the eye movement, usually believed to be in the form of an efference copy. Until recently no plausible neural correspondence had been found to support such a process. However, in the last few years visual cells have been found in the posterior parietal cortex whose firing rate is influenced by eye position. Richard Andersen was one of those responsible for the discovery, and in his symposium presentation (not presented in this volume but see Andersen, 1989) he reviewed this empirical work and also demonstrated with a connectionist model how such cells might encode spatial position.

More time will be needed to evaluate the significance of these findings. The cortical location of the cells suggests their function may be restricted to immediate sensorimotor co-ordination. Elsewhere in the brain, physiologists have failed to find cells encoding eye position signals, despite considerable effort. However cells whose firing faithfully reflects eye velocity are found in several brain areas. This finding formed one starting point of the work of Jacques Droulez, who presented an impressive novel solution to one problem of spatial coding. This problem concerns the spatial programming of saccadic eye movements themselves. The seminal experiments of Mays and Sparks (1980) showed that saccades are planned in a goal-directed way, rather than simply based on a retinal error signal. The usual interpretation given is that saccades are planned in a hypothetical head centred or body centred spatial co-ordinate framework. In his paper (not presented in this volume but see Droulez and Berthoz, 1990) Droulez placed the emphasis rather differently, arguing that the Mays and Sparks experiments reveal spatial memory for the target position. He then showed how such memory could be incorporated into a neural network, and moreover one which uses an eye velocity signal. The model thus has physiological plausibility as well as computational elegance.

Hans Collewijn addresses the question of whether the pathways giving rise to conscious perceptual experience are the same ones used to control the movements of the eyes. In the case of the integration of binocular information the answer is quite clearly that they are not the same. In the two cases of horizontal disparity and of cyclodisparity, the rule appears to be that eye movements respond to absolute values of disparity whereas perceptual experience is influenced only by relative differences in disparity.

A similar dissociation between pathways giving rise to conscious perception and those controlling eye movements is also noted in the case of the saccadic system both by Jan van Gisbergen and Christopher Kennard. Kennard reviews the remarkable phenomenon of blindsight and shows that in some cases accurate saccades can be made to unseen targets in the blind field. Van Gisbergen shows how the saccadic system uses a representation of the visual input in which spatial coding is highly distributed and thus resolution of neighbouring targets is very poor. These properties can be understood with reference to the geometry of retinotopic remapping in the superior colliculus and the use of distributed coding. Van Gisbergen's paper goes on to consider the implications of recent experimental work demonstrating how tightly the rotational characteristics of the eyeball are restricted to two, rather than three, degrees of freedom (Listing's law).

Study of system malfunction produced by brain damage has always been a rich source of information in neuroscience. Christopher Kennard gives examples of this approach in the case of vision and eye movements. As well as the issue of blindsight previously mentioned, a pleasing convergence between physiological and clinical studies may be noted in the analysis of visual motion by cortical area MT.

Andersen R.A. (1989) Visual and eye movement functions of the posterior parietal cortex. *Annual Review of Neuroscience* 12, 177-403.

Droulez J. & Berthoz A. (1990) The concept of dynamic memory in sensorimotor control. In Humphrey D.R. & Freund H.J. (Eds) *Motor Control: concepts and issues*. Chichester: Wiley.

Mays L.E. & Sparks D.L. (1980) Saccades are spatially, not retinotopically, coded. *Science* 208, 1163-1165.

The effect of cortical lesions on visuo-motor processing in man

C. Kennard

Department of Neurology, The Royal London Hospital,
London E1 1BB, United Kingdom

INTRODUCTION

The importance of vision as the major sensory input in higher primates is not only reflected by the fact that about half of all afferent fibres projecting to the brain are from the eye but also by the sophistication of the neural systems controlling eye movements. The development of a specialised fovea made necessary rapid or saccadic eye movements to shift the fovea from one object of interest to another in the visual field. A further system is necessary to ensure that should the object move so it is kept steady on the fovea. The vestibulo-ocular and optokinetic systems cope with movements of the head and body to ensure visual clarity. Although the coordinated eye-head fixation of an object in the peripheral visual field may require an interaction between all of these systems, each type of eye movement is largely subserved by an independent system of neural connections. This eases somewhat the task of elucidating their underlying neurophysiological mechanisms. A considerable amount of the research into oculomotor systems has concentrated solely on the production of eye movements; however most eye movements depend on an appropriate visual input and so visual processing is also extremely important in understanding the neural basis for eye movements. Although relatively independent research has taken place into visual processing and oculomotor mechanisms the area of visuo-motor transformations is a rapidly expanding field of interest.

What then has the investigation of human patients with neurological disease, and cortical lesions in particular, to offer in relation to visuo-motor processing? To the monkey physiologist able to record the activity of single neurones in response to specific visual stimuli as well as the oculomotor output, or the oculomotor behaviour of monkeys following very focal, well defined brain lesions experiments in man must seem relatively crude. However, until the recent advent of positron emission tomographic (PET) brain scanning with functional activation (e.g. Lueck et al., 1989) study of patients with brain lesions was the only method for defining the neural basis for human behaviour and cognition. As Damasio and Damasio (1989) have recently stated, "the lesion method (in man) is only as good as the finest level of cognitive characterisation and anatomical resolution it uses". However, despite the limitations of the method, when carefully applied in man it can provide useful information concerning cortical localisation of function. In this chapter the application of this method is applied to two areas of visuomotor processing; firstly, in relation to the cortical localisation of visual motion processing and and to the cortical pursuit eye movement output. Secondly, the result of lesions of the striate cortex on oculomotor strategies to locate objects presented in the blind hemianopic field will be described. This involves a consideration of patients with blindsight and of subcortical saccadic processing. Finally, the evidence that saccadic training techniques can be used to reduce the size of visual scotomas resulting from occipital lobe lesions in man will be reviewed.

CORTICAL LOCALISATION OF VISUAL MOTION PROCESSING AND PURSUIT OUTPUT

It has long been recognised by clinicians that cortical lesions in the parieto-occipital lobes often results in an impairment of smooth pursuit during tracking of a target moving towards the side of the lesion (Troost et al., 1972b; Sharpe et al., 1979). Despite a number of studies attempting to localise the "pursuit centre" more precisely investigators have always recorded smooth pursuit in these patients in response to targets moving sinusoidally or at constant velocity. However, during pursuit eye movements in these situations the location of the visual stimulus on the retina is not controlled, resulting in activation of either hemisphere. To fully understand the mechanisms and localisation of motion processing and pursuit initiation it is necessary to present moving stimuli in one or other visual hemifield to ensure activation of only one hemisphere. By using a step-ramp stimulus in rhesus monkeys considerable advances in our understanding of these processes have been achieved (summarised in Wurtz et al., 1990). In particular amongst the multiple visual areas in the extrastriate cortex two areas were located in the superior temporal sulcus, the middle temporal (MT or V5) area and the medial superior temporal (MST) area. Area MT appears to be selectively related to visual motion processing, containing directionally selective cells, whereas area MST similarly appears to be related to movements of the eye as well as to motion of a visual stimulus. Lesions of these two areas on monkeys give rise to characteristic deficits of smooth pursuit eye movements in response to step-ramp stimuli as will be discussed below.

Using step-ramp stimuli to control the retinal location of the visual stimuli we sought, in patients with focal lesions, to determine the human homologue of areas MT and MST by attempting to identify similar disturbances of pursuit eye movements in patients with focal lesions to those found after lesions of MT and MST in monkeys (Thurston et al., 1988). Two sets of visual stimuli were used, a predictable target movement in a triangular waveform with constant velocity and non-predictable step-ramp stimuli. In the latter the target stepped into the right or left hemifield and then moved at a constant velocity either centrifugally or centripetally. In a series of 20 patients 5 had smooth pursuit asymmetry while tracking a predictable target. (Fig.1). These same patients showed a characteristic response to step-ramp stimuli. This was a unidirectional tracking deficit (toward the side of the cerebral lesion) in response to stimuli presented into either visual hemifield. This response is typical of the disturbance found in monkeys with lesions of area MST. However, in addition monkeys with such lesions also show a bidirectional deficit in the hemifield contralateral to the side of the lesion which was not found

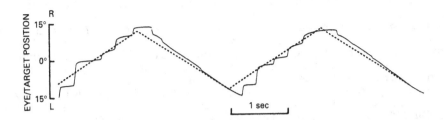

Fig.1. Smooth pursuit of a predictable target moving at a constant velocity of 15 degrees/sec by a patient with a right hemisphere lesion. This shows impaired smooth pursuit when the target moves from left to right. Target position is indicated by the dotted line, eye position by the solid line. R = right; L = left (from Thurston et al, 1988 with permission).

in our patients. This specific type of abnormal response was shown by one patient (Fig. 2). He was unable to track, with pursuit or saccades, targets moving in either direction in the visual hemifield contralateral to the side of the lesion. This response is very similar to the abnormality detailed in monkeys with lesions of MT, and the lesion in this patient was lower (Brodmann area 19 and 37; occipito-temporal junction) than the lesions of the patients with uni-

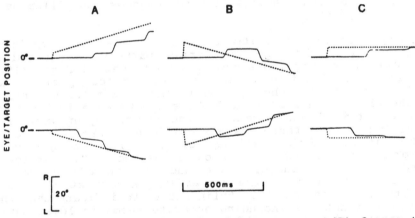

Fig.2. Responses to (A) foveafugal step-ramps, (B) foveapetal step-ramps, and (C) steps in a patient with a left hemisphere lesion. Target position is indicated by the dotted line, eye position by the solid line. Note that saccades made to step-ramps presented in the right visual hemifield are more dysmetric than saccades made to left hemifield step-ramps. Saccades to steps are equally accurate in the right and left hemifields. R = right; L = left (from Thurston et al., 1988, with permission).

directional deficits which were located more superiorally (Brodmann areas 39 and 40). It appears, therefore, that the human homologue of area MT lies at the occipitotemporal junction and this has recently been confirmed by studies using PET (Cunningham et al., 1989).

THE EFFECTS OF STRIATE CORTICAL LESIONS ON SACCADIC LOCALISATION

As has previously been mentioned, the primary role of the saccadic system is to redirect the eyes so that the image of an object of interest present in the peripheral visual field is brought onto the fovea. The spatial characteristics of a visually elicited saccade are, therefore, computed from visual information about the eccentricity and direction of the target. This is the "WHERE" system which behavioural and physiological studies have shown is different to the "WHEN" system which initiates the process (Findlay, 1983). These studies have recently indicated that the "where" system involves visual information passing from the striate and extrastriate cortex to the posterior parietal cortex, and onward to the superior colliculus. Of considerable interest in this regard has been the effect of lesions of the retrogeniculate visual pathways on saccadic localisation. Such studies provide further information on visual processing in non-geniculostriate visual pathways.

Lesions of the geniculostriate pathway, especially the striate cortex, in man were traditionally considered to result in complete and permanent visual loss in an area topographically related to the area damaged (Holmes, 1918). Although such lesions in non-human primates left some degree of visual function, in man before the 1970's only primitive visual functions were known to still exist in the hemianopic field. These included control of pupil size (Magoun, et al., 1935), the presence of the blink reflex (Edinger & Fischer, 1913; Levinsohn, 1913) and a preserved sensitivity to changes in illumination (Brindley, et al., 1969).

Recent interest in the possibility of residual visual function in such patients with lesions of the geniculostriate pathway, originated with the first properly controlled study by Pöppel, Held & Frost (1973). Using the forced choice technique, they studied four patients with lesions of the occipital lobe. A small visual stimulus was briefly moved in the blind visual field and the subjects were instructed to move their eyes towards the target. Although totally unaware of the presence of the target, there was a significant correlation between the target positions and the amplitudes of the corresponding saccades, at least for eccentricities out to about 25 degrees. Similar results were found in another subject, who had had an almost complete resection of the right calcarine cortex (Weiskrantz et al., 1974). These authors introduced the term "blindsight" to indicate that these residual visual functions including accurate saccadic localisation, as well as other forms of visual discrimination, were not consciously perceived (for review see Weiskrantz, 1986).

Accurate saccadic localisation of a stimulus in the blind field is, however, by no means a universal finding. It was only found in 20 per cent of a group of 25 patients with retrogeniculostriate lesions (Blythe, et al., 1987). An interesting group of subjects, from the point of view of localising the visual input providing spatial information to the saccadic system, are those in whom hemidecortication (or so-called hemispherectomy) has been performed. As a consequence, all the striate and extrastriate visual areas are removed unilaterally. Such operations, although rare, are usually carried out in the treatment of severe intractable epilepsy. Gassel and Williams (1963) investigated the eye movements of three patients with hemispherectomy but did not clearly separate the results from a much larger group of patients with simple homonymous field defects. In a careful quantitative study of a single patient, Troost, Weber & Daroff (1972), found that although saccades in both directions were of equal maximum angular velocities, refixation saccades to targets located in the hemianopic field were grossly inaccurate. This is in contrast to two patients studied by Perenin & Jeannerod (1978) in whom there was a good correlation between saccadic amplitude and target positions in the hemianopic field, an observation also made in five similar patients (Sharpe, et al., 1979). Somewhat surprisingly, the latency of the refixational saccades to 20- and 40- degree eccentric targets, located in either the hemianopic or normal visual fields, were equal. Only for 15- degree targets were latencies significantly longer to targets presented in the blind field. This disparity between the latencies to the peripheral and more central targets in the hemianopic field, suggests different modes in the two regions.

These observations in hemispherectomy patients suggest that spatial localisation and initiation of saccades in the blind field may be performed by the superior colliculus which receives a direct retinal input. This is supported by experiments in monkeys which show that either the striate cortex or superior colliculus is sufficient for visual guidance of saccadic eye movements, but joint lesions to both structures eliminate target detection for saccadic guidance (Mohler & Wurtz, 1977).

As has previously been mentioned, accurate saccadic localisation in the hemianopic field is not always observed, and this is particularly so if the various studies using patients with focal lesions (usually due to stroke) of the retrogeniculate pathways are examined. This is rather surprising since saccadic localisation is considered to be one of the main functions of the superior colliculus in primates (Schiller and Sandell, 1983). Pöppel (1977) has suggested that the reason for such a poor response is analogous to the Sprague effect found in cats (Sprague, 1966). The superior colliculus receives both a facilitatory and inhibitory input from the ipsilateral striate cortex and contralateral colliculus respectively. Damage to the striate cortex would, therefore, result in reduced facilitation of the ipsilateral superior colliculus so reducing its inhibitory input on the contralateral colliculus. This in turn results in increased activity in the contralateral colliculus, thereby increasing its inhibitory drive on the ipsilateral colliculus, which reduces its functional capacity.

Another factor which may be of relevance in determining the presence or absence of accurate saccadic localisation in the blind field in such patients is the age at which the cerebral incident occurred. Most of the reported hemispherectomy patients had their operation when children and the latest age was eighteen years. In the study of Blythe, Kennard and Ruddoch (1987), all the subjects who showed absent accurate saccadic localisation in their blind hemifield received their cerebral lesion at twenty years of age or over. In three of the five with good saccadic localisation the occipital lesion occurred when they were less than ten years of age. It is well recognised that the brain is capable of considerable plasticity in these early years so this may well be a factor.

This variability of patients with retrogeniculate lesions to perform accurate saccadic localisation does raise the possibility of the use of training programmes to enhance localisation. Before considering this it is necessary to mention the compensatory oculomotor strategies used by the majority of patients to search for targets presented in their hemianopic field. In a study of such strategies Meienberg et al., (1981) found that the majority of patients moved their eyes step by step (stairstep strategy) until the target appears at their visual field border and can be seen and fixated. Although this is a safe strategy, it is rather time consuming. An alternate strategy, less frequently used, was for the patient to try and "catch" the target by making a large saccade which overshot the target but brought it into the seeing hemifield. There was then a subsequent refixation , usually a drift, to foveate the target. This overshoot strategy is obviously more efficient but, what is not known, is the incidence of these two different strategies and whether or not patients change from one strategy to the other with the passage of time.

Since such different strategies are performed naturally it may be possible to train a patient to improve their saccadic localisation. In a series of experiments carried out over the past ten years Zihl and co-workers have tried to restore visual function in patients with retrogeniculate lesions. In the first set of experiments aimed at improving visually guided eye movements Zihl (1980) trained three patients to discriminate different positions of targets presented briefly in the blind region of their visual field. The targets were located between 10 and 45 degree eccentricity in the blind field and some control sessions using 'blank' trials were also given. Target localisation was poor in all three patients in the first session and they seemed to prefer a specific saccadic amplitude irrespective of the target location. After 9-15 sessions, over a three week period, improvement was shown in one case to near perfect correspondence between target and saccade amplitude. A subsequent study of three different patients was performed. On this occasion two patients were given feedback about the extent of variation of their saccades (but not about the accuracy) and encouraged to shift their final eye position if they did not do so in a block of trials (Zihl & Werth, 1984). There was a clear improvement in saccadic accuracy using this method. A third patient, who was not given any feedback on saccadic variability, eventually developed accurate saccades but after a far greater number of trials. Zihl & von Cramon (1980) have also shown improved saccadic localisation after practice in which the patient indicates by voluntary blinking the detection of a light target presented in the perimetrically blind field.

It would appear from all these observations that the majority of patients with retrogeniculate lesions tend either to make few or constant amplitude saccades to stimuli presented in their blind field. The saccadic performance of such patients can be enhanced by specific training to detect these targets but this improvement can be hastened if during this training feedback is given to the subject about their on-going performance. This raises the interesting question of why there is no spontaneous improvement in these patients even though they use their visual functions, may make saccades into the blind hemifield and may be consciously aware of a vague non-localisable sensation when a stimulus appears in this hemifield. As Zihl (1980) has suggested, these patients may have difficulty using visual stimuli without perceptual experience. The systematic practice used in these experiments might facilitate the use of a capacity to detect and localise targets without the patients' conscious awareness. An important factor in these studies must be linked with an attention mechanism and its enhancement during retraining.

IMPROVEMENT OF VISUAL FUNCTION IN PATIENTS WITH POSTGENICULATE LESIONS BY TRAINING PROCEDURES

It has been known for some time that following lesions of the geniculostriate pathway there may be some spontaneous recovery of visual function (Pöppelreuter, 1917; Riddoch, 1917; Hine 1918). To some degree the aetiology of the lesion determines the extent of this recovery. Following vascular infarction there may be some initial recovery within the first 7-14 days but subsequently there is a poor prognosis for further recovery (Gloning, et al., 1962; Haerer, 1973), whereas traumatic damage (mostly due to gunshot wounds) is often followed by considerable recovery (Hine 1918; Teuber, et al., 1960).

Although several studies of natural recovery have been carried out, systematic attempts to restore lost vision after occipital lobe lesions have rarely been reported until recently. Preobrazhenskaya (cited by Luria, 1967) trained hemianopic patients mainly in reading and reported an enlargement of the peripheral visual field, which resulted in an improved reading performance, a visual function often causing considerable difficulty to such patients. Non-human primate experiments, previously mentioned, showed after striate cortex lesions that systematic training resulted in the restoration of detection and localisation of light stimuli (Cowey, 1967; Mohler & Wurtz, 1977). These experiments led Zihl and von Cramon (1979; 1985) to study similar training techniques in man. Using a psychophysical method in twelve patients in which light difference thresholds were repeatedly determined at the border of the visual field defect they showed an improvement in contrast sensitivity and an increase in the size of the visual field (Zihl and von Cramon, 1979). This improvement was confined to the trained visual field area and showed interocular transfer indicating its central nature. In further studies (Zihl, 1981; Zihl & von Cramon, 1985) they used a saccadic training technique, already described, in which patients with homonymous visual field defects were instructed to make saccadic eye movements to the "guessed" position of light stimuli briefly presented in the perimetrically blind regions. This systematic training lead, in the majority of their 55 patients, to an enlargement of the visual field, in a quarter of cases ranging from 10 to 48 degrees. This improvement could not be attributed to spontaneous recovery since in periods without systematic training there was no observable visual field increase. In most of their patients the newly recovered part of the field included colour and form perception, and appeared to remain unchanged even after the training sessions had ended. Symptomatically, many of the patients reported improvements in their reading performance and in their avoidance of obstacles located on the affected side.

Further analysis of their cases suggests that patients with closed head injury are less likely to show a recovery of colour and form compared to cases after vascular infarction. The degree of cortical damage is also of importance in determining outcome, with a worse outcome in patients with marked necrosis as determined by the degree of hypodensity (<15HU) on the CT scan. The final factor, possibly of critical importance, in determining the effect of these training techniques appears to be the gradient of sensitivity between the intact and the "blind" field. Where the gradient is shallow rather than sharp, training is much more likely to lead to restoration of vision.

The mechanisms proposed for this partial restoration of visual function are unclear. Zihl and von Cramon (1979) propose that training induces selective attention of the defective field so increasing neuronal activity in areas of the striate cortex surrounding the damaged area. They suggest that the superior colliculus (Goldberg & Wurtz, 1972) and the parietal lobes (Lynch, et al., 1977) may be involved in these selective attention mechanisms.

The only reported attempt at replicating the results of Zihl and von Cramon has been by Balliet, Blood and Bach-y-Rita (1985) who studied a group of 12 patients with homonymous hemianopia or quadrantanopia resulting from occipital lesions (vascular infarction). They carried out a similar training programme mainly using the determination of visual thresholds at the border of the visual field defect but also using saccadic localisation in three patientss. They failed to find any significant visual field increases in their hemianopic patients, and concluded that the results of Zihl and von Cramon may have been due to artifacts resulting from "any combination" of (a) large stimuli variability in perimetric testing, (b) changes in detection strategies with practice, (c) compensatory eccentric fixation and (d) reliance on the patients gross subjective impressions. A subsequent rebuttal of these criticisms (Zihl & von Cramon, 1986) has left this important topic in a state of uncertainty. As Zihl and von Cramon (1985, 1986) have pointed out, systematic visual training resulting in an enlarged visual field is only possible to obtain in patients who fulfil certain criteria relating, for example, to the visual gradient at the border of the field defect and the completeness of the neuronal damage in the striate cortex. However, patients with field enlargement did report functional improvement. It is perhaps more realistic for the majority of patients to aim at altering their oculomotor strategies when undertaking specific tasks, such as reading.

REFERENCES

Balliet, R., Blood, K.M.T. & Bach-y-Rita, P. (1985): Visual field rehabilitation in the cortically blind? J.Neurol. Neurosurg. Psychiat., 48, 1113-24.
Blythe, I. M., Kennard, C. & Ruddock, H.H. (1987): Residual vision in patients with retrogeniculate lesions of the visual pathways. Brain 110, 887-905.
Brindley, G.S., Gautier-Smith, P.C. & Lewin, W. (1969): Cortical blindness and the functions of the non-geniculate fibres of the optic tracts. J.Neurol. Neurosurg. Psychiat., 32, 259-64.
Cowey, A. (1967): Perimetric study of field defects in monkeys after cortical and retinal ablations. Quart. J. Exp. Psychol., 19, 232-45.
Cunningham, V.J., Deiber, M.P., Frackowiak, R.S.J., Friston K.J., Kennard, C., Lammertsma, A.A., Lueck, C.J., Romaya, J. & Zeki, S. (1989): The motion area (area V5) of human visual cortex. J. Physiol. 105P.
Damasio, H. & Damasio, A.R. (1989): Lesion analysis in Neuropsychology. Oxford: Oxford University Press.
Edinger, L. & Fischer, B. (1913): Ein mensch ohne Grosshirn. Pflüg. Arch. Gesamte Physiol. Menschen Tiere. 152, 535-61.
Findlay, J.M. (1983): Visual conformation processing for saccadic eye movements. In. Spatially orientated behaviour. Eds. A.Hein & M. Jeannerod, pp. 281-303. New York: Springer.
Gassel, M.M. & Williams, D. (1966): Visual function in patients with

homonymous hemianopia: oculomotor mechanisms. Brain, 86, 1-36.

Gloning, L. Gloning, K. & Tschabitscher, H. (1962): Die occipitale Blindheit auf vasculärer Basio. Albrecht von Graefe's Arch. Ophthalmol., 165, 138-77.

Goldberg, M.E. & Wurtz, R.H. (1972): Activity of superior colliculus in behaving monkeys. II. Effect of attention on neuronal responses. J. Neurophysiol. 35, 560-574.

Haerer, A.F. (1973): Visual field defects and the prognosis of stroke. Stroke., 4, 163-8.

Hine, M.L. (1918): The recovery of fields of vision in concussion injuries of the occipital cortex. Br. J. Ophthalmol., 2, 12-25.

Holmes, G. (1918): Disturbances of vision by cerebral lesions. Br. J. Ophthalmol., 2, 353-84.

Levinsohn, G. (1913): Der optische Blinzel-reflex. Z. Gesamte Neurol. Psychiatr., 20, 377-85.

Lueck, C.J., Zaki, S., Friston, K.J., Deiber, M.P., Cope, P., Cunningham, V.J., Lammertsma, A.A., Kennard, C. & Frackowiak, R.S.J. (1989): The colour centre in man. Nature 340, 386-389.

Luria, A.R. (1963): Restoration of Function after Brain Injury. Oxford: Pergamon Press.

Lynch, J.C., Mountcastle, V.B., Talbot, W.H., Yin, T.C.T, (1977): Parietal lobe mechanism for directed visual attention. J. Neurophysiol. 4, 362-389.

Magoun, H.W., Ranson, W.S. & Mayer, L.L. (1935): The pupillary light reflex after lesions of the posterior commissure in the cat. Am. J. Ophthalmol., 18, 624-30.

Meienberg, O., Zangemeister, W.H., Rosenberg M., Hoyt, W.F. & Stark, L. (1981): Saccadic eye movement strategies in patients with homonymous hemianopia. Ann. Neurol., 9, 537-44.

Mohler, C.W. & Wurtz R.E. (1977): Role of striate cortex and superior colliculus in visual guidance of saccadic eye movements in monkeys. J. Neurophys., 40,, 74-94.

Perenin, M.T. & Jeannerod, M. (1978): Visual function within the hemianopic field following early cerebral hemidecortication in man. I Spatial localisation. Neurophsychologia., 16, 1-13.

Pöppel, E. (1977): Midbrain mechanisms in human vision. In. Neurosciences Research Program Bulletin, Vol. 15, Neuronal Mechanisms in Visual Perception, eds; E. Pöppel, R. Held, & J E Dowling, pp. 335-45. Boston: MIT Press.

Pöppel, E., Held, R., & Frost, D. (1973): Residual visual function after brain wounds including the central visual pathways in man. Nature (Lond.), 243, 295-6.

Pöppelreuter, W. (1917): Die Psychischen Schadigungen durch Kopfschreb im Kriege 1914-1918, Bard 1: Die Stornugun der Niederen und Hoheren Sehleisturgen durch Verletzugen des Okzipitalhirns. Leipzig: L. Voss.

Riddoch, G. (1917): Dissociation of visual perceptions due to occipital injuries with especial reference to appreciation of movement. Brain, 40, 15-57.

Schiller, P.H. & Sandell, J.H. (1983); Interactions between visual and electrically elicited saccades before and after superior colliculus and frontal eye field ablations in the rhesus monkey. Exp. Brain Res., 49, 381-92.

Sharpe, I.A., Lo, A.W. & Rabinovitch H.E. (1979): Control of the saccadic and smooth pursuit systems after cerebral hemidecortication. Brain 102, 387-403.

Sprague, J.M. (1966): Interaction of cortex and superior colliculus in mediation of visually guided behaviour in the cat. Science, NY., 153, 1544-7.

Teuber, H.L., Battersby, W.S. & Bender, M.B. (1960) Visual Field Defects after Penetrating Missile Wounds of the Brain. Cambridge Mas: Harvard University Press.

Thurston, S.E., Leigh, R.J., Crawford, T., Thompson, A. & Kennard, C. (1988): Two distinct deficits of visual tracking caused by unilateral lesions of cerebral cortex in humans. Ann. Neurol. 230, 266-273.

Troost, B.T., Weber, R.B., & Daroff, R.B., (1972a); Hemispheric control of eye movements. I. Quantitative analysis of refixation saccades in a hemispherectomy patient. Arch. Neurol. 27, 441-448.

Troost, B.T., Daroff, R.B., Weber, R.B., & Dell'Osso, L.E. (1972b): Hemispheric control of eye movements. II Quantitative analysis of smooth pursuit in a hemispherectomy patient. Arch. Neurol. 27, 449-452.

Weiskrantz, L., Warrington, E.K., Sanders, D.M. & Marshall, J. (1974): Visual capacity in the hemianopic field following a restricted occipital ablation. Brain, 97, 709-28.

Wurtz, R.H., Komatsu, H., Yamasaki, D.S.G. & Dûrstater, M.R. (1990): Cortical visual motion processing for oculomotor control. In. Vision and the Brain, eds. B. Cohen & I. Bodis-Wollner, pp 211-231. New York: Raven Press.

Zihl J. (1980): 'Blindsight': improvement of visually guided eye movements by systematic practice in patients with cerebral blindness. Neuropsychologia, 18, 71-77.

Zihl J. (1981); Recovery of visual functions in patients with cerebral blindness. Exp. Brain Res., 44, 159-69.

Zihl, J., & von Cramon, D. (1979): Restitution of visual function in patients with cerebral blindness. J.Neurol. Neurosurg. Psychiat., 42, 312-22.

Zihl, J., & von Cramon, D. (1980): Registration of light stimuli in the cortically blind hemifield and its effect on localisation. Behav. Brain Res. 1, 287-298.

Zihl, J., & von Cramon, D. (1985): Visual field recovery from scotoma in patients with postgeniculate damage; a review of 55 cases. Brain, 108, 335-66.

Zihl, J., & von Cramon, D. (1980): Visual field rehabilitation in the cortically blind. J. Neurol. Neurosurg. Psychiat. 49, 965-967.

Zihl. J., & Werth, R. (1984): Contributions to the study of 'blindsight'. I. Can stray light account for saccadic localisation in patients with post geniculate field defects. Neuropsychologia; 22, 1-11.

Binocular eye-movements and depth-perception

H. Collewijn, J. van der Steen and L. J. van Rijn

Department of Physiology I, Faculty of Medicine, Erasmus University Rotterdam, P.O. Box 1738, NL 3000 DR Rotterdam, The Netherlands

In this paper, we shall discuss some aspects of the relation between binocular eye movements and the perception of distance and depth. This topic has a long history, often marked by heated controversies among leading visual scientists. One reason for us to venture into this field is that it has been heavily dominated by the psychophysical approach, while oculomotor behavior has been more often inferred than adequately measured (for a recent review, see Collewijn and Erkelens, 1990). We believe that only the combined study of eye movements and perception, which has been rarely seriously attempted, can lead to the solution of lingering questions and, hopefully, new insights. Specifically, it will be emphasized that in discussions of the role of disparity, relative disparity has to be sharply distinguished from absolute disparity, and that the latter can be only reliably determined by the accurate measurement of eye positions.

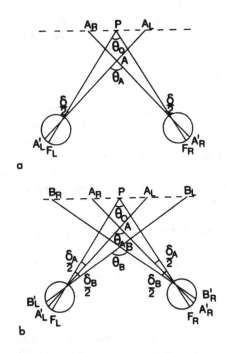

Fig. 1. The geometry of absolute and relative disparity; see text.

THE GEOMETRY OF DISPARITY

The geometric relations of horizontal vergence and disparity will be crucial in our treatment, and therefore we shall begin to review the optical conditions at the hand of Fig. 1. All our experiments were done with stimuli projected on a translucent, flat screen, placed 145 cm in front of the subject. Left and right half-images of 1 or 2 targets were projected apart and optically separated by red and green filters in front of the 4 projectors and the eyes. No other visual targets or frames other than the projected ones were visible. The position of each image could be controlled by galvanometers and mirrors in the horizontal, vertical and torsional direction. Binocular eye movements were also measured in these 3 dimensions by the scleral coil - magnetic field technique. Fig. 1a shows a situation in which a single target is present, without any other visual target or frame of reference. The eyes are converged at an imaginary point P in the plane of the screen, with the lines of sight subtending an ocular vergence angle Θ_O. The targets A_R and A_L, on the other hand, are placed in such a way, that they subtend a larger target vergence angle, Θ_A. The disparity δ_A of target A is defined as the difference angle $\Theta_A - \Theta_O$. This is called an absolute disparity; this is also the parameter typically described as a major trigger feature for certain classes of visual cortical neurons. Obviously, absolute disparity is affected by eye movements, and vice versa. As a matter of fact, absolute disparity is a main factor in the control of ocular vergence: discrepancies between the vergence of the eyes and the vergence of an attended target will induce vergence eye movements, which will reduce this discrepancy.

The presence of only one target is, however, a singular case. In a normal situation we are usually exposed to several targets at the same time. Figure 1b shows a situation in which two targets, A and B, are projected on the screen. The absolute disparity δ_B of target B is, of course, $\Theta_B - \Theta_O$, but in addition, we are dealing now with the relative disparity δ_{AB}, which is the difference $\delta_A - \delta_B$, which is equal to the difference in target vergences, $\Theta_A - \Theta_B$. The salient point of relative disparity is that it is unaffected by eye movements.

DISPARITY AND THE PERCEPTION OF DISTANCE AND DEPTH

What is the role of absolute and relative disparities in the perception of depth? This fundamental question has often been obscured, as few investigators have explicitly distinguished between these parameters. Most studies on stereopsis or motion-in-depth have involved reference targets, fixation points, or visual frames. These were often deliberately introduced in order to maintain constant ocular vergence. The results of such studies actually relate to relative disparity and relative depth, but the results have been typically described as if they related to absolute disparity. As a result, visual scientists (see e.g. Nelson, 1975) have often been puzzled by the apparent discrepancy between the hyperacuity of stereopsis (seconds of arc) and the relative sloppiness of oculomotor control (minutes of arc at best; worse during head movements).

Theoretically, the effort of ocular vergence and/or the absolute disparity of a single target provide adequate neural information for the estimation of the **absolute** distance of a single target. However, careful experiments in the past have shown that such capacity is very limited, and the general consensus is now that angles of binocular parallax induce, at best, some perception of **relative** distances (ranking order) for solitary targets, presented successively, as long as distances are not much longer than arm's length (for review see Collewijn and Erkelens, 1990). Even under the best circumstances, however, the perception of such relative distances on the basis of ocular vergence and absolute disparity for single targets turns out to be far inferior to the perception of relative distance of simultaneously presented, multiple targets.

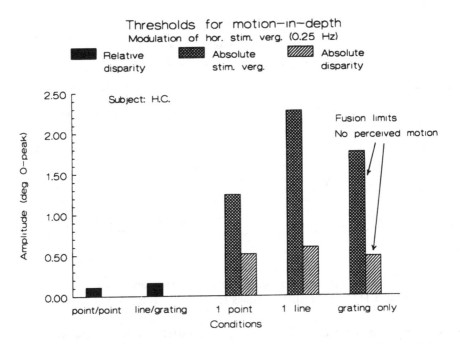

Fig. 2. Thresholds of modulation of absolute or relative disparities for the perception of motion in depth, for several types of visual patterns (see text).

There are two practical ways to test the perception of relative distances of a single target: a) absolute disparity can be modulated continuously as a function of time; or b) targets with different disparities can be shown discontinuously, in succession. The latter condition has been called "sequential stereopsis" (Enright, 1989, 1990). As we shall show, both of these approaches lead to compatible results for distinct, small targets. However, for spatially extended targets it turns out that changes in absolute disparity are not appreciated as changes in distance.

We previously explored these issues in collaboration with Erkelens and Regan (Erkelens and Collewijn 1985a,b; Regan et al., 1986). The most elementary stimulus configuration used in those experiments was a single dot, the target vergence of which was sinusoidally modulated at frequencies of 0.25 to 2.0 Hz. This dot could also be shown in close conjunction with a small, stationary reference target. The latter condition resembled traditional stereograms used in studies of (motion in) depth. The general result was, that for each subject a range of stimulus oscillation amplitudes could be found for which clear motion in depth was experienced when the stationary reference mark, and thus a change in relative disparity, was present, but for which any perception of such motion ceased immediately, when the reference mark was removed, and thus only absolute disparity was available. For the single-dot target, this effect amounted to a substantial elevation in threshold. Some perception of motion-in-depth could be restored by increasing the amplitude of the modulation of absolute disparity by a factor 2-7.

A recent replication of this experiment is shown in Fig. 2. Relative disparity changes between two points, or between a line and a grating, induced a compelling sensation of (relative) motion in depth, with low thresholds (Fig. 2, solid black bars). Modulation of target vergence for a single point or line induced a (less compelling) sensation of (absolute) motion in depth at much higher

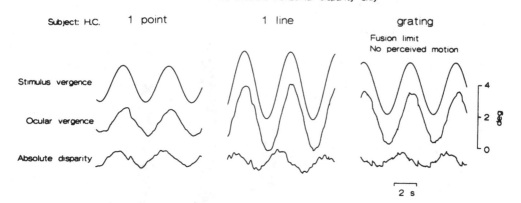

Fig. 3. Modulations of stimulus vergence, ocular vergence and their difference (absolute disparity) in subject H.C. at threshold settings for motion in depth or fusion.

thresholds of target vergence modulation (Fig. 2, cross-hatched bars). These results would be trivial, if the absence of a reference mark had resulted in nearly perfect tracking of the stimulus vergence by ocular vergence, but as we showed previously (Regan et al., 1986) and illustrate here again in Fig. 3, this explanation is inadequate. Although vergence movements were made, their gain was less than unity and their phase lag was up to 50 deg. These errors, in combination, were large enough to leave large residual modulations of disparity (Fig. 2, hatched bars and Fig. 3). Thus, for a point or line target, the threshold for the perception of motion in depth is much higher for absolute disparity changes than for relative disparity changes.

For large target configurations, the difference between the perceptual effects of absolute and relative disparities is even more dramatic. It turns out, that fluctuations in absolute disparity of large stimulus configurations induce no perception of motion in depth whatsoever. For example, we (Regan et al., 1986) tested a random dot pattern (overall size 30x30 deg) in which a central diamond (10x10 deg) could be moved with respect to the surround in such a way as to change the relative disparity between center and background. As expected, this introduced vivid perception of relative motion in depth in every subject. This relative motion was equally well induced by modulating the absolute disparity of either the surround or the central figure; the change in relative disparity was perceived in each of these cases as a relative motion of the central figure relative to the surround. However, when center and background were oscillated as a whole, or when the half-images of a Julesz's type stereogram (with a fixed relative disparity between the central diamond and the surround) were oscillated in counterphase, any perception of motion-in-depth of these patterns as a whole was completely absent, even when the amplitude of the oscillation of target vergence was increased to the limit of fusion. At this point, absolute disparity modulations (after accounting for ocular vergence tracking) amounted to 1-2 degrees, and exceeded the thresholds for the perception of motion-in-depth from relative disparity by a factor of 16-38. Further enlargement of the oscillation just led to loss of fusion, and no threshold for motion in depth could be set for these extended patterns. A recent replication of such experiments for a grating of vertical bars (total diameter 30 deg) is shown in Fig. 2 and 3 (right panels).

To dissociate ocular vergence and absolute disparity even further, Regan et al. (1986) used feedback of ocular vergence to target vergence, in order to clamp the absolute disparity of a Julesz stereogram as a whole at a constant level, and alternated this in steps between 0.6 deg crossed and 0.6 deg uncrossed. Thus, absolute disparity changed in steps of 1.2 deg, while ocular vergence, resulting from this disparity change, followed a triangular course (as first described by Rashbass and Westheimer, 1961) and varied over a range of about 20 deg. Again, we questioned our subjects about any particular perceptions of motion. No motion-in-depth was ever perceived, either in synchrony with the steps in disparity or with the triangular changes in ocular vergence. There was, however, a clear perceptual correlate of vergence of a very different nature. Concomitant with convergence, the apparent size of the stereogram became smaller, while the image appeared to expand again during divergence. These effects were very robust and strong; the factor of shrinkage for a convergence of 25 deg was estimated as about 3, although there was, of course, no significant change in the size of the retinal image. Such effects have been known since Wheatstone's description in 1852 (see e.g. Heinemann et al., 1959; Komoda and Ono, 1974) and are often described as convergence-micropsia. They are most easily interpreted as a size-constancy effect: when an object gets nearer, the eyes converge and at the same time the retinal image expands. Nevertheless, the apparent size of the object remains constant. In our experiment, convergence increased, while retinal image size stayed constant. Central scaling effects interpreted this as a **shrinkage** of the object. Strange enough, there was no simultaneous perception of **approach** of the object.

Thus, it is clear that vergence signals are used by processes of spatial perception, but apparently they are not interpreted in terms of depth or motion-in-depth. We believe that these results are crucial in understanding the robustness of fusion and stereopsis under natural circumstances. In collaborative work with R.M. Steinman, we have shown in recent years that neither gaze stabilization of the eyes apart, nor the yoking of the two eyes are perfect during natural head movements (Steinman and Collewijn, 1980; Collewijn, Martins and Steinman, 1981; Steinman et

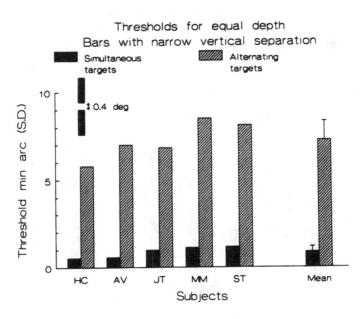

Fig. 4. Disparity thresholds (standard deviations) for equal depth settings of a stereogram with closely adjacent targets, that were shown simultaneously or sequentially.

al., 1985; for a recent review see Collewijn et al., 1990). As a result, ocular vergence may fluctuate over a range of 1-2 deg during natural behavior. As shown by Steinman et al. (1985), such fluctuations in vergence, and thus absolute disparity, do not disrupt stereopsis of a Julesz's type random-dot pattern, even during violent head shaking (see also Patterson and Fox, 1984). Motor control cannot produce the degree of precision that is characteristic for stereopsis and other hyperacuities; the visual system solves this problem by using difference signals (as first proposed by Westheimer, 1979), rather than absolute signals: relative disparities are, by definition, unaffected by eye movements, and can be calculated as long as vergence maintains the eyes within the fusional range, which is typically 1-2 degrees for extended patterns.

SEQUENTIAL STEREOPSIS

We recently followed up these previous observations with some studies of sequential stereopsis. We showed the elements of a two-bar stereogram either simultaneously (with relative disparity) or alternatingly (sequential absolute disparities), and asked subjects to set the absolute disparity of one of the bars for equal apparent distance of the two bars. The standard deviations of these settings are shown for 5 subjects in Fig. 4.

This task was experienced as easy for two, simultaneously shown, vertically aligned bars, separated by a narrow gap, with standard deviations (thresholds) on the order of 1 min arc. Sequential presentation of the bars (with zero time interval) elevated these thresholds roughly by a factor of 10. Lower thresholds could probably have been obtained by training and selecting subjects and using finer stereograms, as suggested by earlier experiments by Foley (1976), Westheimer (1979), Fendick and Westheimer (1983) and Enright (personal communication, 1990), but we were mainly interested in the differences in thresholds, not in the absolute, minimum levels of thresholds. Comparable results were obtained for the same bars, separated by a short (1 deg) horizontal gap.

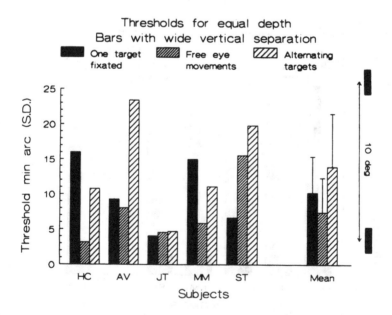

Fig. 5. Thresholds for equal depth settings of the same stereogram as in Fig. 4, but with the bars vertically separated by 10 deg. See text for further explanation.

Thus, the sequential sampling of absolute disparities of two adjacent targets is definitely inferior as a source of depth information to the simultaneous sampling of their absolute disparities, which probably allows a much more accurate computation of the difference in absolute disparities, that is, the relative disparity.

Sequential sampling is also a characteristic of the successive fixation by saccadic eye movements of targets that are separated by some distance . In the past, the question whether such eye movements improve the precision of depth estimates of targets that are not closely adjacent has been raised a number of times (Wright, 1951; Ogle, 1953, 1956, 1962) and taken up recently by Enright (personal communication, 1990). The issue is in fact complex, because several factors are interacting: 1) the decrease of stereo-acuity as a function of eccentricity; 2) the degrading of stereo-acuity with successive instead of simultaneous sampling; 3) the fluctuations in vergence associated with saccades; 4) the perceptual fading of a target in the periphery during steady fixation.

We experimented with a pair of bars, separated vertically or horizontally by 10 deg. Subjects were instructed to set the disparity of one bar for equal distance of the two bars while (1) they fixated one target continuously, or (2) looked freely at either target, or (3) the targets were presented alternatingly (with eye movements free). As pointed out by Fendick and Westheimer (1983), this condition does not differentiate between effects of eccentricity and effects of lateral or vertical separation between the targets, but this confusion is unavoidable in a task involving sequential fixation of the targets.

The results for vertically separated targets are shown in Fig. 5. The results for a similar experiment with the targets separated horizontally by 10 deg were comparable. With steady fixation of one of the bars, stereo-acuity for the 10 deg eccentric target was, in our conditions, inferior to the foveal stereoacuity for alternating targets. We did, in general, find an improvement of stereo-acuity for targets that were 10 deg apart when eye movements were allowed, except in one of our five subjects, who showed an opposite effect. At the average, performance with free saccades between distant, simultaneously presented targets was comparable to performance with closely adjacent targets that were presented sequentially. This seems a reasonable outcome, because successive presentation near the fovea occurs in either of these situations, with the additional assumptions that post-saccadic vergence is reproducible, and that foveal, sequential stereopsis is superior to stereopsis at the particular eccentricity tested. For sequentially presented, non-adjacent targets, performance was worse than with free eye movements between the simultaneously presented targets. Such alternating presentation of distant targets might lead to poorer discrimination of distance, due to saccadic latencies and inaccuracies, which would introduce a significant time gap between the foveation of the two targets.

THE PERCEPTION OF SLANT

So far, we have discussed depth perception only in terms of horizontal vergences and point-to-point disparities. A type of depth perception that may be special, and show a specific relation to torsional eye movements, is the perception of tilt-in-depth, briefly called "slant". A contour that is inclined in the sagittal plane, will be imaged as tilted in opposite lateral directions on the two retina's, and perceived as slanted. The mechanism of this kind of stereopsis is by no means completely clear. In fact, the psychophysical literature on slant-perception is quite meagre, compared to the literature on stereopsis related to simple horizontal disparity.

There are two binocular mechanisms that could serve slant perception: the usual point-to-point horizontal disparities, discussed until now, and a second mechanism, first proposed by Blakemore,

Fiorentini and Maffei (1972) and often called "orientation disparity". This second mechanism would rely on the direct detection of orientation angles, which are also a major trigger-feature of visual cortical neurons. The detection of orientation disparity, or "cyclodisparity", would further require classes of neurons that are sensitive to differences in orientation of contours in the two eyes. A large number of neurophysiological studies has been devoted to the demonstration of such units, but the psychophysical and oculomotor aspects of slant-perception based on orientation disparity have not been recently explored.

Continuing our line of thinking in absolute and relative disparities, we can state that a single, slanted vertical contour, without a background, contains a vertical gradient of absolute, horizontal, position disparities, and therefore, in principle, a gradient of relative, horizontal, position disparity. Thus, a vertical contour contains its own internal reference, and the perception of slant should, in principle, be independent of the presence of additional frames of reference. The same is true for orientation disparity, which is, by definition, an angular difference between the orientation of a similar contour in the two eyes. Both of these types of absolute disparity, and their derived relative disparities, interact with cyclovergence eye movements: cyclodisparity will drive cyclovergence, and cyclovergence will reduce cyclodisparity. This is a potential weakness of cyclo-stereopsis, because it will be vulnerable to imperfections in the binocular coordination of ocular torsion, in the same sense that horizontal absolute disparities are vulnerable to imperfections of horizontal yoking.

Also in this case, a parameter which is independent of eye movements could be the more significant cue to slant-perception. This would be relative cyclodisparity, defined as cyclodisparity with respect to a background; this parameter is insensitive to cyclovergence. The question is then,

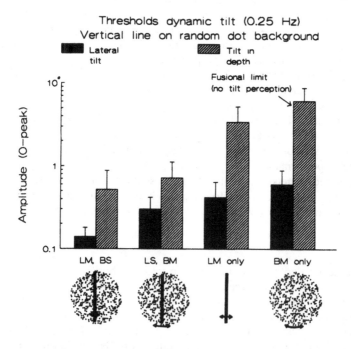

Fig. 6. Threshold settings for the perception of dynamic lateral tilt (induced by conjugate cycloversion of the target) and dynamic slant (induced by cyclovergence of the target). Target conditions depicted at bottom.

whether slant is better observed relatively than absolutely, and whether there are clear indications suggesting a contribution of orientation-disparity in addition to point-to-point disparity.

Before looking at some results, it is useful to examine the geometry of slant-perception.

Slant angles are projected as opposite, lateral tilts on the retina's. This projection leads to considerable reduction of the actual angles under most circumstances, because the distance to the slanting object will usually be much larger than the interocular distance.

Let α be the angle of slant of the target with respect to the vertical; 2β the angle of target cyclovergence, and 2ϕ the angle of horizontal target vergence; then it can be shown that:

$$\tan \beta = \sin \phi \tan \alpha \qquad [1]$$

As a result of this relation, which was derived in almost the same form by Ogle and Ellerbrock (1946), the tilt angles on the retina will be less than one tenth of the slant angles for object distances larger than 65 cm. Thus, we may expect that the perceptual thresholds for slant will be essentially higher than for lateral tilt. The angles discussed in the following always relate to cyclovergence and cyclodisparity; not to the corresponding, much larger, slant angles.

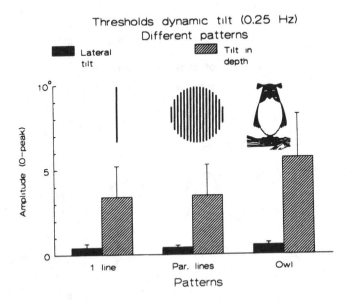

Fig. 7. Threshold settings for the perception of dynamic lateral tilt (induced by cycloversion of targets) and dynamic slant (induced by cyclovergence of targets). Means and standard deviations (bars) for three subjects.

ABSOLUTE AND RELATIVE CYCLODISPARITY

To distinguish between the sensitivity to absolute and relative disparities, we projected a single vertical line and a random dot pattern (diameter about 28 deg) alone, or superimposed on each other. Either the line or the random dot pattern could be oscillated in cyclovergence or, for comparison, in yoked cycloversion. The oscillation frequency was 0.25 Hz. We preferred to measure sensitivity to slant motion rather than to static slant angles, because the latter are known to be affected by adaptive processes tending to adjust the perception of tonically slanted contours to apparently vertical (Ogle and Ellerbrock 1946; De Valois and De Valois, 1988). We determined the perceptual thresholds for lateral tilt and slant in 3 subjects. The mean thresholds and standard deviations (bars) are shown in Fig. 6. During modulation of relative cyclodisparity, the line was perceived as moving in slant, irrespective of whether the line (left columns in Fig. 6) or the background was actually moving (Fig. 6, 2nd columns from left), at thresholds on the order of 0.5 - 1 deg of cyclovergence motion. When the background was removed, and only absolute cyclodisparity of the line was present, thresholds for the perception of slant-motion increased by a factor of about 7 (Fig. 6, 3rd columns from left). When the random-dot background was presented alone, modulation of its absolute cyclodisparity was ineffective in inducing convincing slant perception of the pattern as a whole in any of our subjects, within fusional limits (Fig. 6, rightmost columns). It should also be noted that angular thresholds for slant perception were, under all conditions, markedly higher than for the perception of yoked, lateral tilt. In addition, perception of lateral tilt was veridical, in the sense that subjects recognized whether the line or the background was oscillated.

These observations suggest that: 1) slant perception is more easily induced by relative cyclodisparity than by absolute disparity; 2) absolute cyclodisparity induces slant perception only if the stimulus pattern contains clearly oriented, vertical contours.

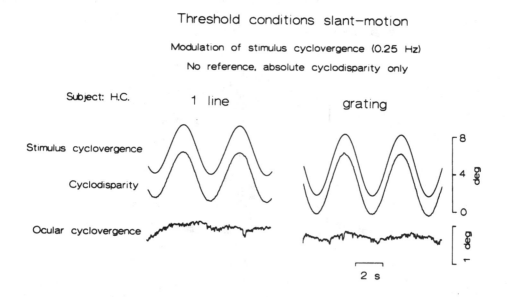

Fig. 8. Modulation of cyclovergence of stimulus and eyes, and of (absolute) cyclodisparity at threshold settings for the perception of slant motion in subject H.C. The stimulus consisted of 1 vertical line or a vertical grating.

These tendencies were supported by further experiments. Slant-motion was perceived not only for a single vertical line, but also for a vertical grating, and an owl-caricature on the basis of absolute cyclodisparity, as shown in Fig. 7. Note again the high thresholds, especially compared to the thresholds for lateral tilt. This shows that the tilting of the contours on the retina is easily perceived, but that tilts that are opposite on the two retina's tend to be cancelled out. Comparable observations have been made several years ago for horizontal vergence and version (Erkelens and Collewijn, 1985a).
Perception of slant-motion was absent or marginal for patterns that lacked clear orientation, such as the random-dot pattern shown before. Also a Julesz stereogram, or even parallel rows of dots were weak in inducing slant perception on the basis of absolute disparity.

However, it is certainly not true that un-oriented patterns such as the random-dot configuration cannot be seen as slanted. When a smaller, second random-dot configuration was superimposed on the large random-dot pattern, and the latter (forming now a background) was oscillated in cyclovergence, slant motion of the smaller configuration was immediately and strongly perceived with a threshold as low as for relative motion of the vertical line on the same background, as shown before in Fig. 6. In this case, a vertical gradient of point-to-point **relative** horizontal disparities was apparently effective in inducing slant, whereas a similar vertical gradient of **absolute** horizontal disparities was completely ineffective. These observations are compatible with the following hypotheses:

1) In patterns that lack clear (vertical) orientation, slant is only induced by the presence of a vertical gradient of **relative** horizontal disparities; absolute horizontal disparities are ineffective, even when they form a gradient;
2) In patterns that are clearly oriented, slant can be perceived on the basis of absolute cyclodisparities of vertical contours. Thresholds are about an order of magnitude higher than for relative disparities; it seems feasible that this slant perception is based on the detection of changes in orientation by oriented visual neurons.

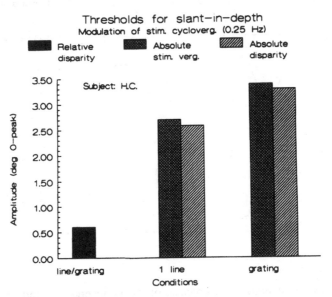

Fig. 9. Thresholds for the perception of slant motion in subject H.C. from relative cyclodisparity (line on grating) or absolute cyclovergence and disparity (line or grating alone).

CYCLOVERGENCE EYE MOVEMENTS

We did record the binocular torsional eye movements elicited by cyclorotation of stimulus patterns. Our recordings show very clearly that ocular cyclovergence and cycloversion responses are generally induced under the conditions of our experiments. The most salient points are, that the magnitude of these torsional eye movements shows large idiosyncratic differences; that even in subjects with relatively large responses compensation is far too incomplete to substantially reduce cyclodisparity with respect to stimulus cyclovergence, that patterns with many contours are more effective than a single line, and lastly, that there is no parallel between the oculomotor and perceptual responses.

As an example, Fig. 8 shows absolute stimulus cyclovergence, ocular cyclovergence and absolute cyclodisparity (the difference between these signals) for one subject at threshold settings for slant motion of a single vertical line or a vertical grating, without any frame of reference. A very small but systematic cyclovergence response was elicited by the vertical grating, and a barely perceptible response by the single vertical line. This subject (H.C.) had low gains of cyclovergence. As a result, the changes in cyclodisparity were almost identical to the changes in cyclovergence of the stimulus. On the other hand, the perceptual thresholds of this subject were slightly lower than those of the other two formally tested subjects.

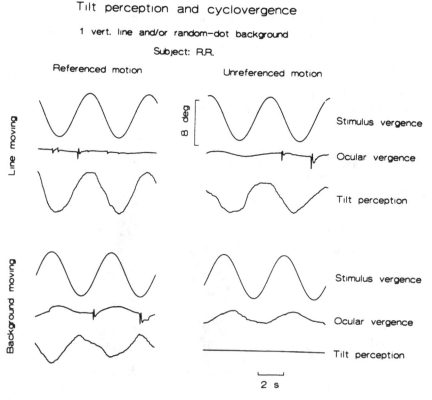

Fig. 10. Recordings of stimulus cyclovergence, ocular cyclovergence, and tilt-in-depth (slant) perception in a subject (R.R.) with good cyclovergence, for the stimulus combinations shown in Fig. 6.

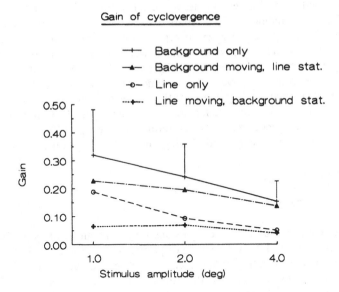

Fig. 11. Gain of cyclovergence (eye movement/stimulus movement) as a function of stimulus amplitude, for the stimulus combinations shown in Fig. 6. Frequency 02. Hz. Means and S.D. for 5 subjects.

These thresholds of subject H.C. are shown in Fig. 9, in which the threshold for the perception of slant motion on the basis of relative disparity (a line superimposed on a grating) is compared to thresholds for a similar perception on the basis of changes in absolute stimulus vergence and cyclodisparity (which are nearly the same) for the line or grating presented alone. These thresholds are comparable to those shown for the corresponding conditions in Fig. 6 (lines) and Fig. 7 (grating). We may conclude that the thresholds for changes in relative cyclodisparity of a vertical line moving on a vertical grating or on a random-dot pattern were similar, and relatively low. The thresholds for changes in absolute cyclodisparity of a single line and a vertical grating, both oriented patterns, were also similar, but much higher.

Other subjects made larger torsional eye movements. Figure 10 illustrates these for subject R.R. for the stimulus condition of a single vertical line and a random dot background, apart or in combination, as shown in Fig. 6. Fig. 10 shows stimulus vergence, ocular vergence and perceived slant (signalled by a joystick) for the 4 situations. The stimulus amplitude remained fixed at 8 deg (peak-to-peak), and does not reflect threshold values. For the line moving upon the background, ocular vergence was nearly absent, but slant perception was strong. In contrast, cyclovergence of the background alone induced relatively strong cyclovergence, but no perceived slant. Thus, changes in absolute cyclodisparity of a non-oriented pattern are detected by the nervous system and used to control cyclovergence. This same signal, however, it is not interpreted as a slant motion of an object.

Average values for gain of cyclovergence (eye movement/stimulus movement) for the pattern combinations shown in Fig. 6 are shown for 5 subjects in Fig. 11. Gain is shown as a function of stimulus amplitude (zero-to-peak), for a fixed stimulus frequency of 0.2 Hz. Gain tends to decrease as a function of stimulus amplitude, but less than proportionally, so that the largest stimulus amplitudes do elicit the largest cyclovergence movements. Mean gain was highest for the motion of the large random-dot pattern alone, but even in this condition it was only about 0.3, with a substantial spread among subjects (bars in Fig. 11 represent standard deviation of intersubject means). These relatively large cyclovergence movements were not associated with any perception of slant motion. Cyclovergence movements were smaller when a stationary vertical line was superimposed on the moving random-dot background. On the other hand, slant motion was strongly perceived in this condition. Even cyclovergence of the vertical line alone induced some significant ocular cyclovergence, while perceptual thresholds for slant motion were relatively high in this condition (Fig. 6). Finally, the line moving on the stationary random-dot pattern elicited virtually no ocular cyclovergence, although in terms of slant perception this situation was equivalent to the one with the background moving and the line stationary.

GENERAL DISCUSSION

Absolute and relative disparities

Our present results reinforce our earlier conclusions (Erkelens and Collewijn, 1985a, b; Regan et al., 1986) on the role of absolute and relative disparities (Fig. 1) in binocular vision. Absolute disparities are the adequate stimulus for the control of vergence eye movements, which will be driven to a position of correspondence for the attended targets. Horizontal absolute disparities are reasonably well compensated by the horizontal vergence system, although substantial errors will occur in a dynamic situation (Fig.3). Absolute cyclodisparities do induce cyclovergence, but the gain of this reflex is too low (Figs. 8, 11) to achieve any significant stabilization of retinal contours in a dynamic condition. Uniform, absolute horizontal disparities caused by such errors in vergence are not interpreted as changes in distance or depth, and do not disrupt stereopsis, as long as fusional limits are not exceeded. Primary sources of such overall non-correspondence of the two retinal images are head movements and imperfections in the coordination of binocular eye movements (for review see Collewijn et al., 1990). In our experiments on stereopsis, this kind of errors is best mimicked by the modulation of absolute disparities of large stimulus configurations.

Modulation of the horizontal absolute disparity of extended patterns such as random-dot stereograms and gratings (Fig. 2) does not induce any perception of motion in depth within the fusional range (or beyond it). We can now extend this conclusion to cyclodisparity. Modulation of the absolute cyclodisparity of visual configurations that lack distinct vertical contours, such as random-dot patterns, does not induce any perception of slant motion (Fig. 6) within fusional limits. For configurations containing vertical contours, slant motion can be perceived on the basis of changes in absolute cyclodisparity, but the thresholds are very high (on the order of several degrees; Fig. 7). We can exclude that these high thresholds were caused by the inability of the visual system to detect the associated retinal image tilts, because **conjugated** cycloversion of the same patterns, in the same sessions, were perceived as distinct lateral tilt motion (in the frontal plane) at thresholds that were an order of magnitude smaller (Fig. 7) than those for the perception of slant motion (in the sagittal plane).

Relative horizontal disparities, i.e., differences in the absolute disparities of two, simultaneously visible, targets, induce stereoscopic depth at exceedingly low thresholds, and modulation of such relative disparity induces a compelling sensation of motion in depth (see Erkelens et al., 1985a, b; Regan et al., 1986). Similarly, relative cyclodisparity induces a compelling perception of slant motion. The threshold for the observation of slant motion of a vertical bar, moving in cyclodisparity upon a stationary background was about 0.5 deg (Figs. 6, 9), while a comparable perception without the random-dot background (absolute cyclodisparity) had a threshold of about

3.5 deg (Fig. 6). Even in the case of relative motion, the perception of the cyclorotation in phase of the bars seen by the right and left eye as tilt in the frontal plane had a much lower threshold than the perception of out-of-phase tilting as slant in the sagittal plane (Fig. 6, left columns).

The perceptual neglect of absolute horizontal disparity

There is something very peculiar about the difficulty in perceiving slant on the basis of absolute cyclodisparity in random-dot patterns, because cyclodisparity of any pattern with a significant vertical extent does imply the presence of a vertical gradient of absolute horizontal disparity. Differencing of such absolute disparities along the vertical meridian should provide a gradient of relative disparity, which should induce the perception of a slanted surface. That a vertical gradient of **relative** disparity does indeed induce the perception of slant, is convincingly shown by the fact that relative cyclodisparity between two uncorrelated random-dot patterns is equally effective in inducing the perception of slant, as relative cyclodisparity between a vertical bar and a random-dot pattern. One possible reason for this result could be a computational constraint. It seems conceivable, that the disparity-processing system would be able to interpret a vertical gradient of locally given relative, horizontal disparities as a slant in the sagittal plane, but that the system would be unable to compute such a vertical gradient of relative disparity directly from a gradient of absolute disparity. However, this constraint seems unlikely because of the exquisite sensitivity of stereopsis to relative disparities between vertically adjacent line segments. It seems more plausible that there is an inhibitory effect of the surround, when large stimulus configurations are subject to uniform modulations of **absolute** disparity.

A parallel can again be drawn between absolute horizontal and cyclo-disparities. Regan et al. (1986) experimented with transitions between a single dot and a random-dot pattern. Modulation of the absolute horizontal disparity of a single dot induced perception of some motion in depth, at a high threshold (see Fig. 2 and 3 for a replication of this phenomenon). This sensation was abolished when the single dot was closely surrounded by randomly distributed, similar dots, which were moved en bloc with the central dot. However, some perception of motion in depth of the central dot alone, relative to the surrounding dots, was restored when the central dot was separated from the surrounding dots by a blank annulus with a width of at least 1-2 deg (Regan et al., 1986, experiment nr. 5). Although the mechanism of this inhibitory surround-effect remains unknown, we may postulate that a similar effect occurs for a uniform field of absolute cyclodisparity. It will remain for future experiments to determine, if blank zones around a central vertically oriented dot configuration amidst a surrounding dot configuration will restore the perception of slant.

Orientation disparity

A more fundamental question is, whether cyclodisparity can indeed be reduced to a gradient of horizontal disparities along the vertical meridian, or should rather be regarded as a composite signal, which truly reflects a cyclorotational vector, computed by synthesizing horizontal and vertical disparities, or by a totally different mechanism. An obvious neural encoding of the cyclorotation of retinal images is at hand in the form of the orientation-tuning of neurons in the visual cortex. Orientation disparity as the basis of slant perception, based on cortical orientation detection, was proposed originally by Blakemore et al. in 1972 (for a recent review, see De Valois and De Valois, 1988). Orientation detection is essentially a monocular mechanism, which can be elaborated independently of the detection of horizontal or vertical disparity. Binocular synthesis of orientation sensitivity may include the tuning of binocular neurons to specific differences in orientation of similar contours in the two eyes. This would constitute a representation of absolute cyclodisparity, or orientation disparity.

Our observations support a role of absolute orientation disparity, independently of position disparity, in the perception of slant. The pertinent evidence is, that modulation of absolute cyclodisparity of patterns containing vertical contours does lead to the perception of slant-motion (albeit at rather high thresholds), whereas similar motion of random dot patterns, which lack oriented contours, does not induce perception of slant motion. Thresholds are roughly similar for the perception of slant of a single line or a grating (Fig. 7, 9). This suggests that there is little interaction between the signals related to orientation changes of parallel contours.

Thus, there is reasonably good evidence that the visual system can use absolute orientation disparities in driving cyclovergence, as well as in perceiving slants. Whether in addition a **relative** cyclodisparity signal is computed, in parallel to the processing of horizontal position disparities, is not clear from the present experiments. The presence of a reference pattern dramatically lowers the threshold for slant-perception of a vertical line, but this effect is quantitatively similar for a random dot or a vertical grating as background (Figs. 6 and 9). This suggests that, in the presence of a background, slant perception is primarily mediated by relative, horizontal position disparity, which is intrinsically more sensitive than absolute orientation disparity as a cue to the perception of slant. Orientation disparity would then mainly serve as a back-up system for the perception of slant, in the case that no reference is available. In view of the high thresholds in terms of projected tilt in the frontal plane, and the relation given in eqn. [1], unreferenced slant of distant objects would have to be very large to be detectable by this mechanism alone, in the absence of secondary cues such as perspective, size, shadowing etc. We are unaware of relevant literature to this issue, except for the old report by Ogle and Ellerbrock (1946), which is somewhat hard to interpret in the light of the probable presence of frames of reference and other secondary cues.

The role of eye movements in depth-perception

There are two sides to the interaction between eye movements and stereopsis. On the one hand, eye movements assist binocular processes by bringing the two retinal images into correspondence by vergence movements, and by binocularly foveating significant features of visual targets. On the other hand, imperfections in oculomotor control are potentially disruptive to stereopsis, because they have effects on disparity in several ways. The crucial issue here is, that stereopsis at its best is much more finely tuned than any type of binocular eye movement.

The insensitivity of depth-perception to rather large but uniform absolute disparities of extended configurations may be interpreted as a protection of spatial vision against oculomotor imperfections. Under any normal conditions, i.e., in humans that are moving freely around, considerable errors in horizontal binocular alignment will occur continuously (see Steinman et al., 1982). Nevertheless, stereopsis is not noticeably impaired by free behavior, or even by violent head motion (Steinman et al., 1985; Collewijn et al., 1990, Patterson and Fox, 1984), nor is it disrupted or modified significantly by experimentally imposed absolute horizontal disparities (see Regan et al., 1986). With regard to errors of cyclodisparity, matters are slightly more complicate. For targets that are separated vertically by some distance, cyclovergence errors will affect the absolute disparities of the different targets not uniformly, but as a function of vertical position. Therefore, the relative disparity between vertically separated targets will change as a result of changes in cyclovergence (Enright, 1990). The magnitude of this effect will be proportional to the variation in cyclovergence. Recent work of Enright (1990), based on measurements of ocular torsions in video-images of the two eyes, gives an estimate of the standard deviation of cyclovergence, in the course of repeated fixations, of about 17 min arc. Our current impressions from binocular 3-D measurements with scleral coils at high resolution (Van Rijn and Van der Steen, in preparation) suggest that this order of magnitude is correct. Earlier coil measurements (Collewijn et al., 1985) also showed that torsional eye movements induced by oscillation of the head in roll were largely conjugate. Assuming 17 min arc as a reasonable estimate of the S.D. of cyclovergence, this would lead to an S.D. of 2.6 min arc of the relative, horizontal disparity in a stereogram in which the elements are separated vertically by 10 deg. On the other hand, such torsional variability would have only negligible effects on the relative, horizontal disparity of targets that are horizontally separated by a similar distance. The same instability of cyclovergence would, of course, lead to an S.D. of orientation disparity of 17 min arc. The question is, how seriously errors of this magnitude affect stereopsis.

Under our present conditions (Fig. 5), the S.D. of disparity settings for equal depth of two bars, that were vertically separated by 10 deg, was usually considerably larger than 2.6 min arc when one of the two targets was steadily fixated. Moreover, these thresholds were not markedly different for targets separated horizontally by 10 deg. Thus, it seems unlikely that instability of cyclovergence was a limiting factor in our conditions. Likewise, Enright (1990) found for a similar experiment with vertically separated targets an S.D. for disparity settings that was larger than the 2.6 min arc "best possible" minimum, at least during steady fixation. With regard to orientation disparity, our thresholds for absolute orientation disparity to be perceived as slant were an order of magnitude larger than Enright's (1990) estimate of cyclovergence instability (Figs. 6, 9). In summary, the evidence does not suggest that instability of cyclovergence is large enough to compromise stereoacuity.

On the other hand, there is now clear evidence that saccadic eye movements between targets, that are separated by some distance, may improve stereoacuity, as shown before by Wright (1951), Ogle (1962) and Enright (1990), who coined the term "sequential stereopsis". We confirm this effect (Fig. 5). The effect is complicated by the trade-off between a number of factors. Stereoacuity degrades with increasing target separations and eccentricity; stereoacuity also degrades when the elements of a stereogram are viewed successively instead of simultaneously. The latter point was demonstrated in detail by Westheimer (1979), who found that stereoacuity for fine, foveal stereograms, viewed under optimal circumstances by highly experienced subjects, as measured by accurate psychometric methods, increased by a factor of about 10 when targets were presented successively, rather than simultaneously. Sequential stereoacuity thresholds were about 1 min arc (see also Foley, 1976). Using less refined methods, we found a similar ratio (Fig. 4) for closely adjacent targets, although the absolute level of our thresholds was higher. In the sequential condition, there is successive sampling of absolute disparities, which can be compared only across time. It would be hard to imagine how sequential stereopsis achieved by saccades could ever be better than sequential stereopsis for foveal targets, and indeed we found comparable average stereoacuity values for sequentially presented foveal targets and sequentially fixated peripheral targets (Figs. 4, 5). A same effect was described by Enright (1990), with the important difference that, with eye movements of about 10 deg, he found stereoacuities as low as 1 min arc, which is as good as has ever been demonstrated for successive target presentations (Westheimer, 1979; Foley, 1976). Our values were higher; this may be related to details of the experimental conditions. In any case, the trends are similar to Enright's findings. Although these results appear to argue for some role of eye movements in improving stereopsis for non-adjacent targets, the present results, obtained for isolated and small targets, may not necessarily be representative for every-day conditions. As most natural targets are embedded in a rich visual context, it will be important to investigate in how far eye movements enhance depth-estimates in normal conditions, in which relative, rather than absolute disparities will be sequentially sampled.

REFERENCES

Blakemore, C., Fiorentini, A. & Maffei, L. (1972): A second neural mechanism of binocular depth discrimination. *J. Physiol. (Lond.)* 226, 725-749.

Collewijn, H. & Erkelens, C.J. (1990): Binocular eye movements and the perception of depth. In *Eye movements and their role in visual and cognitive processes, Reviews of oculomotor research, Vol. 4*, ed. E. Kowler, pp. 213-261. Amsterdam: Elsevier.

Collewijn, H., Martins, A.J. & Steinman, R.M. (1981): Natural retinal image motion: origin and change. *Ann. NY Acad. Sci.* 374, 312-329.

Collewijn, H., Van der Steen, J., Ferman, L. & Jansen, T.C. (1985): Human ocular counter roll: assessment of static and dynamic properties from electromagnetic scleral coil recordings. *Exp. Brain Res.* 59: 185-196.

Collewijn, H., Steinman, R.M., Erkelens, C.J. & Regan, D. (1990): Binocular fusion, stereopsis and stereoacuity with a moving head. In *Vision and visual dysfunction, Vol. 10A: Binocular vision*, ed. D. Regan, pp. 121-136. London: MacMillan.

De Valois, R.L. & De Valois, K.K. (1988): *Spatial vision*, pp. 305-307. New York: Oxford University Press.

Enright, J.T. (1989): Convergence during human vertical saccades: probable causes and perceptual consequences. *J. Physiol. (Lond.)* 410, 45-65.

Enright, J.T. (1990): Stereopsis, cyclotorsional "noise" and the apparent vertical. *Vision Res.* 30, 1487-1497.

Erkelens, C.J. & Collewijn, H. (1985a): Motion perception during dichoptic viewing of moving random-dot stereograms. *Vision Res.* 25, 583-588.

Erkelens, C.J. & Collewijn, H. (1985b): Eye movements and stereopsis during dichoptic viewing of moving random-dot stereograms. *Vision Res.* 25, 1689-1700.

Fendick, M. & Westheimer, G. (1983): Effects of practice and the separation of test targets on foveal and peripheral stereoacuity. *Vision Res.* 23, 145-150.

Foley, J.M. (1976): Successive stereo and vernier position discrimination as a function of dark interval duration. *Vision Res.* 16, 1269-1273.

Heineman, E.G., Tulving, E. & Nachmias, J. (1959): The effect of oculomotor adjustments on apparent size. *Am. J. Psychol.* 72, 32-45.

Komoda, M.K. & Ono, H. (1974): Oculomotor adjustments and size-distance perception. *Percept. Psychophys.* 15, 353-360.

Nelson, J. (1977): The plasticity of correspondence: after-effects, illusions and horopter shifts in depth perception. *J. Theor. Biol.* 66, 203-266.

Ogle, K.N. (1953): The role of convergence in stereoscopic vision. *Proc. Phys. Soc.* 66B, 513-514.

Ogle, K.N. (1956): Stereoscopic acuity and the role of convergence. *J. Opt. Soc. Am.* 46, 269-273.

Ogle, K.N. (1962): Spatial localization through binocular vision. In *The Eye, Vol. 4*, ed. H. Davson, pp. 316-319. New York: Academic Press.

Ogle, K.N. & Ellerbrock, V.J. (1946): Cyclofusional movements. *Arch. Ophthalmol.* 36, 700-735.

Patterson, A. & Fox, R. (1984): Stereopsis during continuous head motion. *Vision Res.* 24, 2001-2003.

Rashbass, C. & Westheimer, G. (1961): Disjunctive eye movements. *J. Physiol. (Lond.)* 159, 339-360.

Regan, D., Erkelens, J.C. & Collewijn, H. (1986): Necessary conditions for the perception of motion in depth. *Invest. Ophthalmol. Vis. Sci.* 27, 584-597.

Steinman, R.M. & Collewijn, H. (1980): Binocular retinal image motion during active head rotation. *Vision Res.* 20, 415-429.

Steinman, R.M., Cushman, W.B. & Martins, A.J. (1982): The precision of gaze. *Human Neurobiol.* 1, 97-109.

Steinman, R.M., Levinson, J.Z., Collewijn, H. & Van der Steen, J. (1985): Vision in the presence of known natural retinal image motion. *J. Opt. Soc. Am. A* 2, 226-233.

Westheimer, G. (1979): Cooperative neural processes involved in stereoscopic acuity. *Exp. Brain Res.* 36, 585-597.

Wheatstone, C. (1852): Contributions to the physiology of vision. Part the second. On some remarkable, and hitherto unobserved, phenomena of binocular vision. *Phil. Trans. R. Soc.* 142, 1-17.

Wright, W.D. (1951): The role of convergence in stereoscopic vision. *Proc. Phys. Soc. B* 64, 289-297.

The parsing of optic flow by the primate oculomotor system

F. A. Miles, U. Schwarz and C. Busettini

Laboratory of Sensorimotor Research, National Eye Institute,
Building 10 Room 10C101, Bethesda, MD 20892, USA

Eye movements function to improve vision, in part by preventing excessive slippage of the image(s) on the retina. Tasks dependent on good visual acuity are seriously impaired if retinal images are allowed to drift at speeds in excess of a few degrees per second and, should the image of interest move, then the oculomotor system will track it. Motion of the observer poses a serious threat to the stability of retinal images and is dealt with chiefly by labyrinthine reflexes that generate compensatory eye movements. The primate labyrinth has two types of receptor organs sensing head movements, the semicircular canals and the otoliths, that respond selectively to rotational and translational accelerations, respectively, and provide the input for two vestibuloöcular reflexes that compensate selectively for rotational and translational disturbances of the head. These labyrinthine reflexes, which we shall refer to as the rotational and translational vestibuloöcular reflexes (RVOR and TVOR), operate open-loop insofar as they produce an output, *eye* movement, that does not influence their input, *head* movement. One serious consequence of this is that if these open-loop reflexes fail to compensate completely, which is not uncommon, then the eye movements will not completely offset the head movements and the image of the world on the retina will tend to drift at such times. However, such retinal image slip activates visual tracking mechanisms that operate as closed-loop negative feedback systems to produce eye movements which reduce the slip. In studies of these visual backup systems it has been usual to consider only rotational disturbances of gaze and to examine the associated visual compensatory mechanisms by rotating the visual surroundings around the stationary subject, the ensuing ocular following being termed *optokinetic nystagmus* (OKN). Only recently have translational visual stimuli been introduced, though studies using conventional optokinetic stimuli may also be relevant here. Thus, the primate optokinetic response has two components and we have recently suggested that, in the real world, only one of these deals mainly with rotational disturbances, the other being more concerned with translational problems.

THE TWO COMPONENTS OF THE PRIMATE OPTOKINETIC RESPONSE

In the traditional optokinetic test situation the stationary subject is usually seated inside a cylindrical enclosure with vertical patterned walls that can be rotated about a vertical axis. Usually, the experiment starts with the subject in the dark to allow the cylinder time to reach the desired constant speed, at which point the lights are turned on for a period generally lasting many seconds. When the lights come on the subject tracks the continuously moving walls of the cylinder with his eyes (slow phases), necessitating regular saccades (quick phases) to reset/recenter the eyes: OKN. If the cylinder rotates rapidly enough, the slow phases do not immediately reach the speed of the cylinder and the development of the response shows two distinct episodes that are generally thought to reflect two distinct mechanisms: **1)** An initial rapid rise in slow-phase eye speed during the first few hundred milliseconds that generally leaves the eyes somewhat short of the cylinder speed; this has been termed the "direct" component of OKN by Cohen *et al* (1977) but we shall refer to it simply as the *early* component (OKNe). 2) A subsequent gradual increase in slow-phase eye speed extending over a period of perhaps half a minute during which the eyes reach an asymptotic speed more nearly approaching that of the cylinder; this has been termed the "indirect" component by Cohen *et al* (1977) but we shall refer to it as the *delayed* component (OKNd). Cohen and his coworkers attributed OKNd to the gradual charging up of a central velocity storage integrator (Cohen *et al.*, 1977; Matsuo & Cohen, 1984; Raphan *et al.*, 1977; Raphan *et al.*, 1979). When the lights are finally extinguished, the sequence of events is reversed: there is an initial rapid drop in eye speed reflecting loss of OKNe, followed by a roughly exponential drop extending over a period of many seconds reflecting the more gradual loss of OKNd, perhaps due to discharge of the proposed velocity storage integrator (optokinetic afternystagmus, OKAN). This OKAN can be followed by a further period of nystagmus in the opposite direction (OKAAN) before the eyes finally come to rest.

We have recently suggested that only OKNd evolved as a backup to the RVOR to help compensate for rotational disturbances of gaze, OKNe evolving as a backup to the TVOR to help compensate for translational problems (Miles, in press; Miles *et al.*, 1989; Miles *et al.*, in press). The major distinguishing features of these two proposed visual tracking mechanisms are listed in Table 1. While the decomposition of head movements into rotational and translational components by the labyrinth is complete and unequivocal - at least in primates - the decomposition of optic flow into rotational and translational components by the optokinetic system is incomplete and has only indirect supporting evidence. Before considering this evidence it is necessary to discuss some general aspects of the optic flow experienced by the moving observer and of its analysis by the optokinetic system.

Table 1. The major features of the two visual tracking systems which we hypothesize evolved to deal with rotational and translational disturbances of gaze.

THE ROTATIONAL MECHANISM	THE TRANSLATIONAL MECHANISM
1. *Delayed* component of OKN: - Long time-constant. - Strong after-nystagmus. - Sensitive to low acceleration/speed.[a] - Centripetal/centrifugal asymmetry.[c]	1. *Early* component of OKN: - Short time-constant. - Weak after-nystagmus. - Sensitive to high acceleration/speed.[ab] - No centripetal/centrifugal asymmetry?
2. Backup to canal-ocular reflex (RVOR)? - Sensitive to gain of RVOR[d] - Insensitive to gain of TVOR?	2. Backup to otolith-ocular reflex (TVOR)? - Insensitive to gain of RVOR[d] - Sensitive to gain of TVOR.[e]
3. Organized in canal planes.[f]	3. Organized in otolith planes?
4. Helps to stabilize gaze against global disturbances? - Cannot utilize motion parallax? - Insensitive to disparity? - Input drive from entire visual field?	4. Helps to stabilize gaze on the local depth plane of interest? - Can utilize motion parallax.[b] - Sensitive to disparity.[g] - Input drive from binocular field only.[b]
5. Pretectum/Accessory Optic System.[h]	5. Direct Cortico-Pontine-Cerebellar System.[j]
6. In all animals with mobile eyes?	6. Only in animals with good binocular vision?

[a] Cohen, Matsuo & Raphan (1977); Zee, Tusa, Herdman, Butler & Gucer (1987). [b] Miles, Kawano & Optican (1986). [c] Van Die & Collewijn (1982); Naugele & Held (1982); Westall & Schor (1985); Ohmi, Howard & Eveleigh (1986). [d] Lisberger, Miles, Optican & Eighmy (1981). [e] Schwarz, Busettini & Miles (1989). [f] Soodak & Simpson (1988); Simpson, Leonard & Soodak (1988); Leonard, Simpson & Graf (1988); Graf, Simpson & Leonard (1988). [g] Howard & Gonzalez (1987). [h] Schiff, Cohen, Buttner-Ennever & Matsuo (1990). [j] Kawano (in press).

OPTIC FLOW

In discussing the visual motion experienced by the moving observer it is usual to consider the retinal image as distributed over the surface of a sphere and created by projection through a single point at the center of that sphere. Thus, the center of the sphere can be regarded as the vantage point. Neglecting for the moment any compensatory eye movements, pure rotation of the observer results in a rigid rotation of the spherical optic array, the flow pattern resembling circles of latitude: see Fig. 1A. The direction and speed of flow at all points are dictated entirely by the observer's own motion and, in principle, the retinal image motion can be entirely eliminated by appropriate compensatory eye movement: the angular velocity of the eye in the orbit must simply match the angular velocity of the observer's head in space.[1] In contrast, pure translation of the observer creates a flow pattern resembling circles of longitude with lines emerging from one pole ahead and disappearing into another behind: see Fig. 1B. In this situation, the pattern (or direction) of flow again depends solely on the observer's motion, but the *speed* of the flow at any given point also depends on the *viewing distance* at that location. In fact, to a first approximation retinal image speed is inversely proportional to the viewing distance during lateral translation of the observer's eye, i.e., motion with a component orthogonal to the line of sight (Schwarz *et al.*, 1989). Thus, translation causes complex image shear as the nearby objects move across the field of view more rapidly than the distant ones (motion parallax) and a given compensatory eye movement can only eliminate the retinal image motion of objects stationed at a particular viewing

---

1 Actually, the eyes lie some distance in front of the axis of rotation of the head and so undergo some translation during normal head turns. In fact, the RVOR takes account of this (Biguer & Prablanc, 1981; Blakemore & Donaghy, 1980; Gresty & Bronstein, 1986; Gresty *et al.*, 1987; Hine & Thom, 1987; Viirre *et al.*, 1986).

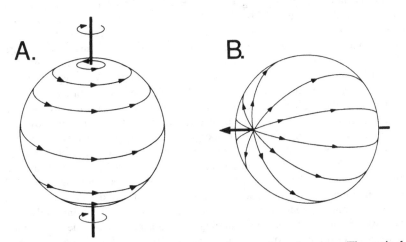

Figure 1. The optic flow pattern experienced by the moving observer. The retinal image is assumed to be distributed over the surface of a reference sphere whose center is the vantage point. We here neglect compensatory eye movements and consider the path taken by the images when the observer rotates about the vantage point (A) or moves in a straight line (B). In both A and B the pattern (or direction) of the flow is solely determined by the observer's motion; however, in B (but *not* in A) the speed of flow at any given point also depends (inversely) on the viewing distance at that location. (From Miles, in press.)

distance. Clearly, normal head movements generally have both rotational and translational components and we are suggesting that, in primates, any residual shifts of gaze are dealt with by the two components of the optokinetic response operating largely independently, the one to derive the global flow vector associated with rotational disturbances of gaze and the other to derive the local flow vector associated with the object of regard - as distinct from its surroundings - during translational disturbances of gaze.

SOME EVOLUTIONARY CONSIDERATIONS

In thinking about the primate optokinetic system's parsing of optic flow patterns it is useful to consider the evolutionary steps that might have led to its development. Two points are critical for understanding the hypothesis that we wish to develop. Firstly, in our view, the two components of the primate optokinetic response are the product of sequential - rather than parallel - evolution. Secondly, evolutionary processes addressed the *specific* problems of normal everyday conditions and did not solve the *general* problem, i.e., did not attempt a true mathematical decomposition of optic flow into rotational and translational components. In our view, the latter was neither necessary nor possible.

All animals with mobile eyes appear to have both canal-ocular and optokinetic reflexes, and we suggest that the latter originally evolved in lateral-eyed animals as a backup to the former, dealing almost exclusively with rotational disturbances of gaze. This primordial optokinetic system is still evident in contemporary primates as OKNd and developed special features that rendered it largely blind to the commonest translational disturbances of gaze - those associated with forward locomotion. Indeed, we suggest that the primordial visual system relied on the motion parallax associated with locomotion to discern the 3-D structure of the surrounding world. Further, we assume that up to this point the oculomotor rôle of the otolith organs was limited to maintaining the orientation of the eyes with respect to gravity during head tilts, providing the tonic drive for *sustained* counter-rolling of the eyes to supplement the *dynamic* counter-rolling produced by canal-ocular reflexes. This is still the primary rôle of the otolith input in lateral-eyed animals like the rabbit (Baarsma & Collewijn, 1975). However, as foveal vision emerged - together with frontal vision, vergence eye movements and stereopsis - we suggest that the challenge now was to stabilize binocular gaze on the depth plane containing the object of regard and translational disturbances could no longer be tolerated. We propose that the "new" labyrinthine and visual reflexes - the TVOR and OKNe - evolved primarily to deal with this problem. Of course, the mechanisms compensating for rotation - the RVOR and OKNd - were still required and still in place, and no doubt these older systems were improved upon during subsequent evolution. However, the uncompensated translation confronted the system with a radically different challenge. At some point - perhaps linked to the increasing emergence of manual feeding, which dispensed with one major reason for prolonged head tilting - the need for tonic counter-rolling largely disappeared, leaving the oculomotor system free to evolve a new use for otolith information - to compensate for linear accelerations of the head, i.e., the TVOR. However, to be optimally effective, the output of the TVOR should be inversely proportional to the viewing distance, necessitating range-finding mechanism(s). One cue to viewing distance - the accommodative effort used to focus the eyes on the image(s) of interest - had long been available to the system and a new one emerged with the development of frontal vision - the vergence effort used to align the two eyes on the object of regard. Recent experiments on man and monkeys indicate that the output of the TVOR is appropriately modulated with viewing distance and uses both vergence and accommodation cues, though other cues such as perspective, size, overlay, texture and so forth may also be involved (Baloh *et al.*, 1988; Berthoz *et al.*, 1987; Bronstein & Gresty, 1988; Buizza *et al.*, 1979; Israël & Berthoz, 1989; Paige, 1989; Schwarz *et al.*, 1989; Vidic *et al.*, 1976). The task confronting OKNe, the new visual backup to the TVOR, was to distinguish the object of regard from its surroundings, a task perhaps made even more difficult once the TVOR (and OKNe) begin to compensate and, thereby, create a whole new swirling pattern of optic flow as the foreground and background images go their separate ways.

We think that such complex visual motion necessitated cortical processing. Thus, ocular compensation for translation added a further dimension to an already complex situation and, in our view, played a critical rôle in refining the visual input driving OKNe.

OKNd: BACKUP TO THE RVOR?

The delayed component of the primate optokinetic response (OKNd) is probably homologous to the classic optokinetic reflex that seems to be ubiquitous in animals with mobile eyes and has been studied intensively in rabbits. Recordings from single neurons assumed to mediate OKN in the rabbit suggest that the RVOR and OKN in this species share the same frames of reference, facilitating their orderly summation: individual cells can have very large visual receptive fields and respond best to global rotations about axes that generally coincide with the rotational axes that are optimal for activating particular semicircular canals (Graf et al., 1988; Leonard et al., 1988; Simpson et al., 1988; Soodak & Simpson, 1988). For example, the anterior canals are oriented 45° to the parasagittal plane and the head rotations that are best for activating, say, the right anterior canal involve forward pitch combined with rightward roll. If during such head movements the signals emanating from this canal were compromised then the subject would see upward motion in the right anterior visual field and downward motion in the left posterior field: this is exactly the pattern of motion preferred by one class of visually driven cells.

The rabbit's oculomotor system is relatively insensitive to both the visual and the vestibular consequences of forward locomotion (Baarsma & Collewijn, 1975; Collewijn & Noorduin, 1972), and it is generally supposed that this is desirable in this lateral-eyed animal: divergence of the eyes would disrupt the images of the scene ahead, reducing the animal's visual capabilities in the very region of the field where potentially interesting new images are being sought (Howard, 1982). However, the mere fact that the visual receptive fields mediating OKN are organized to respond optimally to rotations does not render them totally insensitive to such translation. The rabbit's eyes are not perfectly yoked but even if they were, i.e., received the same pooled drive signals from the entire (binocular) visual field and responded conjugately, differences in the three-dimensional arrangement of objects on either side of the animal would cause asymmetrical optic flow speeds that might be expected to result in a net optokinetic response towards the side of the nearer objects. Any internal imbalance in the sensitivity to motion emanating from the two eyes would have a similar effect.

Significantly, the rabbit's optokinetic system has at least three special features that operate specifically to reduce its sensitivity to the nasal-to-temporal motion created by the animal's forward motion. Firstly, nasal-to-temporal motion generally has only a suppressive effect on the activity of the neurons in the optokinetic pathway so that the sole drive here is the withdrawal of the resting maintained discharge, which is often low (Collewijn, 1975; Simpson et al., 1988; Soodak & Simpson, 1988). The effect here is akin to that of a rectifier. Secondly, these neurons are insensitive to motion in the lower visual field (Collewijn, 1975; Simpson et al., 1981; Simpson et al., 1979), thereby excluding a major potential source of translational contamination since objects in the lower visual field are likely to be the ones most near and hence their retinal images most sensitive to translation. Thirdly, these neurons are insensitive to high speeds. As Dr. J. I. Simpson has pointed out to us (personal communication), this is appropriate for a visual backup to the RVOR, since the latter functions sufficiently well that any residual disturbances of gaze due to uncompensated rotation result only in low-speed retinal slip, hence high-speed slip must generally emanate from translational disturbances of gaze.

Unfortunately, such single-unit data are not available for monkeys. However, some years ago it was shown that optically induced changes in the gain of the monkey's RVOR are associated with parallel changes in OKNd but, significantly, *not* in OKNe (Lisberger et al., 1981), leading to the suggestion that OKNd shares brain stem circuitry with the RVOR and that the two combine to compensate selectively for rotational disturbances of the observer. This concept is illustrated in the block diagram in Fig. 2A, in which the RVOR and OKNd share the variable gain element, G, underlying adaptive gain control in the RVOR. (Note also the shared velocity storage integrator.)

Frontal-eyed animals such as ourselves, monkeys and cats, mainly see an expanding flow field during forward movement (centrifugal optic flow), and no eye movement can compensate entirely for this. However, we assume that the fovea takes precedence, and the compensatory eye movement required to maintain a stable foveal image depends where the animal directs its gaze in relation to the pole of the flow field. [2] If gaze is aimed directly at the pole of the flow field, pure foveofugal retinal image motion will result and no compensatory eye movements are called for to stabilize the foveal image. Once more, however, differences in the three-dimensional arrangement of objects to either side of the lines of sight can cause asymmetric optic flow speeds that might tend to deviate the eyes. Once again the system seems to deal with this potential problem by means of selective directional asymmetries: when tested monocularly with stimuli confined to one hemifield, human OKN is weaker when the motion is centrifugal than when it is centripetal (Naegele & Held, 1982; Ohmi et al., 1986; Van Die & Collewijn, 1982; Westall & Schor, 1985). There is also evidence that primate OKNd - like the rabbit's optokinetic system - is insensitive to high accelerations/speeds (Zee et al., 1987). Given the effectiveness of the

---

2　The eye movements generated by the TVOR in response to forward motion are dependent on the gaze angle, operating to increase any eccentricity of the eyes with respect to the direction of heading. Thus, if the observer's gaze is directed downwards during forward motion then his/her compensatory eye movements will be downward, while if gaze is directed to the right of the direction of heading then the compensatory eye movements will be rightward, and so forth (Paige et al., 1988).

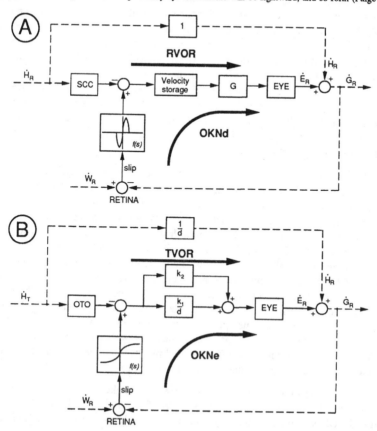

Figure 2. Block diagrams showing the proposed linkages between the visual and vestibular reflexes operating to stabilize gaze. **A:** The open-loop RVOR and the closed-loop OKNd generate eye movements, $\dot{E}_R$, that compensate for rotational disturbances of the head, $\dot{H}_R$. These reflexes share i) a velocity storage element, which is responsible for the slow build-up in OKN and the gradual decay in RVOR with sustained rotational stimuli, and ii) a variable gain element, G, which mediates long-term regulation of RVOR gain. SCC, semicircular canals. The element, $f(s)$, indicates that the visual input is sensitive to low slip speeds only. (After Miles, Schwarz & Busettini, in press.) **B:** The open-loop TVOR and the closed-loop OKNe generate eye movements that compensate for translational disturbances of the head, $\dot{H}_T$, which affect gaze in inverse proportion to the viewing distance, d. These reflexes share i) a variable gain element, k_1/d, which gives them their dependence on proximity, and ii) a fixed gain element, k_2, which gives them an offset. We hypothesize that velocity saturation in the feedback path from the retina, $f(s)$, which was described by Miles *et al* (1986), tends to offset the influence of the variable gain element on the optokinetic response under normal viewing conditions: Since retinal slip speeds will tend to vary inversely with viewing distance, ocular following will tend to show increasing saturation with near viewing. Thus, under everyday conditions OKNe may show little dependence on viewing distance. OTO, otolith organs. Dashed lines represent physical links: $\dot{H}_T$, head velocity in linear coordinates; $\dot{H}_R$, $\dot{E}_R$, $\dot{G}_R$ and $\dot{W}_R$, velocity of head, eyes (in head), gaze and visual surroundings, respectively, in angular coordinates. (From Schwarz *et al.*, 1989.)

primate RVOR, it seems likely that rotational disturbances of gaze are generally minor - and therefore within the range of OKNd - while translational ones (at least the worst) are beyond its range and therefore ignored.

Data such as the above lead us to suggest that the classical optokinetic system - manifest in primates as OKNd - originally evolved to deal selectively with rotational disturbances of gaze. Note, we are *not* suggesting that the rotational component of optic flow is extracted by a true mathematical decomposition. Rather we envisage a system that was originally organized to respond optimally to rotational flow patterns *and* developed a number of special features to reduce the likelihood of a response to translation, especially forward locomotion. These special features are critical and, in animals like the rabbit that lack significant translational compensatory mechanisms, we further suggest that the visual system may actually depend upon the optic flow associated with forward motion to map out the three-dimensional layout of the world to either side perhaps using relative-motion detectors such as those described in pigeon tectum (Frost & Nakayama, 1983; Frost *et al.*, 1981).

OKNe: BACKUP TO THE TVOR?

The early component of the primate optokinetic response (OKNe) has no homology in lateral-eyed animals and, we suggest, evolved to deal with the visual stabilization problems of frontal-eyed animals. We have recently shown that OKNe shares the TVOR's dependence on proximity, consistent with the notion that the two systems are synergistic, sharing central pathways and compensating selectively for translational disturbances of the observer (Schwarz *et al.*, 1989). In these experiments, the visual scene was back-projected onto a tangent screen in front of the stationary animal and laterally translated. The animal's initial visual tracking responses - which we assume to be mediated by the same mechanisms as OKNe - were inversely related to the viewing distance. (Note that this dependence on proximity was truly neuronal and not due simply to the optics of the situation. Obviously, reducing the viewing distance normally increases the size and speed of the retinal image of a given moving object but, in reducing the viewing distance in these experiments, care was always taken to scale the speed and size of the moving stimulus on the screen so that the **retinal stimulus was always the same**.) In our proposed scheme for explaining these findings (see Fig. 2B), the TVOR and OKNe share two gain elements: a variable one (k_1/d, where k_1 is a constant and d is the estimate of target distance), which gives the dependence on proximity, and a fixed one (k_2), which accounts for the slight offset in our data. The variable gain element effectively allows the TVOR to receive inputs encoded in cartesian coördinates [translational velocity of the head ($\dot{H}_T$)] and to respond with outputs encoded in polar coördinates [rotational velocity of the eyes ($\dot{E}_R$)]. In fact, as Viirre et al (1986) have pointed out, we are dealing with vectors and the correct setting of the gain in three dimensions requires the computation of a cross product. It is significant that the two models in Fig. 2 depicting the RVOR and TVOR, with their respective visual backups, OKNd and OKNe, have a strong structural similarity.

To be effective, a visual backup to the TVOR must be able to single out the depth plane of interest and deal with the motion parallax associated with translation, which is a rich source of potentially conflicting motion cues for a visual stabilization mechanism. One option for the system here would be to abandon the global analysis of flow fields and concentrate solely on the local flow in the region of particular interest. In fact, it is known that human subjects can direct their attention to a restricted part of the visual field and respond selectively to the optic flow in that region regardless of whether it is foveal or extra-foveal (Cheng & Outerbridge, 1975; Dubois & Collewijn, 1979; Howard et al., 1989; Murasugi et al., 1986; Van Den Berg & Collewijn, 1987). Indeed, the need to concentrate from time to time upon selected elements of the shifting

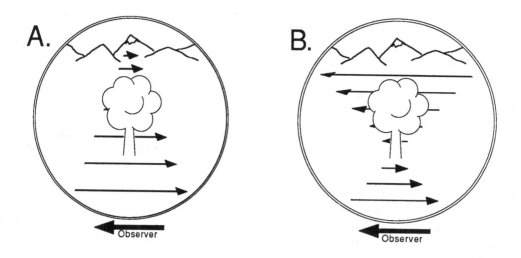

Figure 3. The effect of compensatory eye movements on the optic flow associated with lateral translation of the observer. **A:** The observer makes no compensatory eye movements so that the flow pattern pivots around infinity. **B:** The observer generates compensatory eye movements, presumably due to the combined operation of the TVOR and visual stabilization mechanism(s), so that the flow pattern now pivots around the tree in the middle-ground. The length of the arrows infers the speed of the local flow. (From Miles, in press.)

scene on the retina may have provided the initial pressure to evolve the so-called *pursuit* system. In this scheme, the pursuit system is an attentional focussing mechanism that spatially filters visual motion inputs. It is not difficult to imagine how, once evolved, such a mechanism could also be used to track small moving targets - even across textured backgrounds, the *sine qua non* of the primate pursuit system. However, while granting that subjects may opt to respond solely to local flow in some circumstances, it seems that they respond to more global flow in others - thanks to a second visual tracking mechanism, OKNe, that we suggest evolved to deal with translational problems in the binocular field. We have already seen that viewing distance is a critical parameter governing the optic flow associated with translations - but not rotations - and that OKNe shares the TVOR's sensitivity to this parameter, consistent with the notion of their structural and functional linkage. Additional support for this view of OKNe comes from evidence that it can utilize relative depth cues, such as motion parallax and disparity, to aid in the decoding of translational optic flow and, thereby, to help in stabilizing the eyes on the frontal depth plane of interest. In our further discussion of the visual stabilization mechanisms that we propose deal with translational disturbances of gaze, emphasis will be on the case in which the poles of the flow field are seen only by the far peripheral retina - as during lateral motion. The expanding/contracting flow near the poles can be safely ignored here and we shall not further consider the case in which the observer looks towards the polar region - as during forward motion.

<u>Sensitivity to Relative Depth Cues: Motion Parallax.</u>

Optical geometry dictates that when his vantage point moves the passive observer experiences retinal image motion that is inversely related to viewing distance (Schwarz *et al.*, 1989). This optical lever effect is especially evident when looking out from a fast moving train: the scene appears to pivot around the most distant objects (the mountains in Fig. 3A, for example). If the observer compensates for the motion by rotating his eyes then the optical lever will now pivot about some intermediate point (the tree in Fig. 3B, for example). This complicates the flow pattern yet further, the images of objects beyond this imaginary optical fulcrum now reversing their motion on the retina. If, as seems likely, the observer fails to compensate fully for his own motion then the image of interest (such as the tree in Fig. 3B) will drift back across his central retina while the image of the distant background will be swept forwards across his peripheral

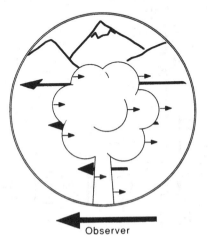

Figure 4. The optic flow experienced by the translating observer who attempts to stabilize his gaze on the object of regard. The situation resembles that in Fig. 3B except that compensation is incomplete and the object of regard has been positioned in the foreground. The net result is antiphase motion in the central and peripheral regions of the visual field. (From Miles, in press.)

retina: see Fig. 4. Interestingly, there is evidence from studies of both monkeys and humans that the optokinetic system can utilize motion parallax cues such as these to improve its performance. Thus, concurrent motion in the central and surround regions that is opposite in direction actually improves the tracking of the central motion, an effect termed *antiphase enhancement*; conversely, surround motion that is in the same direction as that at the center degrades tracking performance, an effect termed *inphase suppression* (Guedry Jr *et al.*, 1981; Hood, 1975; Miles *et al.*, 1986; Ter Braak, 1957; Ter Braak, 1962). Significantly, these effects due to concurrent motion in the peripheral retina are evident at short latency (<100 ms) and hence we assume are characteristic of the early component of the optokinetic response (Miles *et al.*, 1986). It is apparent from this that *en masse* motion is not the optimal stimulus for OKNe, and that antiphase motion in the surround (due to motion parallax) could actually help - rather than hinder - the moving observer attempting to stabilize an object off to one side.

Miles *et al* (1986) have suggested a preliminary model of the tracking system responsible for the early component of the monkey's optokinetic response with provision for antiphase enhancement and inphase suppression. The basic scheme consists of a negative feedback tracking system that is *driven* by foreground images moving in the central retina and *modulated* by background images moving in the peripheral retina. In a further elaboration of this model, the retinal slip signals are assumed to be decoded by directionally selective motion detectors that each respond solely to either rightward or leftward motion (Miles, in press; Miles *et al.*, in press). In this scheme, the processed slip signals that emanate from the central retina - and provide the principal drive to the tracking system - pass to separate gain elements for rightward and leftward motion. Similarly processed inputs emanating from the peripheral retina are assumed to exert their modulatory influence on these gain elements. Note that both the object of regard and the background object(s) are generally stationary, their images moving only because the observer moves, so that all of the visual inputs are *reafferent* (Holst, 1954; Holst & Mittelstaedt, 1950). In such a system, textured backgrounds help the moving observer to stabilize his/her eyes on nearby stationary objects of interest.

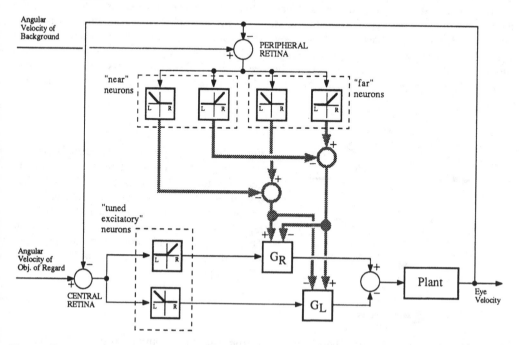

Figure 5. Hypothetical model of OKNe that can parse the optic flow associated with lateral translational disturbances using neurons sensitive to horizontal binocular disparity. The primary

drive comes from objects in the plane of fixation, whose images fall in the central retina and whose motion is signalled by "tuned excitatory neurons". Support for this component of the model comes from the experiments of Howard & Gonzalez (1987). Proper decoding of the motion parallax in the peripheral retina depends on the system's ability to determine whether objects are beyond or nearer than the plane of fixation/stabilization. Antiphase enhancement and inphase suppression result from image motion in the peripheral retina due to objects beyond the plane of fixation. This motion is signalled by "far" neurons. Inphase enhancement and antiphase suppression result from image motion in the peripheral retina due to objects inside the plane of fixation. This motion is signalled by "near" neurons. (From Miles, in press.)

Sensitivity to Relative Depth Cues: Disparity.

In discussing the decoding of motion parallax by the system mediating OKNe we have so far assumed that the peripheral retina always sees images of objects that are *beyond* the object of regard. Under these conditions, it is appropriate that enhancement occurs when motion in the center and surround are antiphase. Of course, in everyday life the surroundings can be closer to the observer than the object of regard, in which case, their retinal images would move in the *same* direction as those at the center and, to be appropriate here, enhancement would have to occur when the motion in the surround was *in phase* with that at the center. In order to deal with this situation, feedback signals from the peripheral retina would have to undergo a sign inversion so that when the surroundings were distant there would be antiphase enhancement and when the surroundings were near there would be inphase enhancement. It is not yet known if the optokinetic system can decode motion parallax in this way, though Miles (in press) has suggested a scheme for achieving this which uses disparity-sensitive neurons that have been recorded in the monkey's visual cortex: see Fig. 5. In this model, the main drive to the visual stabilization mechanism comes from "tuned excitatory neurons" (or "tuned inhibitory neurons") that are sensitive only to stimuli in and around the plane of fixation, while the modulating inputs come from "far" and "near" neurons (Poggio & Fischer, 1977; Poggio & Talbot, 1981): the outputs of the "far" neurons are organized to provide antiphase enhancement and inphase suppression while the outputs of the "near" neurons are organized to provide the converse (inphase enhancement and antiphase suppression). The suggestion here is that the proper decoding of motion parallax might be achieved through neurons that are tuned for binocular horizontal disparity. It has been shown recently that human OKN is much better when the moving visual scene is binocularly fused than when disparate (Howard & Gonzalez, 1987). Further, the changes in nystagmus with disparity were sufficiently brisk to suggest to us that they were mediated by OKNe, which dominates human optokinetic responses (Cohen *et al.*, 1981; Koenig & Dichgans, 1981). This study supports the idea that the ocular stabilization mechanism can respond selectively to the motion of objects in the plane of fixation and ignore the motion of objects that are nearer or further. It remains to be seen if motion signals from outside the plane of fixation can make a more positive contribution to visual stabilization such as hypothesized above. [3]

NEURAL MEDIATION

Recent studies on monkeys using single unit recordings and focal chemical lesions indicate that early ocular following is mediated by a pathway that includes the middle superior temporal (MST) area of cortex and the dorsolateral pontine nucleus (DLPN) of the brain stem, which projects to the oculomotor centers via the floccular lobes of the cerebellum (Kawano, in press). Significantly, the antiphase pattern of retinal image motion that is so effective in driving OKNe is also a good stimulus for some neurons in the middle temporal (MT) area of the monkey's cortex (Allman *et al.*, 1985; Tanaka *et al.*, 1986), a region that is known to project to the MST area (Maunsell & Van Essen, 1983b; Ungerleider & Desimone, 1986). In light of our earlier discussion, it is interesting that many neurons in MT show sensitivity to disparity (Maunsell & Van Essen, 1983a). There is evidence that OKNd is mediated by a separate pathway that includes the nucleus of the optic tract (NOT) of the pretectum and the dorsal terminal nucleus (DTN) of the accessory optic system: for recent review see Schiff *et al* (1990).

REFERENCES

Allman, J., Miezin, F., and McGuinness, E. (1985): Stimulus specific responses from beyond the classical receptive field: Neurophysiological mechanisms for local--global comparisons in visual neurons. *Ann. Rev. Neurosci.* **8**, 407-430.

Archer, S. M., Miller, K. K., and Helveston, E. M. (1987): Stereoscopic contours and optokinetic nystagmus in normal and stereoblind subjects. *Vision Res.* **27**, 841-844.

Baarsma, E. A. and Collewijn, H. (1975): Eye movements due to linear accelerations in the rabbit. *J. Physiol.* **245**, 227-247.

Baloh, R. W., Beykirch, K., Honrubia, V., and Yee, R. D. (1988): Eye movements induced by linear acceleration on a parallel swing. *J. Neurophysiol.* **60**, 2000-2013.

Berthoz, A., Israël, I., Viéville, T., and Zee, D. S. (1987): Linear head displacement measured by the otoliths can be reproduced through the saccadic system. *Neurosci. Lett.* **82**, 285-290.

Biguer, B. and Prablanc, C. (1981): Modulation of the vestibulo-ocular reflex in eye-head orientation as a function of target distance in man. In *Progress in Oculomotor Research*, ed. A. F. Fuchs, W. Becker,, pp. 525-530. New York: Elsevier North Holland.

Blakemore, C. and Donaghy, M. (1980): Co-ordination of head and eyes in the gaze changing behaviour of cats. *J. Physiol.* **300**, 317-335.

Bronstein, A. M. and Gresty, M. A. (1988): Short latency compensatory eye movement responses to transient linear head acceleration: a specific function of the otolith-ocular reflex. *Exp. Brain*

---

3 Human ocular following can be elicited by moving stereoscopic contours that lack monocular motion cues (Archer *et al.*, 1987; Fox *et al.*, 1978). However, responses are not obligate and are observed only when the subject perceives the moving contour, perhaps representing an instance of "pursuing the perceptual rather than the retinal stimulus" (Steinbach, 1976).

Res. **71**, 406-410.

Buizza, A., Léger, A., Berthoz, A., and Schmid, R. (1979): Otolithic-acoustic interaction in the control of eye movement. *Exp. Brain Res.* **36**, 509-522.

Cheng, M. and Outerbridge, J. S. (1975): Optokinetic nystagmus during selective retinal stimulation. *Exp. Brain Res.* **23**, 129-139.

Cohen, B., Henn, V., Raphan, T., and Dennett, D. (1981): Velocity storage, nystagmus and visual-vestibular interaction in humans. *Ann. NY. Acad. Sci.* **374**, 421-433.

Cohen, B., Matsuo, V., and Raphan, T. (1977): Quantitative analysis of the velocity characteristics of optokinetic nystagmus and optokinetic after-nystagmus. *J. Physiol. (Lond.)* **270**, 321-344.

Collewijn, H. (1975): Direction-selective units in the rabbit's nucleus of the optic tract. *Brain Res.* **100**, 489-508.

Collewijn, H. and Noorduin, H. (1972): Conjugate and disjunctive optokinetic eye movements in the rabbit, evoked by rotatory and translatory motion. *Pflügers Arch.* **335**, 173-185.

Dubois, M. F. W. and Collewijn, H. (1979): Optokinetic reactions in man elicited by localized retinal stimuli. *Vision Res.* **19**, 1105-1115.

Fox, R., Lehmkuhle, S., and Leguire, L. E. (1978): Stereoscopic contours induce optokinetic nystagmus. *Vision Res.* **18**, 1189-1192.

Frost, B. J. and Nakayama, K. (1983): Single visual neurons code opposing motion independent of direction. *Science* **220**, 744-745.

Frost, B. J., Scilley, P. L., and Wong, S. C. P. (1981): Moving background patterns reveal double-opponency of directionally specific pigeon tectal neurons. *Exp. Brain Res.* **43**, 173-185.

Graf, W., Simpson, J. I., and Leonard, C. S. (1988): Spatial organization of visual messages of the rabbit's cerebellar flocculus. II. Complex and simple spike responses of Purkinje cells. *J. Neurophysiol.* **60**, 2091-2121.

Gresty, M. A. and Bronstein, A. M. (1986): Otolith stimulation evokes compensatory reflex eye movements of high velocity when linear motion of the head is combined with concurrent angular motion. *Neurosci. Lett.* **65**, 149-154.

Gresty, M. A., Bronstein, A. M., and Barratt, H. (1987): Eye movement responses to combined linear and angular head movement. *Exp. Brain Res.* **65**, 377-384.

Guedry Jr, F. E., Lentz, J. M., Jell, R. M., and Norman, J. W. (1981): Visual-vestibular interactions: The directional component of visual background movement. *Aviat. Space Environ. Med.* **52**, 304-309.

Hine, T. and Thorn, F. (1987): Compensatory eye movements during active head rotation for near targets: effects of imagination, rapid head oscillation and vergence. *Vision Res.* **27**, 1639-1657.

Holst, E. von (1954): Relations between the central nervous system and the peripheral organs. *J. Anim. Beh.* **2**, 89-94.

Holst, E. von and Mittelstaedt, H. (1950): Das Reafferenzprinzip. Wechselwirkungen zwischen Zentralnervensystem und Peripherie. *Naturwissenschaften* **37**, 464-476.

Hood, J. D. (1975): Observations upon the role of the peripheral retina in the execution of eye movements. *J. Otorhinolaryngol.* **37**, 65-73.

Howard, I. P. (1982): *Human Visual Orientation*, London: Wiley.

Howard, I. P., Giaschi, D., and Murasugi, C. M. (1989): Suppression of OKN and VOR by afterimages and imaginary objects. *Exp. Brain Res.* **75**, 139-145.

Howard, I. P. and Gonzalez, E. G. (1987): Human optokinetic nystagmus in response to moving binocularly disparate stimuli. *Vision Res.* **27**, 1807-1816.

Israël, I. and Berthoz, A. (1989): Contribution of the otoliths to the calculation of linear displacement. *J. Neurophysiol.* **62**, 247-263.

Kawano, K. (in press): Neural mechanisms of ocular following responses. In *Vestibular and Brain Stem Control of Eye, Head and Body Movements*, ed. H. Shimazu, Y. Shinoda, Tokyo: Springer-Verlag/Japan Scientific Societies Press.

Koenig, E. and Dichgans, J. (1981): Aftereffects of vestibular and optokinetic stimulation and their interaction. *Ann. NY Acad. Sci.* **374**, 434-445.

Leonard, C. S., Simpson, J. I., and Graf, W. (1988): Spatial organization of visual messages of the rabbit's cerebellar flocculus. I. Typology of inferior olive neurons of the dorsal cap of Kooy. *J. Neurophysiol.* **60**, 2073-2090.

Lisberger, S. G., Miles, F. A., Optican, L. M., and Eighmy, B. B. (1981): Optokinetic response in monkey: Underlying mechanisms and their sensitivity to long-term adaptive changes in vestibuloocular reflex. *J. Neurophysiol.* **45**, 869-890.

Matsuo, V. and Cohen, B. (1984): Vertical optokinetic nystagmus and vestibular nystagmus in the monkey: Up-down asymmetry and effects of gravity. *Exp. Brain Res.* **53**, 197-216.

Maunsell, J. H. R. and Van Essen, D. C. (1983a): Functional properties of neurons in middle temporal visual area of the macaque monkey. II. Binocular interactions and sensitivity to binocular disparity. *J. Neurophysiol.* **49**, 1148-1167.

Maunsell, J. H. R. and Van Essen, D. C. (1983b): The connections of the middle temporal visual area (MT) and their relationship to a cortical hierarchy in the macaque monkey. *J. Neurosci.* **3**, 2563-2586.

Miles, F. A. (in press): Visual mechanisms underlying the stabilization of gaze. In *Exploratory Vision: The Active Eye*, ed. T. E. Weymouth, L. T. Maloney, Berlin: Springer-Verlag.

Miles, F. A., Kawano, K., and Optican, L. M. (1986): Short-latency ocular following responses of monkey. I. Dependence on temporospatial properties of the visual input. *J. Neurophysiol.* **56**, 1321-1354.

Miles, F. A., Schwarz, U., and Busettini, C. (1989): Are the two components of the primate optokinetic response concerned with translational and rotational disturbances of gaze?. *Soc. Neurosci. Abstr.* **15**, 783.

Miles, F. A., Schwarz, U., and Busettini, C. (in press): The decoding of optic flow by the primate optokinetic system. In *The Head-Neck Sensory-Motor System*, ed. A. Berthoz, P. -P. Vidal and W. Graf, New York: Oxford University Press.

Murasugi, C. M., Howard, I. P., and Ohmi, M. (1986): Optokinetic nystagmus: the effects of stationary edges, alone and in combination with central occlusion. *Vision Res.* **26**, 1155-1162.

Naegele, J. R. and Held, R. (1982): The postnatal development of monocular optokinetic nystagmus in infants. *Vision Res.* **22**, 341-346.

Ohmi, M., Howard, I. P., and Eveleigh, B. (1986): Directional preponderance in human optokinetic nystagmus. *Exp. Brain Res.* **63**, 387-394.

Paige, G. D. (1989): The influence of target distance on eye movement responses during vertical linear motion. *Exp. Brain Res.* **77**, 585-593.

Paige, G. D., Tomko, D. L., and Gordon, D. B. (1988): Visual-vestibular interactions in the linear vestibulo-ocular reflex (VOR). *Invest. Ophthalmol. Vis. Sci. (Suppl.)* **29**, 342.

Poggio, G. F. and Fischer, B. (1977): Binocular interaction and depth sensitivity in striate and prestriate cortex of behaving rhesus monkey. *J. Neurophysiol.* **40**, 1392-1405.

Poggio, G. F. and Talbot, W. H. (1981): Mechanisms of static and dynamic stereopsis in foveal cortex of the rhesus monkey. *J. Physiol.* **315**, 469-492.

Raphan, T., Cohen, B., and Matsuo, V. (1977): A velocity-storage mechanism responsible for optokinetic nystagmus (OKN), optokinetic after-nystagmus (OKAN) and vestibular nystagmus. In *Developments in Neuroscience, Volume 1*, ed. R. Baker, A. Berthoz, pp. 37-47. Amsterdam: Elsevier/North-Holland.

Raphan, T., Matsuo, V., and Cohen, B. (1979): Velocity storage in the vestibulo-ocular reflex arc (VOR). *Exp. Brain Res.* **35**, 229-248.

Schiff, D., Cohen, B., Büttner-Ennever, J., and Matsuo, У. (1990): Effects of lesions of the nucleus of the optic tract on optokinetic nystagmus and after-nystagmus in the monkey. *Exp. Brain Res.* **79**, 225-239.

Schwarz, U., Busettini, C., and Miles, F. A. (1989): Ocular responses to linear motion are inversely proportional to viewing distance. *Science* **245**, 1394-1396.

Simpson, J. I., Graf, W., and Leonard, C. (1981): The coordinate system of visual climbing fibers to the flocculus. In *Progress in Oculomotor Research*, ed. A. Fuchs, W. Becker, pp. 475-484. Amsterdam: Elsevier North Holland.

Simpson, J. I., Leonard, C. S., and Soodak, R. E. (1988): The accessory optic system of the rabbit. II. Spatial organization of direction selectivity. *J. Neurophysiol.* **60**, 2055-2072.

Simpson, J. I., Soodak, R. E., and Hess, R. (1979): The accessory optic system and its relation to the vestibulocerebellum. In *Reflex Control of Posture and Movement*, ed. R. Granit, O. Pompeiano, pp. 715-724. Amsterdam: Elsevier.

Soodak, R. E. and Simpson, J. I. (1988): The accessory optic system of the rabbit. I. Basic visual response properties. *J. Neurophysiol.* **60**, 2037-2054.

Steinbach, M. (1976): Pursuing the perceptual rather than the retinal stimulus. *Vision Res.* **16**, 1371-1376.

Tanaka, K., Hikosaka, K., Saito, H.-A., Yukie, M., Fukada, Y., and Iwai, E. (1986): Analysis of local and wide-field movements in the superior temporal visual areas of the macaque monkey. *J. Neurosci.* **6**, 134-144.

Ter Braak, J. W. G. (1957): "Ambivalent" optokinetic stimulation. *Fol. Psychiat. Neurol. Neurochir. Neerl.* **60**, 131-135.

Ter Braak, J. W. G. (1962): Optokinetic control of eye movements, in particular optokinetic nystagmus. *Proc. 22th Int. Congr. physiol. Sci,. Leiden* **1**, 502-505.

Ungerleider, L. G. and Desimone, R. (1986): Cortical connections of visual area MT in the macaque. *J. Comp. Neurol.* **248,** 190-222.

Van Den Berg, A. V. and Collewijn, H. (1987): Voluntary smooth eye movements with foveally stabilized targets. *Exp. Brain Res.* **68,** 195-204.

Van Die, G. and Collewijn, H. (1982): Optokinetic nystagmus in man. *Human Neurobiol.* **1,** 111-119.

Vidic, T. R., Barlow, J. S., Oman, C. M., Tole, J. R., Weiss, A. D., and Young, L. R. (1976): Human eye tracking during vertical motion. *Soc. Neurosci. Abstr.* 1062.

Viirre, E., Tweed, D., Milner, K., and Vilis, T. (1986): A reexamination of the gain of the vestibuloocular reflex. *J. Neurophysiol.* **56,** 439-450.

Westall, C. A. and Schor, C. M. (1985): Asymmetries of optokinetic nystagmus in amblyopia: the effect of selected retinal stimulation. *Vision Res.* **25,** 1431-1438.

Zee, D. S., Tusa, R. J., Herdman, S. J., Butler, P. H., and Gücer, G. (1987): Effects of occipital lobectomy upon eye movements in primate. *J. Neurophysiol.* **58,** 883-907.

Current views on the visuomotor interface of the saccadic system

J. A. M. Van Gisbergen, A. J. Van Opstal and A. W. H. Minken

Department of Medical Physics & Biophysics, University of Nijmegen, Geert Grooteplein Noord 21, 6525 EZ Nijmegen, The Netherlands

INTRODUCTION

The function of the saccadic system is to direct the fovea rapidly and accurately on stimuli of interest and to keep it stable in the new position afterwards. It is known from neurophysiological studies (see Sparks and Mays, 1990) and from experiments exploring the effect of local electrical stimulation (Robinson, 1972) and reversible inactivation (Hikosaka and Wurtz, 1985; Lee et al., 1988) that the superior colliculus (SC) in primates plays an important role in the control of these movements. The SC is a sensorimotor center interposed between parts of the visual system and the motor system. Its intermediate position is reflected in the presence of visual neurons in its upper layers and of many visuomotor neurons in its deeper layers which respond with an initial sensory-locked response if a visual stimulus stimulates their receptive field, followed by a fierce movement-related burst in cases where the stimulus attracts a saccade. All layers have a topographical organization which still betray a visual origin. The part of the saccadic system transforming sensory messages into appropriate motor commands subserving the requirements implicit in its function (see above), of which the SC is an essential part, is faced with various requirements and constraints (see also Vilis and Tweed, 1990):

1) *Target selection.* Directing gaze on stimuli of interest implies that target selection is required in all but the simplest situations. Since the SC visual system has a rather low spatial resolution (Cynader and Berman, 1972) and is poorly equipped for feature recognition and spectral analysis (Marocco and Li, 1977), the question arises how the system can cope with several simultaneous potential targets and how it performs when target selection requires recognition of subtle features. So far, very little is known about how this process might operate at a physiological level (see below).

2) *Spatio-temporal recoding.* Even in the case of a single target, the quite different coding prevalent in the visual and the motor system calls for a major signal transformation in which the SC, being at the watershed of both systems, may be assumed to play an

important role. Visual signals are delivered in the format of topographically organized neural maps which, reflecting the inhomogeneity of the retina, form a distorted representation of stimulus space. Since the system appears able to generate goal-directed eye movements, this distortion is somehow taken into account. On the motor side, the eye ball/eye muscle system has visco-elastic and kinematic properties which require specially designed temporally coded pulse-step signals at the motoneuron level (Robinson, 1970; Tweed and Vilis, 1987). It is generally agreed that the high frequency burst (pulse) in agonist motoneurons serves to overcome the effect of viscous forces, thus ensuring that the movement is maximally fast in order to minimize the time lost for vision as a result of retinal image motion, whereas the step signal is required to keep the eye stable in the new position against the restoring elastic forces.

3) Use of internal feedback. Apart from this required recoding, transforming the representation of the target in distorted eye-centered coordinates (map-type coding) to the head-coordinate-based signals at the level of motoneurons (in firing-rate code), the system somehow has to solve the problem that the requirement of fast movements is incompatible with the use of external (visual) feedback to guide them. Since saccades are fast and the visual system has long processing delays, external feedback cannot be used to guide saccades directly. Accordingly, a trade-off between speed and accuracy is not usually thought to occur with saccades (but see Abrams et al., 1989), unlike in the case of arm movement control where this is an established principle (Fitt's law). Instead, there are indications that the saccadic system makes use of internal feedback based on efference copy signals. This is a viable option in the oculomotor system where, unlike the situation in the control of arm movements, external loads are invariant. The notion of internal feedback was worked out in the context of a specific 1D quantitative model for the first time by Robinson (1975). It was used later to explain the behaviour of so-called saccadic burst cells (Van Gisbergen et al., 1981), which are held responsible for the pulse component in the firing patterns of motoneurons, by assuming that the burst cells are driven by the difference (motor error) between a desired eye position signal and an efference copy of eye position. In the model, the latter signal was obtained by mathematically integrating the pulse signal and also provided the step signal in motoneurons. Later work by Sparks and Porter (1983) has shown that at least one type of internal feedback signal ends more upstream, before or at the level of the SC, and led to the idea that the SC codes motor error (desired saccadic displacement) in a map code. Recently, the problem of whether the internal feedback model can be generalized in order to explain the 3D control of saccadic eye movements has attracted much attention. It has become clear from detailed studies that when a subject fixates in various directions (with the head upright and steady), the eye does not use all three degrees of freedom for positioning which are theoretically available. In fact, only those eye positions are actually observed which can be reached from a fixed reference position (the primary position) by fixed-axis rotations in a single plane, a fact known as Listing's law. Recently it has become clear that Listing's law is also valid dynamically during saccades, at least in good approximation. Tweed and Vilis (1987) have proposed a 3D model with internal feedback which can generate saccades obeying Listing's law. An important issue, to be discussed later, is whether the SC is within the proposed 3D feedback loop and whether it plays a role in implementing Listing's law. In this paper, we will first review some earlier work on mechanisms of target selection and a model-based description of SC neuron properties. Subsequently, the possible implications of recent work on the 3D control of saccades for the role of the SC and internal feedback signals therein will be discussed.

MECHANISMS OF TARGET SELECTION

In behavioural studies with human subjects making saccades to targets with sudden visual onset, Ottes et al. (1984) compared responses to single and double targets. In these experiments it was found that the saccadic system chooses either one or the the other stimulus when these are sufficiently wide apart. For smaller separations, the eye jumped in between the two stimuli (averaging; see also Findlay, 1982). In this sense the saccadic system has a much lower spatial resolution than the perceptual system. This was the case not only for target separation in eccentricity but also for double stimuli along different meridians at equal eccentricity. In a second series of experiments (Ottes et al., 1985) the possible ambiguity of having two targets was removed by presenting the subject, in some trials, with a green/red target/nontarget pair. The same low spatial resolution (averaging responses) was found when target and nontarget stimulus were sufficiently close together, but only if latency was short. When subjects were instructed to avoid such mistakes, they could avoid averaging, but only if latency was sufficiently long (beyond 300 msec). Short-latency responses made in these experiments still showed the averaging behaviour.
More recently, electrophysiological experiments using monkeys which performed the same target/nontarget selection task (wide stimulus separations only) were set up in order to learn more about the role of the superior colliculus in the target selection process (Ottes et al., 1987). The monkeys performed this task quite well, although incorrect responses (to the red stimulus) were sometimes made, especially in short-latency trials. The initial visual response of most neurons in the superficial and deeper layers was clear but did not discriminate reliably between the green target and the red nontarget stimulus in a way which would allow one to predict the response the monkey was about to make. By contrast, in the deeper layer neurons with a movement-related burst, the presence or absence of a saccadic burst correlated very well to the actual movement irrespective of whether the saccade was correct or incorrect. This indicated that the SC motor discharge was actually controlling the behavioural response. Since the response patterns of the collicular neurons showed a sudden transition from the non-discriminating sensory-related firing pattern to the saccade-related burst, there was no evidence that the target selection process was actually taking place in the colliculus. Rather, it was concluded that the decision where the saccade should go was apparently coming from some other (cortical?) center which could make the required distinction. Several possibilities for the source of this hypothetical command signal (among them the substantia nigra and area 17) are discussed in the original paper.

In a recent series of experiments (Van Opstal and Van Gisbergen, 1990), designed to investigate the neural basis of saccade averaging, monkeys were trained to track single and double-step stimuli. The double step stimuli were designed to elicit compromise saccades, directed in between the first and final stimulus position. Like humans, monkeys made these intermediate responses when latency was in a certain range relative to the first stimulus step. In one type of double stimulus the spot of light first jumped to a certain eccentricity on one meridian and, after a short period, moved rapidly to a final position on a different meridian at equal eccentricity ($\Delta\Phi$ double step). Meanwhile recordings were taken from an SC neuron with saccade related discharges whose movement field (the ensemble of saccade amplitudes and directions associated with saccade related activity; see below) had been explored first with saccades evoked by single-step stimuli. By manipulating the double-step stimulus parameters it was possible to evoke averaging saccades into the movement field, although the stimulus itself was outside this area. It was found that the SC neuron was active also prior and during these averaging saccades. This work shows that the averaging process must take place inside or above the SC. The data are compatible with a single area of movement cell activity in between the two double step stimulus positions.

Interestingly, the burst associated with averaging saccades was typically somewhat weaker than when a similar saccade was elicited by a single step stimulus. Closer scrutiny revealed that the averaging saccades were actually somewhat slower than comparable single step saccades. The latter finding is reminiscent of other reports in the literature (e.g., Berthoz et al., 1986) that the vigour of collicular movement related bursts is related to saccadic velocity. Thus, it is by now quite well established that the SC is not only responsible for determining the metrics of saccades, but also affects its dynamic properties. Just how far this goes is not yet clear and is obviously a very important problem for future work (see also below). At least one thing has become clear: the SC is a major center in the final common pathway for all saccades.

INTERPRETATION OF COLLICULAR MOVEMENT FIELD PROPERTIES

Quite analogous to the receptive field in sensory neurons, the movement field of SC cells (Wurtz and Goldberg, 1972) is not just the summary of a set of measurements but equally a challenge to our understanding of the total system (McIlwain, 1976). Of course, as in the case of sensory receptive fields, if one wishes to interpret (understand) the properties of SC movement fields, one needs a model which can fit the data points quantitatively. Such a model would have to explain the particular asymmetrical shape of SC movement fields, as well as the size differences which are noticable between neurons associated with small and large saccades. The movement field of an SC neuron is a plot of firing rate in the movement-related burst against the saccade parameters amplitude and direction. In all cells the movement field appears to be limited and to have its highest firing rate for a particular saccade (the optimum saccade) which depends upon the neuron's anatomical location in the collicular map. Saccades of the same size with directions on both sides of the optimum saccades show a symmetrical gradual reduction in firing rate. By contrast, the movement field profile along an amplitude cross section of the field is typically asymmetrical (skewed). As saccade size increases from the optimum saccade size onward, one observes a gradual decrease in firing rate whereas a much more abrupt decrease is observed when saccade size decreases. Neurons with larger optimum saccade sizes have typically larger movement fields than those found in the more rostral area of the colliculus which is devoted to small saccades. The fact that SC burst neurons are not specialized in one particular saccade vector but are involved in a rather broad range of saccade sizes and directions makes clear that a considerable percentage of SC neurons must be active for each saccade. What is the shape and the extent of this activity profile in the SC neural map? Is it similar everywhere in the colliculus?

Robinson (1972) has shown, by electrical stimulation, how the locus of activity in the deeper collicular layers is related to the metrical properties of the resulting saccade. From combined electrical stimulation and single unit recording experiments (Schiller and Stryker, 1972), it is known that local electrical stimulation elicits a saccade which is very similar to the optimum saccade of nearby movement cells. An important aspect in Robinson's results is that the representation of motor space on the collicular map (i.e., the motor map) is highly nonhomogeneous. Analogous to the visual system where the foveal and parafoveal visual field is overrepresented, the SC motor map devotes relatively much more space to small saccades. The shape of the motor map, the size and shape of collicular movement fields, the orderly relation between movement field location and anatomical location in the motor map, together with the fact that visually elicited saccades are normally quite accurate, has led us to propose a model in which the following aspects of collicular connectivity and activity patterns are of crucial importance (Van Gisbergen et al., 1987a):

1) The mapping from the retina onto the motor map, determining which cells are recruited in response to a given target position on the retina (afferent mapping). For reasons to be explained below, we used a complex logarithmic mapping in our model.

2) The shape and the extent of the movement cell activity profile occurring when a saccade is made (population activity profile; see below). We assumed that cell density in the motor map is uniform.

3) The connectivity of collicular burst cells with the motor system downstream (efferent mapping), which allows their activity to direct movements of the eye towards the external stimulus which recruited them through the afferent mapping. To account for this, we assumed that the efferent mapping, from motor map to motor space, has to be the inverse of the afferent mapping from visual space onto the motor map. For details we refer to the original paper.

4) A summation rule. We assumed that the total movement produced is the vector sum of the movement contribution of each active cell, which is directed at the center of its movement field.

The results of the afferent mapping fit procedure are worth mentioning here. Using the complex logarithmic mapping function we found that the magnification factor in the direction domain is clearly higher than in the amplitude domain. In other words, the motor map is anisotropic. Using the best-fitting mapping results in fitting the movement field data recorded in trained monkeys, Ottes et al. (1986) found good fit when they assumed that the population activity profile had a circular symmetrical Gaussian shape. When, instead, an isotropic afferent mapping function was used the fit was less good unless the population activity profile was allowed to be asymmetric. Accordingly, the simplest model to explain both sets of data is the combination of an anisotropic logarithmic afferent mapping function and a rotation-symmetrical activity profile in the motor map. The peculiar shape of collicular movement fields (skewed along the amplitude axis, but symmetrical along the direction domain) reflects the distortion resulting when a symmetrical (Gaussian) profile along collicular coordinates is plotted along motor space coordinates, thereby reflecting the inverse logarithmic mapping which operates when the SC map is transformed into movement. If the model were ideally true, all movement fields replotted in collicular coordinates should be rotationally symmetrical and translation invariant (see Ottes et al., 1986).

Recently Van Opstal and Van Gisbergen (1989) have used a somewhat similar type of analysis in a behavioural study designed to investigate the amount, the nature and the possible source of noisy variation in the saccadic system. The basic idea behind this work was that if noise at the level of the afferent mapping stage in the SC would play a marked role, it would be expected to create certain identifiable features in the scatter of saccade vectors to an identical retinal stimulus. Suppose that the center of the population activity profile shows a little bit of noisy variation around the mean position assumed in the model when the target is presented repeatedly at the same position. If this is the case, and the model is correct, one would expect to see more scatter in amplitude than in direction (because the mapping is anisotropic) and more scatter, in an absolute measure, in larger than in small saccades (because the mapping is inhomogeneous). Several aspects of these theoretical predictions were in fact observable in the data. As can be seen in Fig. 1, the scatter clouds are elongated and increase in size with target eccentricity. If the model were to explain all of the noisy variability, the same scatter clouds should become invariant when replotted in collicular coordinates. It was found that this is true within certain limits.

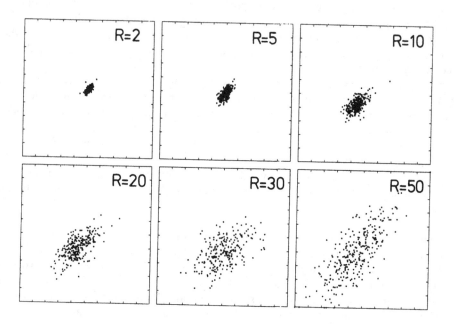

Fig. 1. Saccade endpoint scatter plots, corrected for variability in prior fixation, for six different target eccentricities (R) along the $\Phi = 225$ deg meridian. In all scales one division equals one degree. From Van Opstal and Van Gisbergen (1989).

LISTING'S LAW AND INTERNAL FEEDBACK

Tweed and Vilis (1990a) measured eye movements in humans and monkeys with the 3D eye coil technique described by Ferman et al. (1987a) and represented their data in a head-fixed coordinate system. From their results it appears that it is always possible to define a plane (Listing's plane) such that every eye position could be obtained by rotating the eye from a reference position (called the primary position) to the present position along a fixed axis appropriately oriented in Listing's plane. On this basis (eye position has only two degrees of freedom) they stated that the eye obeys Listing's law both statically during fixations and dynamically during saccades. Ferman et al. (1987b), using a different coordinate system, obtained rather similar results in an earlier study on humans.

But how does the eye rotate during eccentric saccades which do not cross the primary position? Tweed and Vilis (1990a) found that eye position during these saccades again obeyed Listing's law but, to achieve this, the angular velocity axis cannot remain in Listing's plane. Instead, the velocity axis has to tilt out of this plane by an amount which can be computed from the start and endposition of the saccade, provided the movement is executed by a single axis rotation. Tweed and Vilis investigated saccades made at various eccentricities and found that their velocity axis tilted out of Listing's plane by the right amount and furthermore that the orientation of the rotation axis remained roughly stable throughout the saccade. These data, then, suggest that the saccadic control system contributes a torsional rotational signal to the pulse signal of motoneurons to ensure that eye position, controlled by the step, remains in Listing's plane throughout the eccentric movement. Tweed and Vilis (1987) have proposed a neural circuit with internal feedback which can explain the observed tilting behaviour. Since the feedback loop carries a 3D position signal which interacts multiplicatively with the proposed desired position signal (also 3D), the model is quite complex.

While monkey saccades have a reputation of having rather straight trajectories, this is much less the case in humans. The curvature of human saccades has recently been studied quantitatively (Smit and Van Gisbergen, 1990). It was shown that there exists idiosyncratic but consistent direction-dependent curvature, especially in oblique saccades. Dramatically curved saccadic movements have been recorded in a two dimensional double step paradigm (Van Gisbergen et al., 1987b). In these experiments the subject fixates the target on an eccentric position until it jumps to a new eccentric position along a given direction, stays there for some 50-100 msec and then jumps to the final position on a different meridian relative to the original fixation point. Depending upon latency and the precise parameters of the stimulus, the eye may first move to the initial stimulus location, stop there and then make a second saccade to the final position. The more interesting type of response, which is sometimes made, occurs when the eye changes direction in midflight and ends up near the final target position in one continuous curved movement. In earlier work we noticed that eye velocity along the trajectory often had a dip suggesting the possibility that these movements in fact consisted of two overlapping saccades, one in response to the first stimulus movement, the other in reaction to the second movement (two-saccade hypothesis).
Another possible interpretation of these responses is that they are driven by the difference between an internal representation of the target and an internal feedback signal representing eye position (internal feedback hypothesis). So far, neither of the two explanations can be firmly excluded. We wondered whether, by recording the same type of response with an accurate 3D recording technique, new valuable information, relevant for this issue, could be obtained. The main point of interest in these new experiments, of course, is whether the strongly curved movements obey Listing's law. There is no a priori reason to expect that they should since this type of movements has not been investigated so far. If 3D internal feedback as proposed by Tweed and Vilis (1987) is available, the system can continuously adjust the tilting of the rotation axis so that Listing's law is obeyed despite the fact that these movements are not single axis rotations.

According to the two-saccade scheme, the command signals for the two separate saccades are added to yield a combined response which is the sum of the individual time-shifted responses to each of the two steps. However, relying on the results of Tweed and Vilis, it can be shown by using a simple example that in this case the combined response may violate Listing's law, even if each separate saccade does not. Suppose, we combine a saccade from 20 deg left of the primary position with with a 20 deg upward saccade starting from the primary position. In the extreme case, when both component saccades start simultaneously, the resultant saccade will be an eccentric movement from 20 deg at the left to 20 deg upward. This saccade will require a tilted rotation axis to obey Listing's law but the two separate component saccades, in this example, will not. The reason why the sum of saccadic command signals, each of which is adequate to yield a Listing saccade, may nevertheless violate Listing's law is that 3D rotations do not commute.

The experiments to be reported here were performed on five male human subjects. The 3D eye coil (Ferman et al., 1987a) was extensively calibrated before the experiment, using a procedure developed at the Zürich University Hospital. We used a right-handed, head-fixed, Cartesian coordinate system (X,Y,Z) with its origin at the ocular center of rotation (Haustein, 1989). In this system rotations along the X-axis (forward), the Y-axis (leftward) or the Z-axis (upward) yield a deviation in the torsional, vertical and horizontal direction, respectively. In Figs. 2 and 3 eye position is represented by a rotation vector. This vector is oriented as the rotation axis which would rotate the eye from the primary position into the present position and has a magnitude of $tg(\rho/2)$ where ρ is the angle of rotation. Three projection planes (frontal view, top view, side view) can be used to represent the components of each rotation vector (Fig. 2). For example, a fixation at 20 deg left from the primary position will be represented by a rotation vector with components (0, 0, 0.176). When the eye is looking at 40 deg upwards relative to the primary position this yields the components (0, -0.364, 0). During the experiments, subjects faced a screen at 95 cm in front of them with a background illumination of about 1 cd/m². Vision was binocular. The head was stabilized with a head rest and a chin support. The eye position signals were low-pass filtered (-3 dB at 150 Hz), sampled at 500 Hz with a precision of 12 bits and stored on disk. In one type of experiment, subjects were required to refixate the 0.3 deg circular target (5 cd/m²) after it jumped randomly from the straight ahead position to one of 72 different target positions which were arranged at 7 different eccentricities (5-35 deg, at 5 deg intervals) along 12 meridians (every 30 deg). In other experiments single step saccades were elicited towards circular arrays (15 deg diameter; 8 directions, every 45 deg) presented at various starting positions relative to the straight-ahead position.

In the double-step experiments, the stimulus jumped from the initial position (F) at 20 deg left of the straight-ahead position, through an amplitude of 40 deg to a new position in the upper (Φ = 22.5 deg; location A) or the lower (Φ = -22.5 deg; location B) part of the right half field, stayed there for a time (τ) of 50-70 msec and then made a purely vertical movement to the other location (A or B). Thus, FAB double stimuli required a clockwise change in eye movement direction whereas FBA double step trials called for an anti-clockwise change. An experimental sequence contained a total of 20 FAB and FBA double-step trials, 10 (FA and FB) single step controls and 10 other trial types where the target made a single-step movement in various directions towards locations at various eccentricities. In order to manipulate latency, we varied the time (Δ) between fixation spot offset and stimulus onset at the first peripheral stimulus location. In different experiments, Δ could be varied from -50 (gap condition) to +50 msec (overlap condition). In all subjects, except one, we noticed examples of responses with spectacular curvature in the direction of double-step motion.

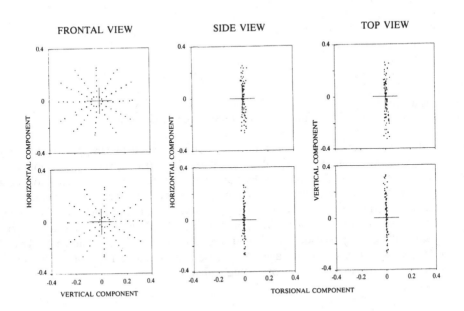

Fig. 2. Eye position represented as rotation vectors (see text), during fixations up to 35 deg from the straight-ahead position. Note very small torsional components (middle and right-hand column).
Data from subjects JK (upper row) and AM (bottom row).

Single-step responses showed a much narrower range of curvature.
Before one can have faith in measurements of eye torsion, it is essential to have some reassurance that coil slippage, which may most easily occur in the torsional direction, was not a major problem. Therefore we used the data from each experiment to evaluate the amount of torsion coil slippage. This could easily be deduced from an overall shift of the cloud of data points from the best-fit Listing plane determined in the first experiment, based on the in-situ calibration. The amount of slippage, from one experiment to the next, had a mean of 0.67 deg across subjects (range: 0.19 - 0.93 deg; n = 5). The cumulative change of coil position in the torsional direction during an entire session had a mean value of 1.92 deg (range in the different subjects: 0.46 - 3.54 deg). The impression obtained is that slippage in each single experiment was quite modest and probably in the range of biological variation with which Listing's law is implemented (see below).

The next step was to check how accurately Listing's law is obeyed in our subjects during fixation and conventional single-step saccades. We first consider the fixation data, obtained from fixation episodes in the single-step tracking experiments (Fig. 2). If Listing's law is strictly valid, the fixation eye positions, which scatter over a wide section of the oculomotor range, should all fit within a single plane (Listing's plane). Although a best-fit plane was obvious in all subjects, there is scatter in the data points, most easily expressed by the standard deviation, which ranged from 0.70 to 1.09 deg in our subjects (Table 1).

subject	fix	sacc	cursacc
JK	1.02	1.05	1.03
AM	0.73	0.70	0.46
JVG	0.78	0.72	0.56
BVDD	0.70	0.66	1.65
JD	1.09	1.02	----

Table 1. Standard deviation (deg) of eye position data during fixation (fix), normal saccades (sacc) and curved double-step saccades (cursacc) relative to best-fit Listing plane in five subjects.

A similar analysis was applied to eye position data obtained when the eye was in saccadic motion in response to single-step stimulation. These data (sacc) show an amount of scatter in the torsion component which does not clearly exceed that of the fixation data (Table 1). On this basis, we can state that our measurements and techniques confirm the earlier conclusions in the literature that Listing's law is valid in reasonable (Ferman et al., 1987b) or even good approximation (Tweed and Vilis, 1990a), not only during fixation, but even dynamically during saccades. We could also confirm the earlier finding by Tweed and Vilis (1990a) that many eccentric saccades can be considered as fixed axis movements whose axis of rotation shows tilting out of Listing's plane.

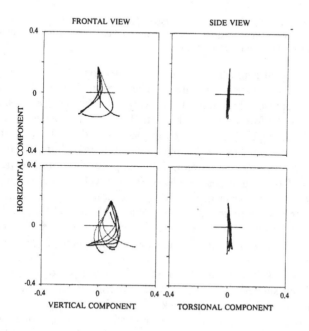

Fig. 3. Curved double-step saccade data from subjects JVG (top) and BVDD (bottom) in frontal and side view. Also the top view (not shown) confirms that the torsional component is quite small.

Reassured that our techniques reproduce the essence of extensive earlier studies performed by other groups, we can now look at the 3D behaviour of saccades with strongly curved trajectories. Since these do not conform to the idealized stereotype of single axis rotations, the torsion velocity component of the rotation axis required to ensure that eye position remains in Listing's plane has no longer a fixed relation to the horizontal and vertical components, but should have an independent time course. Fig. 3 shows a few examples of curved saccades, performing non-single axis rotations, which nevertheless appear to obey Listing's law quite well.

Results of a statistical analysis of the torsional scatter in the eye position data of curved double-step saccades are shown in the right-hand column of Table 1. The fact that the standard deviation does not clearly exceed that in fixation episodes and single-step saccades implies that Listing's law is equally valid for very curved saccades.

DISCUSSION

Saccades towards, through or from the primary position (centric saccades for short) have their rotation axis in Listing's plane (Tweed and Vilis, 1990a) which not only allows eye position to remain in Listing's plane but also brings the eye from the initial to the final position along the shortest rotation path possible. As has been stressed by Tweed and Vilis, adherence to both Listing's law and the principle of shortest path rotation in 3D space is no longer compatible in eccentric saccades. There are infinitely many possibilities to bring the axis of gaze in a single axis rotation from its starting point to its final destination. If, as in centric saccades, the rotation axis of eccentric saccades remains in Listing's plane (no torsional tilt) neither will eye position obey Listing's law nor will the eye execute a shortest path rotation. In other words, in eccentric saccades the two 'requirements' cannot be reconciled. Shortest path rotations require that the rotation axis is perpendicular to the plane defined by the initial and final directions of gaze whereas the amount of tilting required to obey Listing's law (still considering fixed axis rotations) is intermediate between these two extremes.

We were interested to see whether the slight violations of Listing's law noticeable in the double-step responses (Fig. 3) might be due to inappropriate tilting of the angular velocity axis. For each saccade (both straight and curved saccades were included) we took the difference in torsion between saccade onset and offset as the worst-case measure for its violation of Listing's law. This was compared with the predicted torsion violation which would have occurred if the angular velocity axis would have stayed in Listing's plane throughout the movement. While the actual torsion violation was in the order of a few degrees, the predicted error (no tilting of the velocity axis) was much higher and reached values of up to 13 deg in the largest eccentric curved saccades. As the plots in Fig. 4 show, the predicted torsion violation has virtually no relation with the actual results. From this we conclude that if the system did not take special measures during eccentric saccades by appropriate tilting of the velocity axis, much larger Listing violations would have occurred.

Recently, two different theories on how the motoneuron signal, crucial for the implementation of Listing's law, may be generated by the neural control system have been proposed by Tweed and Vilis (1987,1990b) and Hepp (1990), respectively. These models

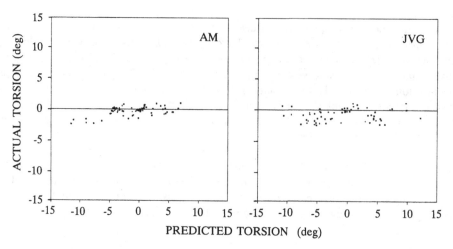

Fig. 4 Comparison of actual cumulative change in torsion from saccade onset to offset and change in torsion predicted if the angular velocity axis would not have tilted out of Listing's plane during the movement. Subjects AM and JVG.

assign quite different roles to the superior colliculus in the generation of the 3D angular velocity signal. Our data show that the brain adopts a control strategy to ensure that saccades obey Listing's law in good approximation. It is of interest to note that saccades with strongly curved trajectories in the horizontal-vertical plane were no exception to this rule. Since it is not easy to see how the Listing behaviour of our curved saccades could be explained otherwise, this result encourages the idea that the system controlling the amount of rotation axis tilting is not preprogrammed and rigid but has access to instantaneous eye position information. In this sense, our results can be reconciled with the 3D Tweed and Vilis (1987) internal feedback model. But based on recent results from the Zürich group (see below), we hesitate to accept certain implications of this model. These concern the implied role of the SC in the implementation of Listing's law and the dimensionality of the putative internal feedback signal.

The fact that the angular velocity axis has to tilt out of Listing's plane during an eccentric saccade, but not during a similar centric saccade, implies that the pulse command signal at the level of motoneurons requires three degrees of freedom. Tweed and Vilis (1990b) have suggested that the SC can supply all required information. In their proposal the SC motor map receives information about the current discrepancy between a 3D signal representing required eye position and a 3D internal feedback signal representing current eye position. With this scheme, the Listing behaviour of our double-step responses can be understood in principle but, apart from the question of whether target position is coded in head coordinates (and in 3D), we have to consider whether the proposition that the SC codes 3D information is realistic.

Recent experiments at the Zürich University Hospital (Van Opstal et al., 1990; Hepp et al., 1990) indicate that the SC code is 2D. This view is based on the fact that saccades evoked by collicular electrical stimulation obey Listing's law, irrespective of the starting position of the eye. These facts are clearly at variance with predictions from Tweed and Vilis' (1990b) colliculus model.

Hepp (1990) has discussed an alternative scheme which delays the precise specification of the angular velocity command signal until a very late stage, downstream of the SC. In his model, the SC supplies information about the desired displacement of the visual axis in the frontal plane. The next stage, the temporal pulse generator (embodied by pontine and mesencephalic burst cells) codes desired visual axis velocity, still a 2D signal. In this view the tilting of the velocity axis is a last-minute addition to the pulse generator signal to prevent the build up of torsion during eccentric saccades. It is reasonable to assume that a 2D internal feedback signal (H/V) would suffice to supply the required eccentricity information. Therefore, if internal feedback plays a role in implementing Listing's law (a realistic possibility in view of our results), there is no compelling reason to assume that either the SC or the feedback signal would have to carry 3D information. Also, it cannot be excluded at this point that inflow information from the eye muscles, rather than efference copy signals, may be used for this purpose.

In conclusion, in the present paper we have outlined a scheme which allows certain properties of collicular saccade-related neurons to be interpreted as the consequence of a specified process whereby a 2D topographical representation of the target is transformed into a goal-directed eye movement. Precise 3D recordings of saccadic eye movements have earlier revealed that, although eye position during saccades has two degrees of freedom, the velocity command signal controlling saccades requires a 3D format which depends critically on eye position. We have demonstrated that although this format becomes very complex in curved saccades, Listing's law is nevertheless also obeyed in these movements and have proposed how the existence of peripheral mechanisms accounting for such accurate 3D control could reconcile these data with the notion of 2D control at the collicular level.

ACKNOWLEDGEMENTS

We thank Jurgen Kemmelings for his essential assistence in the 3D experiments and the data analysis and acknowledge financial support from NWO (Netherlands Organization for the Advancement of Pure Research) and the EC (BRAIN initiative and ESPRIT II project MUCOM 3149).

REFERENCES

Abrams, R.A., Meyer, D.E. & Kornblum, S. (1989): Speed and accuracy of saccadic eye movements: Characteristics of impulse variability in the oculomotor system. *J.Exp. Psychol.: Human Perception and Performance* 15, 529-543.

Berthoz, A., Grantyn, A. & Droulez, J. (1986): Some collicular efferent neurons code saccadic eye velocity. *Neuroscience Letters* 72, 289-294.

Cynader, M. & Berman, N. (1972): Receptive field organization of monkey superior colliculus. *J. Neurophysiol.* 35, 187-201.

Ferman, L., Collewijn, H., Jansen, T.C. & Van den Berg, A.V. (1987a): Human gaze stability in the horizontal, vertical and torsional direction during voluntary head movements, evaluated with a three-dimensional scleral induction coil technique. *Vision Res.* 27, 811-828.

Ferman, L., Collewijn, H. & Van den Berg, A.V. (1987b): A direct test of Listing's law-II. Human ocular torsion measured under dynamic conditions. *Vision Res.* 27, 939-951.

Findlay, J.M. (1982): Global visual processing for saccadic eye movements. *Vision Res.* 22, 1033-1045.

Haustein, W. (1989): Consideration on Listing's law and the primary position by means of a matrix description of eye position control. *Biol. Cybern.* 60, 411-420.

Hepp, K. (1990): On Listing's law. *Comm. Mathemat. Phys.* , in press.

Hepp, K., Van Opstal, A.J., Hess, B.J.M., Straumann, D. & Henn, V. (1990): On the central implementation of saccades in Listing's plane. *Soc. Neurosci. Abstr.* , vol. 16, part 2, p. 1084.

Hikosaka, O. & Wurtz, R.H. (1985): Modification of saccadic eye movements by GABA-related substances. I. Effect of muscimol and bicuculline in monkey superior colliculus. *J. Neurophysiol.* 53, 266-291.

Lee, C., Rohrer, W.H. & Sparks, D.L. (1988): Population coding of saccadic eye movements by neurons in the superior colliculus. *Nature* 332, 357-360.

Marocco, R.T. & Li, R.H. (1977): Monkey superior colliculus: properties of single cells and their afferent inputs. *J. Neurophysiol.* 40, 844-860.

McIlwain J.T. (1976): Large receptive fields and spatial transformations in the visual system. *Int. Rev. Physiol.* 10, 223-248.

Ottes, F.P., Van Gisbergen, J.A.M. & Eggermont, J.J. (1984): Metrics of saccade responses to visual double stimuli: two different modes. *Vision Res.* 24, 1169-1179.

Ottes, F.P., Van Gisbergen, J.A.M. & Eggermont, J.J. (1985): Latency dependence of colour-based target vs nontarget discrimination by the saccadic system. *Vision Res.* 25, 849-862.

Ottes, F.P., Van Gisbergen, J.A.M. & Eggermont, J.J. (1986): Visuo-motor fields of the superior colliculus: a quantitative model. *Vision Res.* 26, 857-873.

Ottes, F.P., Van Gisbergen, J.A.M. & Eggermont, J.J. (1987): Collicular involvement in a saccadic colour discrimination task. *Exp. Brain Res.* 66, 465-478.

Robinson, D.A. (1970): Oculomotor unit behavior in the monkey. *J. Neurophysiol.* 33, 393-404.

Robinson, D.L. (1972): Eye movements evoked by collicular stimulation in the alert monkey. *Vision Res.* 12, 1795-1808.

Robinson, D.A. (1975): Oculomotor control signals. In *Basic mechanisms of ocular motility and their clinical implications*, ed. P. Bach-y-Rita & G. Lennerstrand, pp. 337-374. Oxford: Pergamon Press.

Schiller, P.H., & Stryker, M. (1972): Single unit recording and stimulation in superior colliculus of the alert monkey. *J. Neurophysiol.* 35, 915-924.

Smit, A.C. & Van Gisbergen, J.A.M. (1990): An analysis of curvature in fast and slow human saccades. *Exp. Brain Res.* 81, 335-345.

Sparks, D.L., & Porter, D. (1983): Spatial localization of saccade targets. II.Activity of superior colliculus neurons preceding compensatory saccades. *J. Neurophysiol.* 49, 64-74.

Sparks, D.L., & Mays, L.E. (1990): Signal transformations required for the generation of saccadic eye movements. *Ann. Rev. Neurosci.* 13, 309-336.

Tweed, D. & Vilis, T. (1987): Implications of rotational kinematics for the oculomotor system in three dimensions. *J. Neurophysiol.* 58, 832-849.

Tweed, D.B. & Vilis, T. (1990a): Geometric relations of eye position and velocity vectors during saccades. *Vision Res.* 30, 111-127.

Tweed, D.B & Vilis, T. (1990b): The superior colliculus and spatiotemporal translation in the saccadic system. *Neural Networks* 3, 75-86.

Van Gisbergen, J.A.M., Robinson, D.A. & Gielen, S. (1981): A quantitative analysis of generation of saccadic eye movements by burst neurons. *J. Neurophysiol.* 45, 417-442.

Van Gisbergen, J.A.M., Van Opstal, A.J. & Tax, A.A.M. (1987a): Collicular ensemble coding of saccades based on vector summation. *Neuroscience* 21, 541-555.

Van Gisbergen, J.A.M., Van Opstal A.J., & Roebroek J.G.H. (1987b): Stimulus-induced midflight modification of saccades. In *Eye Movements: from Physiology to Cognition,* eds J.K. O'Regan and A. Levy-Schoen, p. 27-36. Amsterdam: Elsevier Science Publishers

Van Opstal, A.J. & Van Gisbergen, J.A.M. (1989): Scatter in the metrics of saccades and properties of the collicular motor map. *Vision Res.* 29, 1183-1196.

Van Opstal, A.J. & Van Gisbergen, J.A.M. (1990): Role of the superior colliculus in saccade averaging. *Exp. Brain Res.* 79, 143-149.

Van Opstal, A.J., Hepp, K., Straumann, D. & Hess, B.J.M. (1990): Role of superior colliculus in 3D eye movement generation: I. Results of electrical stimulation. *European J. Neurosci.* , in press.

Vilis, T. & Tweed, D. (1990): What can rotational mechanics tell us about the neural control of eye movements? *Proc. Dahlem Konferenzen* , in press.

Wurtz, R.H. & Goldberg, M.E. (1972): Activity of superior colliculus in behaving monkey. III. Cells discharging before eye movements. *J. Neurophysiol.* 35, 575-586.

PART 4

Tacit assumptions in vision research

Andrei Gorea and Michel Imbert

We probably share the opinion that sensory research, in general, and visual research, in particular, has always been guided by more or less explicit assumptions which, to the extent that they could not be experimentally falsified, were regarded as axiomatic. Axioms are not everlasting. With time, most of them get into the realm of falsifiability and can be accepted or rejected as any banal hypothesis. Some of them do not.

Is our current understanding of sensory processes independent of such guiding assumptions, or do we still build our theories while referring (implicitly or explicitly) to one or more untestable propositions?

For example: Has the underlying idea of the *Specific Nerve Energy* doctrine of Johannes Müller become a trivial (and testable; e.g. "labelled detectors") proposition, or do we still refer to it as to an axiomatic base for the interpretation of the sensory and perceptual processes? What is the current status of the *real-world-constraints* (Marr) vs. *direct perception* (Gibson) views? How sure can one be that all anatomically, physiologically or psychophysically revealed specialisations should be correlative and that they must have a "useful" perceptual counterpart (a Darwinian "state of mind"), and, in the same vein, what is the biological significance of the *ideal observer* as a template for estimating visual coding efficiency? To what extent are we still willing to accept the dogma that *early vision* is a *bottom-up* process? Etc.

In the first chapter of this Part, A. Gorea presents a succinct overview of some of the main developments in vision research during the last three decades and argues that Johannes Müller's *Specific Nerve Energy* doctrine still serves as a main (implicit or explicit, i.e. "labelled detectors") reference frame for most of these experimental and theoretical contributions. After introducing over two dozen strategic questions in vision research and discussing some of them (particularly those related to visual attention), B. Julesz (last chapter of this Part) also discusses the current status of Johannes Müller's doctrine and develops his line of argumentation within the framework of present philosophical debates related to the enduring question of "whether machines (algorithms) can ever behave like brains".

S. Klein and C. W. Tyler (second and third chapters of this Part, respectively) address the philosophical question of *duality* as it relates to the understanding of subjective phenomena, in general, and as it applies to visual psychophysics, in particular. S. Klein argues that the subjective-objective split is the only split that reductionists have been unable to bridge and claims that von Neumann's "moveable split" solution, as it was proposed in the context of quantum mechanis, provides the perfect framework for us to make sense of the dualities we meet in *psycho-physics*. C.W. Tyler insists on the reasons why the processing of the brain can never be fully understood and discusses in detail the possibility of misleading interpretations of early visual processing based on the specific assumptions we make on the nature of visual *thresholds* and on the general validity of *Weber's law*.

Another two potentially misleading assumptions related to the interpretation of 3-D objects are discussed by P. Cavanagh and by H. Barlow. P. Cavanagh elaborates on the tacit assumption according to which top-down processing requires that something be "up top" and that this resolves otherwise irresolvable problems. He presents images in which form is carried by shadows and for which the interpretation requires top-down help. While for natural 3-D images the prototype may be selected based on 3-D information, the interpretation of his high contrast images is still possible although they do not provide any direct or indirect 3-D information. Cavanagh proposes that the first step in the process of their recognition consists of a 2-D match that selects the best prototype to guide further interpretation. Such memory stored prototypes may be limited to basic objects - as complex as a face or as simple as a cylinder - and a single prototype might serve for an entire class of objects.

H. Barlow discusses the misleading assumption according to which 3-D representation must be computed before objects are recognized (Marr). He argues that an inference-like process occurs at many stages in sensory information processing and that this imposes two requirements that have been largely overlooked: (a) knowledge of the *prior probability* (unexpectedness) of various items being signalled and (b) *independence* without which it is virtually impossible to determine the prior probabilities of conjunctions of signals. A representation formed on these principles is vastly richer than the 3-D representation usually considered, because it incorporates knowledge of the associative structure in the world around it.

More or less explicit assumptions on the brain as an information processing device, on the modularity of visual processes, the independence of algorithm and hardware, the role of explicit representations, etc., are presented by S. Ullman in the context of present computer-vision research. The author discusses possible challenges to these assumptions as raised by, e.g., connectionism, direct perception (Gibson) and the notion of perception as a "bag of tricks" (Ramachandran). From a rather different perspective, S. Anstis discusses the intrinsic "assumptions" made by the visual system in the processing of visual motion (i.e. spatial redundancy, visual inertia, figure-ground organization, coincidence detection and biological motion) and elaborates on their role of accelerating visual processing.

Thoughts on the specific nerve energy

Andrei Gorea

Laboratoire de Psychologie Expérimentale Associé au C.N.R.S.,
Université René Descartes, 28 rue Serpente, 75006 Paris, France

The *specific nerve energy* (*SNE*) doctrine is derived from the common sense observation that there must be something specific about sensory processing merely because, everybody would agree, sensations are specific[*]. There is also common sense beneath the credo that objects "give off images of themselves, which are carried to the mind by the nerves" (see Boring, 1942, p. 69). At the time when they were spelled out, these two propositions were equally untestable, and for that reason regarded as axiomatic.

Let me stress from the very beginning the fact that a strict definition of what is actually meant by either "specific nerve energy" or "giving off images of itself" has never been provided. In fact, Johannes Müller himself phrases the essence of the former as a mirror image of the latter:

"A sensation is not the conveyance to consciousness of a quality or a state of an external object, but rather the conveyance to consciousness of a quality or state of our nerves, brought about by an external cause." [translated by Bela Julesz from the *Handbuch der Physiologie des Menschen*, 4th edition, Coblenz, Verlag von J. Hoelscher, 1844.]

Although equally reasonable from the "common sense" standpoint, by the end of the 19th century, the *SNE* approach had completely overthrown the "giving off images..." idea. Indeed, all of the evidence provided by anatomy, cytology, electrophysiology and neuropsychology transformed the *SNE* doctrine into a matter of investigation, "proved" correct its consequences and allowed the transfer of its axiomatic status into the realm of current knowledge.

The sensory scientist and more particularly the modern psychophysicist might be aware that the implicit and continuous use of this "current knowledge" in everyday model/theory building is not as obvious as it might appear. It is my claim that such a feeling is not only due to the general difficulty of applying "current knowledge" to specific investigation, but essentially to the fact that the *SNE* doctrine remains a doctrine in that it has never been satisfactorily tested. Paraphrasing Boring, not all the men who discarded this problem as trivial understood it (Boring, 1942, p. 68).

---

[*]In latin, sensation is "qualis", i.e. quality. "Specific sensation" may thus be looked at as tautological.

1 THE *SNE* AS A DEDUCTIVE APPROACH

1.1 A Syllogism

A. The basic concepts of the *SNE* doctrine are rooted in the following two observations: "The same stimulus acting on different nerves gives rise to different qualities [or perceptual states]; different stimuli acting on the same nerve give rise to the same quality. It is the nerve, not the stimulating object, that matters." (Müller, *passim* Boring, p. 71).

B. The *SNE* doctrine cannot be related to specific *receptors* (where a neural pathway begins) since stimulation beyond the receptor site still gives rise to specific perceptual states.

C. *SNE* must then be related to the site where a pathway **ends**, the *sensory center*, or to the neural pathway as a whole (including the "sensory center").

D. Modern and contemporary neuroscience - as initiated by Mountcastle (1957) and extensively developed by Hubel and Wiesel (1977) - basically consists of new experimental and theoretical instances of Müller's point of view.

Points A to C develop the logic beneath the *SNE* doctrine and point D provides experimental support. Altogether, points A to D lead to the apparently obvious conclusion that "... if you cross-connect,(say) the optic and the auditory nerves, you could see tones and hear colors" (du Bois-Reymond, *passim* Boring, p. 78).

Whether logical or intuitive, the above statement is based on the *belief* that "sight is not hearing because the optic fibers are projected on the occipital lobe and the auditory upon the temporal lobes" (*op. cit.*, p. 78). It is crucial to note that, despite point D, this premise is no more testable than the fact that we do *see* colors and *hear* tones, which is a pure matter of convention. In fact points A to D do not tell us what a specific sensation or "perceptual state" is, or what a nervous "ending" (or "site") is meant to stand for.

1.2 Where does deduction stop?

Let us consider the following:

E. If you cross-connect, during a *critical period*, the optic and the auditory nerves you would still see colors and hear tones. There is no *a priori* reason to reject this proposition. In the absence of any available test, it is as axiomatic as its reversed formulation. (However, accepting it will bring the *SNE* concept down in pieces. Or, will it?)

F. Accepting propositions A,B,C and E as true entails that the specificity of sensations does not lie either in the stimulus, or in the receptors, or in the pathways, or in the "sensory center", whatever their precise definition might be. Where does it lie, then?

I see two possible answers to this question. It either lies nowhere: there is no such thing as *specific* sensations or perceptual states. Or, it is related to a specific *conjunction* of the physical characteristics of the stimulus, their sensory receptors, the pathways and the "sensory centers", altogether.

The first alternative presents the advantage of eliminating all reference to the *SNE* concept. However, since it also presents the disadvantage of bringing back the sensory sciences to the deep blur of untestable philosophical propositions, I shall not discuss it here.

The spirit of the second alternative is definitely more biological. If sensation is the *stimulus-brain conjunction*, none of the enumerated "links" is "specific" with respect to sensation. On the other hand, at least in the early stages of development, none of these links should be missing without specificity of sensation being lost. All links are, however, specific in their own right. The photoreceptors respond to light, while the ciliary cells respond to vibration, etc. The optic and auditory tracts are spatially distinct and neurons at different stages are selective to distinct aspects of the stimulus in different physical domains. Depending on the location within the processing hierarchy and on the extent of the critical period, their selectivity is more or less experience dependent.

The inconvenience of the above perspective is that, while offering a general paradigm for conceptualising our general knowledge and deductions, it does not lead to specific models and theories and even less to specific experiments in the field of sensory research. The concept of *SNE* remains inherent to this formulation but it is no longer a matter of test. How could one test the (self-evident and circular) proposition that sensation is "labelled" by the interactions among all the elements, from stimulus to brain, intervening in that particular behavior?

The *SNE* doctrine appears to be rather diffuse in our minds. If identified with the concept of *stimulus-brain conjunction*, it provides an implicit frame for our thinking and, under this acceptation, it is more like a paradigm (Kuhn, 1962). If taken under its more localized version, it is more like a tool. As a tool, however, it requires strict definitions of what we mean by concepts such as "perceptual states" (or specific sensations) and neural "sites" (*viz.* "nervous endings" or "sensory centers").

It is in the rapidly changing ways of using the *SNE* concept, especially as a tool but also as a paradigm, that one may realize the extent to which the choice of the stimuli, the setup of our experiments, the interpretation of our results and the building of our models are pervaded by tacit assumptions the source of which can be, in most cases, traced down to the *SNE* doctrine. I shall try to make this point clear by briefly discussing what I think have been the main ideas within our field of research during the last three decades and their dependency on distinct definitions of the perceptual state-neural site duo.

The reader should be aware that, in this brief historical discussion, I have taken a (psychophysically) biased perspective and chosen to cite only a very limited number of authors. The likelihood of having omitted some basic developments in vision research is thus quite high.

2 PERCEPTUAL STATES AND NEURAL SITES IN THE LAST THREE DECADES OF VISION RESEARCH

2.1 Stimulus-specific (feature) detectors

2.1.1 General considerations. The main idea behind the feature-detector approach - at least as it started in the early fifties - is that there exists a set (to be specified) of distinct, canonical visual mechanisms (sensory centers) whose sensitivity profiles are such that they respond selectively to a set of distinct, canonical visual stimuli (features). It is implicit in this formulation that a given sensation or perceptual state is directly related to the activation of a given mechanism (or *homunculus*). As such, the feature-detector approach represents an orthodox implementation of the *SNE* doctrine.

The immediate problem with such an approach is that mechanisms and stimuli are defined with respect to one another. The actual existence of a canonical mechanism cannot be proven unless we specify the corresponding canonical stimulus and, reciprocally, the specification of a canonical stimulus requires knowledge about the corresponding canonical mechanism. Thus, in the same vein as the paradigmatic *stimulus-brain conjunction* approach, the feature-detector approach apparently misses from the very start its ultimate objective of providing the basis for a one-to-one relationship between the activation of a specific neural site and the experience of a specific sensation (or perceptual state). How was it actually used?

In 1953, Barlow discovered cells in the frog retina responding selectively to *small* objects *moving*, within a restricted velocity range, across their receptive fields. He proposed that *small* + *movement* = *fly* (which is to be eaten; paraphrasing Barlow, 1953, p. 86) and that the cells he described are thus *fly detectors*. This was an *a posteriori* interpretation. Its merit consisted in its biological *meaningfulness* (for frogs). It soon became a matter of dogma to assume (implicitly or explicitly) that all features (i.e. canonical stimuli) and feature-detectors are biologically meaningful. While meaningfulness would be a superb guide for our experimental and theoretical work, it is rarely (if ever) specified *a priori* (see also 2.5.1).

The initial suggestion of Hubel and Wiesel (1959) that orientation is a meaningful attribute of the visual image and that "oriented" units are meaningful relays in visual processing was an *a posteriori* assessment too. Visual scientists were ready, however, to accept meaningfulness in early vision as a key concept in visual research and proceeded for more than a decade as if the assessment of the meaningfulness of the mechanisms they were about to study was an *a priori* endeavour. This assumption started the golden age of the feature-detector approach.

2.1.2 Feature-detectors and hierarchical processing. The meaningful sensory center may be peripheral (for the frog) or, at least in principle, as central as one desires (for more sophisticated species; Barlow, 1972). Correlatively, perceptual states may be as elementary as *movement* and as complex as *small + movement = fly.* Of course, they can be *faces* but also *grandmothers.* As everybody in the field must have realized and acknowledged from the very start of the feature-detector tradition, *grandmothers* and, say, *yellow* are equivalent perceptual states, although grandmothers are frequently colored, old and typically friendly. If this is so, another major problem with the feature-detector approach is its effort to match an apparently hierarchical structure of meaningful sensory centers (a very influencial view since Hubel and Wiesel, 1968) with an apparently unstructured domain of perceptual states.

On the one hand, this convergence scheme (the higher the processing stage, the higher the complexity of the processed visual attribute) is difficult to conciliate with the idea that distinct detectors are the substrate of distinct perceptual states or, at least that distinct perceptual states are *always* related to the activation of distinct mechanisms. How could this be if the neural substrate is just a relay within a hierarchical chain of neural transformations? On the other hand, the ultimate implication of the convergence scheme, namely the existence of a higher up *master homunculus*, is identical with that of the *SNE* doctrine.

During the golden years of the feature-detector approach, few debated the inherent ambiguity of the stimulus/mechanism definition or the meaningfulness of orientation (rather than curvature, for example) and orientation-selective (rather than curvature-selective) units for visual behavior. Instead, "flies" and "fly"-detectors multiplied as mushrooms and pervaded the classical frontiers of early vision.

Similar considerations apply to the psychophysical feature- detector approach where the existence of bar and edge detectors was questioned on grounds that a different stimulus description (*viz.* in the frequency domain) may be more meaningful and certainly more general. The psychophysicists put then more emphasis on an alternative approach based on the idea of *filtering.* However, the basic assumption that filters, as well as feature-detectors, are *labelled* such that they can be directly related to a specific perceptual state, was not abandoned (see 2.1.4). Two reasons for this were that no other paradigm was immediately available and that the labelling concept was extremely fruitful in naming a whole set of *specific perceptual states* and *detectors* on solid (electrophysiological and psychophysical) experimental grounds. Given the above considerations, the extent to which these specific detectors, their visual functions and the underlying experimental evidence are beyond any doubt remains a matter of debate.

2.1.3 Feature-detectors and parallel processing. The view of a strictly hierarchical processing of the visual image coexisted practically from the very beginning with the view that visual information is initially blown into a number of primitives processed in parallel up to some higher (unknown) associative areas (for recent reviews see DeYoe and Van Essen, 1988; Livingstone and Hubel, 1988; Zeki and Shipp, 1988). Within this context, vision research emphasized the idea of "super-flies" related to concepts such as space and time, form and motion, chromatic and achromatic dimensions each of which is presumably processed within parallel pathways (e.g. Livingstone and Hubel, 1988).

The first challenge of the hierarchical view may have been the discovery by Enroth-Cugell and Robson (1966) of the X and Y ganglion cells in the cat retina. X and Y cells were shown to differ in many respects of which their distinct spatial and temporal processing characteristics were of main interest. In a relatively short lapse of time, psychophysical research managed to impose the idea that shape (i.e. spatial information) and what was indifferently referred to as flicker or motion (i.e. temporal and spatio-temporal information) were processed by more or less independent mechanisms. The issue soon became ambiguous both neurophysiologically and psychophysically. The X/Y distinction at higher processing levels became controversial and the status of motion perception (which is inherently spatio-temporal) raised theoretical problems concerning the separability of space and time (see Burt, 1987).

While space and time may be conceptually (and experimentally) difficult to relate to distinct perceptual states, the perception of color, form, motion and depth may be easily regarded as orthogonal and studied independently. This conceptual and experimental facility, I would guess, led to the reinforcement of the generalized parallel processing idea which partly overshadowed the hierarchical processing one. The initial feature-detectors which, in principle, could be selective to any specific combination of visual attributes (like "yellow submarine") were replaced by specific pathways dealing with specific attributes at all levels of complexity. Saying that two pathways are distinct is to say that they carry specific (perceptual) information and thus specific nerve energies.

There are two main objections to this approach. The first is experimental and relates to the increasing number of cross- connections between presumably distinct pathways and to the difficulty of demonstrating their exclusive selectivity to a given stimulus dimension (e.g. DeYoe and Van Essen, 1988; Zeki and Shipp, 1988). The second is theoretical and relates to the integration of attributes processed independently within an unique and meaningful visual object. This integration problem, repeatedly addressed by both neurophysiologists (e.g. Zeki and Shipp, 1988) and experimental psychologists (e.g. Treisman and Gelade, 1980) is far from being solved.

It is interesting to note that there is an integration problem only when one rejects the possibility that the cross-talk among distinct pathways can be captured within the activity of a unique (meaningful) mechanism. Indeed, there is no such a problem if one is ready to accept the existence of a feature-detector (of unspecified complexity) selective within a multidimensional physical space. Such a feature-detector is in fact a neurophysiological replica of the stimulus and, as such, it can be directly associated with a perceptual state (of unspecified complexity). Hence, there is an integration problem only outside of the conceptual frame determined by *SNE* doctrine.

2.1.4 "Identification" and the labelling argument. Recently, Watson and Robson (1981) performed the following experiment. They randomly presented during one of two temporal intervals one out of two spatial frequency patches whose contrast covered the whole threshold range. Observers were asked to specify the interval which contained the stimulus (detection) and to identify the stimulus (identification). They measured the detection/identification performances as a function of contrast for a number of stimulus pairs and found that when the two stimuli in a pair were sufficiently disparate (in spatial frequency), the detection and identification functions of contrast overlapped. Since the system is capable of identifying the stimulus any time it detects it and since, at threshold, the probability of activating more than one (optimal) detector is very low, it follows that this detector *must* be *labelled*. Thus, *all* detectors must be labelled. What "labelled" was meant to specify is unclear, although everybody would probably agree that the "labelling" idea is directly related to that of a specific perceptual state and of a specific nerve energy. If so, is the above described experiment a proof of, or just another way of restating the *SNE* doctrine?

The relationship between identification and detection has been discussed by Helmholtz and its modelling has been shown since to depend on factors such as the underlying detection theory, the linearity of the detection process, the independence of the detectors, etc. (Graham, 1989). It is clear, for example, that the interpretation of the above experiment is critically dependent on the assumption that threshold performances are determined by the activation of an optimal detector. One may, however, doubt whether this assumption will ever be a matter of formal testing. Moreover, the optimal detector is specified psychophysically and formal proofs of its neurophysiological site are missing (see, however, Newsome in this volume). Thus, the application of the *SNE* concept to early vision remains a matter of consensus.

2.2 Linear filters and feature-detectors - Mechanisms

2.2.1 Traditional approach. Linear filters (Campbell and Robson, 1968; Sachs, Nachmias and Robson, 1971) and feature-detectors have been and still are hostile friends. From the *SNE* point of view, they are equivalent to a large extent. The basic and perhaps only difference between them is that, in principle, the filter approach requires a limited number of filters to account for a much larger number of perceptual states. The underlying idea (which can be traced back to Young) is that a perceptual state is related to some specific pattern of activation of a limited set of low-level units. Current understanding of color perception and of discrimination/ identification visual performances in early vision (see para 2.1.4) is heavily dependent on this principle.

The manipulation of the filter concept eventually led to the specification of theoretically *optimal detectors* which, in turn, permitted the specification of the *appropriate stimulus* (Watson, Barlow and Robson, 1983) to be used in the process of testing (psychophysically or electrophysiologically) the existence of the optimal detectors... such as spatial frequency or directionally tuned filters displaying a more or less pronounced even or odd spatial symmetry, with a more or less Gaussian spatial sensitivity weighting function, etc.

This increasingly sophisticated engineering approach also raised problems related to the independence of spatial and temporal processing (see para. 2.1.3), to biological noise and its correlation across distinct detectors and, more generally, to the linear *vs.* nonlinear processing of visual information (Graham, 1989). As Bela Julesz pointed out a while ago, the fact that visual processing is strongly nonlinear necessarily leads us back to the feature-detector approach since nonlinearities are features (or bugs or flies).

2.2.2 *Pyramids*.

It has been proposed (Marr, Ullman and Poggio, 1979) that early vision may be modelled as a parallel, multiple-scale filtering process. Since the representation of physical information is isomorphical to the related percept at any scale of the "pyramid", a perceptually "popping-out" feature is a "popping-out" neuron (or group of neurons) at at least one of these filtering levels.

This "pyramidal" scheme (see Part 1 of this volume) is an obvious extension of the filter approach and it was initially developed to provide higher efficiency coding (but not decoding) primitives (kernels, wavelets, 2-D Gabors, etc.) and algorithms in the luminance domain. It cannot thus account for more than second-order, black-and-white phenomena such as texture discrimination, "pop-out" effects and the like.

In principle (but not yet in practice), the pyramidal approach could be applied at *all perceptual domains*, whether at the same early vision processing stage (such as the chromatic domain), or at a higher processing stage (such as, say, the domain within which we account for shape-from-motion phenomena). At the same processing level, a large population of units would share the same multidimensional tuning space, while others would be more or less one-dimensionally biased. At different processing stages, primitives would differ qualitatively so that the higher the processing level, the more elaborated the coding primitive. Moreover, the multiple-scale processes (at all complexity levels) may be made interactive and the perceptual states may be related to the state of the pyramid(s) as a whole rather than to the activity of some of its (their) layers (see para. 2.5).

If implemented, this architecture of interactive "pyramids-on-pyramids" may develop unexpected behaviors. It dilutes any specific meaning of the perceptual state concept. It also leads to a major problem: What is the scaling metric for higher level pyramids?

2.2.3 *Textons and statistics*.

Bela Julesz never hesitated to identify the texton and the feature-detector concepts (Julesz, 1981). They are both just different names for low-level visual primitives (i.e. "atoms of perception") and, in principle, may be extended to any visual entity independently of its complexity (the "grand-mother" detector). The problem, of course, is of defining what a visual entity is. For the texton theory, *crossings* and *terminators* were important perceptual "atoms". Which brings us to (in this particular case, binary) *statistics*.

The texton story started with the idea that, from a Fourier point of view, two stimuli (textures) with identical power spectra can be discriminated only on the basis of their *spatial phase* characteristics. Black-and-white stimuli with identical power spectra are also identical in terms of their second order statistics (they are *iso-dipoles*) but they are not necessarily identical in terms of their higher order statistics. In the Fourier domain, statistics of an order higher than 2 may always be related to the phase spectrum of the stimulus.

Julesz and col. showed that the iso-dipole texture-pairs they initially used, were not "instantaneously" discriminated and concluded that what they had already coined as the *preattentive* visual system was not sensitive to spatial phase. Within this theoretical framework, discrimination based on spatial phase (or higher-order statistics) requires *scrutiny* (i.e. some kind of ill-defined mind's eye search process). Later on, Julesz and col. found a few iso-dipole texture-pairs readily discriminable. Since, according to them, the first set of experiments had shown that spatial phase information could not be processed without scrutiny, they concluded that discrimination of texture-pairs which do not share 3rd or higher order statistics must be based on the analysis of very *local* and distinct patterns which they called *textons*.

The first logical step having led to the texton concept was not sufficiently validated. First, the phase distortion in the texture- pair was not quantified and it was probably quite small. Besides, whether attentive or preattentive, phase discrimination as such is quite poor to start with. It is hence an error to conclude on the basis of the above experiments that phase information is not processed by early vision. Second, phase information may be regarded as the relevant parameter only to the extent that one has in mind a *global* Fourier analysis.

Since the description level at which these textons could be characterized was not obvious, Julesz and col. "scrutinized" the texture-pairs producing high and low discrimination performances and pinpointed some specific shapes which they called "blobs", "terminators", "crossings", "connections"... In the last few years, a series of papers has demonstrated, however, that all typical and apparently atypical cases of texture discrimination can be accounted for by the parallel processing of the image by a population of local linear filters at different spatial scales. The texton no longer had any reason to exist.

The idea that the visual system might compute and subsequently discriminate visual stimuli on the basis of their n^{th}-order statistics was new and insightful. Conceptually, it can be regarded as one of the first attempts to get rid of the *SNE* doctrine. Statistics must be computed over a large number of units which do not need to be labelled with respect to the dimension along which discrimination takes place and whose correlative perceptual states thus become irrelevant. The perceptual state is related to the statistics themselves.

While computation by the visual system of nth-order statistics did not receive experimental support, it definitely prefigurated the *connectionist* philosophy (see para. 2.5), as well as recent electrophysiological research demonstrating resonant activity in neural populations (see Gray in this volume).

2.3 Matching as a perceptual state

Research in stereopsis (Julesz, 1960, 1971) and motion perception (Reichardt, 1961) led in the early sixties, to the formulation of the concept of *matching* as a direct substrate of perceptual states. The underlying idea was that a given sensation is characterized by the extent to which the activities of a given (rather than of any other) pool of neurons are matched (or cross-correlated) in space (for stereopsis) or in space and time (for motion). Out of the very large number of possible binary matchings, only those which are *globally coherent* (or concordant) are finally selected through global interactions.

This formulation only apparently solves the dilemma introduced by the *SNE* doctrine: relating perceptual states (as well as states of mind) to matching states in the brain does not require, in principle, the use of labelled *primitives (see para 2.5)*. Posing that a given perceptual state depends on the matched activity within a neural population does not exclude that it also depends on the particular neurons involved in the matching process. On the other hand, identical neural populations may give rise to very different perceptual states.

Depth perception is in all respects distinct from motion perception. The underlying matching processes, as modelled, are of a very different kind. But so are the neurons subserving each of the two perceptual states. In contrast, motion and texture perception may be related to very similar matching processes across similar or identical cell populations (Gorea and Papathomas, 1990). The remaining two combinations are also possible. Is thus the specificity of sensations related to the process (of matching) or to its neurophysiological substrate? Whether blunt or dull, this question has no obvious answer. Definitely no more than "Where are the nervous sites of our perceptions?" Hence, the use of the matching concept as the neurophysiological counterpart of a perceptual state is not entirely independent of the *SNE* doctrine.

2.4 The computational approach

In order to build a machine that "sees", one is facing conceptual problems analogous to those encountered in the process of unveiling the nature of biological vision (see Ullman in this volume). Marr's work (1982) is exemplary in having intimately combined these two domains of research.

One of Marr's conceptual contributions to the study of biological vision was to let the neural "matching" process be guided by **real world constraints**. A second contribution was to reverse the perspective of the current theoretical inquiry. Instead of asking "What is it [the visual system] doing?", he asked "What is it supposed to do?" The underlying idea is that descriptions of function should provide information about (neural) substrate. The analysis of the "natural stimulus" and of the "biologically plausible functions" of a system vis-à-vis this "natural stimulus" may stand as a revival of Gibson's (1966, 1979) philosophy and as the ultimate concept behind the *computational* approach.

On the one hand, one may say that Marr's approach stressed the *natural* stimulus-end of the visual process. Vision (but also any other sense and for that matter, experience) is constrained. More than anything, the job of the vision scientist is to realize, inspect, understand and determine how the system reacts to and takes advantage of those constraints.

Constraints are physical in the sense that the physical arrangement in space and time of the visible matter determines the nature of *visual assumptions* concerning the visual meaningfulness of that physical matter (see the chapters by Anstis and by Cavanagh in this volume). This makes the irrefutable point of materiality, namely that vision is the stimulus-visual brain conjunction.

On the other hand, constraints are biological in the sense that *we see what we need*. This makes an ambiguous point. It may have been intended to mean that the system "needs" specific information concerning some vital functions of ours like moving within a sophisticated environment. But it may also mean that the system "needs" something sufficiently well specified to be experimentally evaluated by some master biologist. The point here is that it is equally likely that *we need what we see*.

The distinction between *seeing what we need* and *needing what we see* is crucial when elaborating the concept of *perceptual state*. In the first case, perceptual states are given *a priori*. In the second case, they are physically determined. It is thus the second alternative which leads explicitly to the specification of perceptual states in terms of physical dimensions. But it is also the alternative which objects to the interest of asking "What is the system supposed to do?". This contradiction in the premises of Marr's thought is, of course, inherent to the nature/nurture dilemma. The consequence of which is that the specification of perceptual states remains a paradoxical matter.

It is the inherent implementation of Marr's approach which, despite its formal rigour, brings us back to the *SNE* concept. Whether explicitly or implicitly accepted, processing stages and parallel processes have *meaning*. They are labelled. Knowledge about "out there" is provided *directly* at/within those processing stages and parallel pathways. Of course, in a strictly computational sense, knowledge is a purely decisional matter, but one may argue that sensing and interpreting is also a decisional matter.

The specification of the physical and biological constraints of visual behavior is not necessarily an objective matter. Certainly, the visual system of some diving birds has adapted so as to automatically correct for refraction errors. But is there any *objective* reason explaining why our visual systems did not evolve to process infrared light? Why is our retina inhomogenous? And then, why don't we fly? Etc.

2.5 The connectionist approach - Networks

There is (almost) nothing new under the sun. Most fashionable nowadays, networks are "matching" devices. In addition, the connectionist approach leans heavily on the idea that meaningfulness of neural processing is intrinsically related to the interaction of processing units both with the outer world and among themselves. This implies that the units themselves and their interconnections (networks) are (or must have been at some point) memory devices. The idea that memory is a distributed process may be traced back to William James and is unanimously accepted nowdays.

What is new about the connectionist approach is that it might offer a possible solution to the problem of "high-level-vision". The solution is conceptual and, to the extent that it can be *simulated*, it is objective. The notorious problem with this approach is that it is (notoriously) untestable.

"High-level-vision" is certainly something ill-defined. Being ill-defined may have hidden advantages. For example the advantage of insinuating that the concept of perceptual state is itself ill-defined.

Most will agree that there is more to vision than orientations, disparities, movement detectors and so (see Barlow in this volume). The connectionist approach has just started to face problems such as shape and size constancy, 3-D recovering from 2-D representations (also addressed by the computational approach), etc. Of course, most will also agree that there is more to vision than shape and size constancy... The question is how much more. The question is, *How do we define the scope and represent the complexity of what vision is supposed to account for in our behavior?*

Going beyond early vision is a dominant preoccupation today (see Cavanagh in this volume) and the connectionist approach is simultaneously a consequence of this preoccupation and a means to study (simulate) behaviors related to it. From the standpoint of the present discussion, the connectionist approach appears to dilute the problem of both sensory centers and perceptual states. Accounting for complex visual behaviors such as watching a yellow submarine or visualizing a tempest in terms of specific sensory centers and perceptual states is definitely an uneasy task. In what sense would these two behaviors be qualitatively different?

The *SNE* doctrine is tautological with the concept of specific-neural-sites-distinct-perceptual-states. In that respect, the connectionist approach may be the alternative solution. The "states" of a network, which are difficult to qualify as qualitatively different, are perceptual states. An untestable solution... Unless, contrary to traditional modelling and experimentation, simulation is to be accepted as scientific proof.

3 CONCLUSION

Things (and thoughts) can be indefinitely more confusing. Consider this. When you listen to a complex tone, you may, especially if you are well trained, pick up some of its components. Recent papers suggest that this kind of selectivity indicates that the specific underlying filters do have "direct access to perception" (e.g. Welch, 1989). Would those scientists agree on the reciprocal, viz. that "direct access to perception" necessarily implies the existence of specific filters, mechanisms and what more? Probably not, if you consider that "direct access to perception" of a yellow submarine does not imply the existence of a yellow submarine specific detector...

It is consensually accepted that *access to perception* refers to a sensorial (visual) entity. It is generally implied that if the neural substrate of a sensorial entity is itself a neural entity (namely that it may be spatially localized in the cortical space) specific for analyzing a given physical (or otherwise conceptual) dimension of the stimulus, that neural entity has direct access to perception.

The unanimously shared conviction that we do have direct access to Gabor-patches (as visual primitives), to oriented edges (by the virtue of zero-crossings), to red (but also to yellow) etc., is puzzling. How is that anatomically possible? If cells in V4 code *color as seen* (i.e. respect color constancy - Zeki, 1980), what about visual behavior accounted for by the activity of CGL, color-opponent cells? Through what path do the latter access perception?

What shall we think about the perceptual status of a feature-detector, if the only evidence we have about its materiality is obtained *via* stimulation with an *exclusive* class of stimuli, whether defined along a physical or otherwise conceptual dimension?

Orientation and spatial frequency specific detectors exist "beyond any doubt" and their stimulation is positively assumed to account for the capacity of "identifying" our own orientation- and frequency-related sensations. However, all evidence is against the slightest capacity of visually identifying the harmonic components of a square-wave grating.

Suppose that we have a metric for ordinating faces. Are we sure that selective adaptation, masking, subthreshold summation and the like experiments with faces *would not* provide results equivalent to those obtained with sinusoidal gratings? What would our conclusions be?

*

None of the insights provided by the theoretical (and conceptual) approaches of these last decades has been proven definitely wrong. What we know about vision today is what all of them taught us.

We (think we) know that our visual system is built up of (spatial and spatio-temporal) orientation detectors, face and hand detectors, and also of more or less narrowly tuned (chromatic, but also color, spatial and temporal frequency, disparity, etc.) filters and of more or less specific (X-Y, magno-parvo, luminance-chrominance, etc.) pathways... We also have the firm conviction that all these detectors and filters and pathways, all of which *must* have perceptual meaning and thus direct access to perception, interact within rather huge networks whose states also have perceptual meaning, presumably at a higher complexity level...

A few might think they even know that perceptual meaning is a perfectly useless concept. Like the ether, say. But, a "unifying" theory of vision making the economy of this concept has not as yet been proposed. The *SNE* doctrine is the doctrine of perceptual meaning. As such, it could never be formulated as a question to be answered experimentally. It is a state of mind.

The *SNE*'s paradigmatic nature may be looked at in a different way. If we mix all the required ingredients: feature-detectors, one-and multidimensional filters, matching devices, pyramids-on-pyramids, parallel pathways and distributed processing *plus* a rich ecological visual environment (in Gibson's (1979) sense) and if we let it be, this artificial system *must* develop a perceptually meaningful behavior identical to that of our visual brain. It seems to me that this unescapable conclusion is rooted in the philosophy according to which understanding the visual brain cannot go beyond this isomorphical, but also circulary, explanation (see chapters by Klein and by Tyler in this volume).

Visual behavior is meaningful. Meaningfulness does not require consciousness. The purpose of studying visual behavior is to uncover the neural substrate of visual meaningfulness as defined at a given moment. Sooner or later, this is achieved either when the neural substrate (or a model of it) appears to match the meaningful behavior as defined, or when we manage to redefine meaningfulness such that it matches a given substrate. The problem of the appropriate stimulus matching a sensory entity as *experienced* is nowadays as intact as it has ever been.

ACKNOWLEDGEMENT. I am particularly grateful to Bela Julesz, Patrick Cavanagh, Christopher Tyler, Stanley Klein, Horace Barlow, Shimon Ullman and Maggie Shiffrar for their constructive comments on earlier versions of this paper.

REFERENCES

Barlow H.B. (1953) Summation and inhibition in the frog's retina, *J. Physiol (London)* **119**, 69-88.

Barlow H.B. (1972) Single units and sensations: a neuron doctrine for perceptual psychology? *Perception* **1**, 371-394.

Boring E.G. (1942) *Sensation and perception in the history of experimental psychology*, New York, D. Appleton-Century Company.

Burt P.J. (1987) The interdependence of temporal and spatial information in early vision, In *Vision, brain and cooperative computation* (Eds. M.A. Arbib & A.R. Hanson), Cambridge, MIT Press.

Campbell F.W. & Robson J.G. (1968) Application of Fourier analysis to the visibility of gratings, *J. Physiol (London)* **197**, 551-566.

DeYoe E.A. & Van Essen D.C. (1988) Concurrent processing streams in monkey visual cortex, *Trends Neurosci.* **11**, 219-226.

Enroth-Cugell C. & Robson J.G. (1966) The contrast sensitivity of retinal ganglion cells of the cat, *J. Physiol (London)* 187, 517-552.

Gibson J.J. (1966) *The senses considered as perceptual systems*, Boston, Houghton Mifflin.

Gibson J.J. (1979) *The ecological approach to visual perception*, Boston, Houghton Mifflin.

Gorea A. & Papathomas T.V. (1990) Texture segregation by chromatic and achromatic visual pathways: an analogy with motion perception, *J. Opt. Soc. Am. A* 7

Graham N.V.S. (1989) *Visual pattern analyzers*, New York, Oxford University Press.

Hubel D. & Wiesel T.N. (1959) Receptive fields of single neurones in the cat's striate cortex, *J. Physiol. (London)* 148, 574-591.

Hubel D. & Wiesel T.N. (1968) Receptive fields and functional architecture of monkey's striate cortex, *J. Physiol. (London)* 195, 215-243.

Hubel D. & Wiesel T.N. (1977) Functional architecture of macaque monkey visual cortex, *Proc. R. Soc. London* B. 198, 1-59.

Julsez B. (1960) Binocular depth perception of computer-generated patterns, *Bell Syst. Tech. Jour.* 39, 1125-1162.

Julesz B. (1971) *Foundations of cyclopean perception*, Chicago, University of Chicago Press.

Julesz B. (1981) Textons, the elements of texture perception and their interaction, *Nature* 290, 91-97.

Kuhn T.S. (1962) *The structure of scientific revolutions*, The University of Chicago, 1st edition; 1970, 2nd edition.

Livingstone M. & Hubel D. (1988) Segregation of form, color, movement and depth: anatomy, physiology and perception, *Nature* 240, 740-749.

Marr D. (1982) *Vision*, San Francisco, Freeman & Co.

Marr D., Ullman S. & Poggio T. (1979) Bandpass channels, zero- crossings, and early visual information processing, *J. Opt. Soc. Am.* 69, 914-916.

Müller J. (1844) *Handbuch der Physiologie des Menschen*, 4th edition, Coblenz, Verlag von J. Hoelscher.

Reichardt W. (1961) Autocorrelation, a principle of evaluation of sensory information by the central nervous system, In *Sensory coding* (Ed. W.A. Rosenbluth), New York, John Wiley.

Sachs M.B. Nachmias J. & Robson J.G. (1971) Spatial-frequency channels in human vision, *J. Opt. Soc. Am.* 61, 1176-1186.

Treisman A. & Gelade G. (1980) A feature integration theory of attention, *Cognitive Psychol.* 12, 97-136.

Watson A.B. & Robson J.G. (1981) Discrimination at threshold: labelled detectors in human vision, *Vision Res.* 21, 1115-1122.

Watson A.B., Barlow H.B. & Robson J.G. (1983) What does the eye see best? *Nature* 302, 419-422.

Welch L. (1989) The perception of moving plaids reveals two motion-processing stages, *Nature* 337, 734-736.

Zeki S. (1980) The representation of colours in the cerebral cortex, *Nature* 284, 412-418.

Zeki S. & Shipp S. (1988) The functional logic of cortical connections, *Nature* 335, 311-317.

The duality of psycho-physics

Stanley A. Klein

School of Optometry, University of California
Berkeley, CA 94720, USA

This is another paper on the mind-body problem and on the question of whether computers can ever become conscious and have feelings. Dozens of papers and books are written on this topic every year. The great philosophers from Plato through Descartes and Kant wrote and worried on it. So why another paper? Four reasons:

1. This conference on visual perception is in the land of René Descartes, who gave the first clear statement of the mind-body problem, and of Alain Aspect, the physicist whose recent experiments (to be discussed) cast doubt on an objective world that exists without being observed. It seems entirely appropriate to honor and connect the contributions of these two Frenchmen.
2. This particular session of the conference is about tacit assumptions in our models of vision. We should ask what assumptions about consciousness and the homunculus we need for our models. Is it possible to truncate the infinite regress often associated with the homunculus? How are our subjective impressions linked to our objective button pushes? It is good to get the tacit assumptions out on the table.
3. The name of our field, "psychophysics," entitles us to be the "chosen field" to worry about the connection between psyche and physics.
4. Last, but not least, I have a novel solution for the mind-body or the subjective-objective duality, related to a little-known aspect of quantum mechanics.

Section 1 of the paper attempts to show that there is a problem worth worrying about. Section 2 introduces the laws of quantum mechanics. Section 3 provides a solution to the mind-body problem based on a little-known quirk of quantum mechanics.

Section 1. Is there a problem? Most scientists, being materialists, would say that there is no mind-body problem; they would say that the mind is reducible to the operations of the brain. So the first task of this essay is to convince the reader that there is a mind-body problem. I will begin with definitions of what I mean by mind and body. Then a brief history of the mind-body problem is presented to help us understand how this state of affairs came about. Part of this history is an examination of how the success of reductionism tempts one to believe that the mind can be understood in terms of the brain. Finally I will present arguments for why my mind is not reducible to the physical operations of my brain. Once we have established the non-equivalence of mind and brain we are back to the classic problem of how the mind and brain interrelate. It is to this question that I offer a novel answer in Section 3.

This morning as I was writing this paper I asked my wife whether she thought a robot could ever have feelings or be conscious. She thought it a dumb question because without the slightest doubt she answered, "NO!" She feels there is something very special about our emotions and thoughts that will always be missing from robots. Those readers who agree with my wife can skip Section 1 and proceed with Section 2 on quantum mechanics. Section 1 is for those people whose scientific training has given them too great a faith in reductionism, and for those whose science fiction reading (Hofstadter & Dennett, 1981) convinced them that it is no big deal for robots to have humor, emotions and self-consciousness.

A hint that the mind might cause trouble for reductionists is that so many respectable scientists and philosophers are still writing and arguing about it. The same issue is found under many guises: Will computers ever gain consciousness? Can algorithmic machines solve nonalgorithmic problems? What is the substrate for human consciousness? Can one have free will in a deterministic body? How does one identify a thought such as "I am happy" with brain mechanisms? The surprising point of this essay is that these questions might have answers.

1a. Definition of body. By "body" I mean the brain as a machine. The essential property of a machine is that its operations can be understood in terms of the presently known laws of physics, chemistry and biology, similar to how the eye is understood (or is expected to be understood).

It is possible to define body differently. Recently Searle (1980, 1990) and Penrose (1989) made a distinction between a machine that is equivalent to an algorithm (a universal Turing machine) and a machine with "meat" (as allowed in our definition) in which the processing is hardware dependent. The definition of body as algorithm is associated with what Searle calls the Strong AI position. Searle believes Strong AI to be wrong and Section 1d offers a suggestion on how Strong AI can be modified to appease him. Another view of body is proposed by Penrose who believes that changes in the laws of physics are needed to account for brain operation. Section 1e will examine Penrose's unique view of the brain's machinery. I will argue, however, that these different definitions of what is meant by "body" are irrelevant to the mind-body problem.

Definition of mind. The aspect of mind to be considered in this paper is subjective <u>feelings</u>. By feelings I mean the type of feelings I feel when I feel strong feelings. The point of this circular definition is that subjective states don't like to be reduced to words. Haugeland (1985) does an excellent job of classifying feelings and pondering what it would mean for a computer to have feelings. I focus on feelings rather than thoughts and awareness (consciousness) since feeling a sharp pain is clear and vivid. Defining mind in terms of thoughts and awareness would have led us astray into abstract philosophical discussions of intentionality. There is nothing abstract about a strong feeling. We all know the feeling of pain. Subjective feelings are sufficient to produce a mind not reducible to a brain, which is all that is needed for this essay. In a strict sense I am the only entity with feelings, but as shall be discussed in connection with Turing's "imitation game", there are circumstances in which, through empathy, I might become convinced that another entity also has subjective feelings. Since this definition of mind focuses on subjectivity, the mind-body problem is one version of the subjective-objective problem: How can we reconcile the observer's subjective world with the objective world of the observed? Those who have studied quantum mechanics will notice a familiar ring to this way of stating the problem. Section 3 will discuss the important distinction between mind as a carrier of feelings vs. mind as observer.

1b. A statement of the mind-body problem. Science seeks to understand *you* as a machine, to explain *your* every action, including why *you* scream when badly hurt, squirm when tickled and why *you* say *you* feel happy (or frustrated) when *you* contemplate philosophy. On the other hand, as pointed out by philosophers of mind, *my* subjective feelings while in pain or being tickled, or while contemplating philosophy, are not equivalent to the associated neural impulses. The mind-body problem arises because of the paradox of being able to totally reduce *your* (observed) brain to neural events, while not being able to do the same reduction to *my* (observer's) brain. Quantum mechanics, with its sophisticated duality, provides a novel solution to this dilemma.

1c. History of mind-body problem. Galileo (1564 - 1642), the father of modern science, is a good starting point. He showed that combining reason with experiments produced a powerful tool for learning about nature. However, he got into trouble with the church because, by not explicitly removing mind from his domain of study he was encroaching on the province of the church. The first to make a clear separation of mind from nature was Descartes (1596 - 1650), the first modern philosopher and the first modern mathematician. Cartesian duality is usually discussed in the context of how nasty it is because of the difficulty in figuring out how the physical and the mental halves of the duality interact. An excellent analysis of Cartesianism is given by Lockwood (1989). What is often forgotten in discussions of Descartes' contribution is that his duality provided the underpinning for the scientific revolution that was to follow. By getting the mind out of the operation of the physical world, Descartes' metaphysics paved the way for Newton and followers to develop a purely mechanistic universe with minimal complaints from the church. It was Newton (1642 - 1727) who had the greatest impact on removing mind and magic from the operation of nature (although Newton himself did believe in magic). Before Newton, the heavens were still assumed to operate according to aesthetic principles, such as Kepler's (1571 - 1630) proposal that the planets move according to "musical harmonies." After Newton, the motions of all bodies, both heavenly and earthly, moved in straight lines unless affected by simple forces such as gravity. A mind wasn't needed to keep the system running, and it would keep running even if no minds observed it. The success of this mechanical reductionist approach to explaining nature is well known to us.

Reductionism. The materialist worldview is that everything in nature can be understood in terms of the physical world. Complicated phenomena can be explained in terms of simpler underlying mechanisms. The scientific explorations of 300 years following Newton have produced a worldview in which the realm of the physical has been constantly expanding. The success of the reductionist program has left little room for mind. It is sometimes claimed that quantum mechanics has drastically changed the materialist worldview. Not so. The world of quantum mechanics still operates by precisely stated principles and forces and there is no room for influences outside its narrow framework. Quantum mechanics does not, for example, tolerate a "mind-force" working outside the physical realm that is nevertheless able to influence the probabilities of events.

It may surprise some people that even though reductionism places limits on possible explanations of events, it is not opposed to duality, but is in fact one half of a duality. To make this point I reprint the last paragraph of Descartes' "Treatise of Man" (Descartes, 1664):

> "I desire you to consider, further, that all the functions that I have attributed to this machine, such as ... waking and sleeping; the reception by the external sense organs of light, sounds, smells, tastes, heat, and all other such qualities; the imprinting of the ideas of these qualities in the organ of common sense and imagination; the retention or imprint of these ideas in the memory; the internal movements of the appetites and passions; and finally, the external movements of all the members that so properly follow both the actions of objects presented to the senses and the passions and impressions which are entailed in the memory--I desire you to consider, I say, that these functions imitate those of a real man as perfectly as possible and that they follow naturally in this machine entirely from the disposition of the organs--no more nor less than do the movements of a clock or other automaton, from the arrangement of its counterweights and wheels."

This dramatic affirmation of reductionism by the father of duality emphasizes that a person can believe in both duality and reductionism.

Less than 100 years ago there was still a belief that the reductionist program might be halted. It was thought that there might be something special about life that couldn't be reduced to chemistry. Biologists and physicists were searching for a new fundamental force that they thought was needed to explain life. However, within the past 50 years it has become clear that the life-force can be understood in terms of DNA and proteins, using the same laws of chemistry that govern inanimate matter. It has been found that biology can be reduced to chemistry, which can be reduced to physics and quantum mechanics.

When it comes to explaining consciousness and mind, there is again the possibility that the reductionist program will fail. This time the possibility of failure is greater than ever before for one paramount reason. The reductionist program has collapsed in another arena: quantum mechanics. Since, as I will argue, the quantum duality is split along a similar subjective-objective cut as the mind-body problem (remember our definition of mind as subjectivity) it is possible that it is the same duality (the theme of this essay). Possibly the clever way in which the quantum duality both is and is not reductionistic might apply to the mind-body duality.

In order to clarify my views of the mind-body problem it is useful to compare them to alternative viewpoints. Sections 1d and 1e will examine the points of view of Searle (1980, 1990) and Penrose (1989). Both of these authors have generated a good deal of discussion recently.

1d. Strong AI: an aside on Turing's imitation game. Strong AI believes that both mind and body will some day be understood in terms of a complex algorithm, independent of the hardware implementing the algorithm. Searle (1980, 1990), however, argues that algorithms can not be minds. He makes the analogy to how a person doesn't understand Chinese if all he knows is how to follow an algorithm written in English that specifies how to respond with Chinese letters to a question given in Chinese letters. That is, minds can understand, algorithms can't.

The Strong AI program seeks to convince us that a sufficiently clever algorithm can have feelings by playing Turing's imitation game. However, I now will argue that the usual version of the imitation game would not work for me.

The way a robot would go about convincing me about its feelings is that we would spend a few days living together. I am a skeptical person, but I am open-minded and might be convinced. I do, for example, believe that dogs have feelings. I have seen the looks on their faces and their wagging tails. I know that if I felt happy and excited and had a tail I would wag it just as a dog does. Most important for me is that I know that dogs are made of the same kind of stuff as I am. A robot on the other hand, would have great trouble convincing me that it had feelings. I can easily imagine a futuristic robot with a modulated voice (such as HAL in the movie 2001 or C3PO in Star Wars) that could evoke my empathy even better than could a dog. However, I don't think that either HAL or C3PO would convince me that they truly had feelings. I'd be skeptical and worry that they were just programmed to imitate feelings and to respond as would a feeling person. I, in agreement with Searle, will never be convinced that an algorithm can have feelings.

Does this mean that no robot could convince me that it has feelings? No. Here's what a robot could do to convince me. After conversing for a few days and getting angry with each other, I would finally realize that one of the reasons that I doubted its feelings is that for me feelings involve sensations in my stomach, muscles and nerve endings throughout my body. At that point, the robot, would hand me a screwdriver and allow me to inspect some of its innards. It would show me the flask in its chemistry panel where dozens of chemicals were being mixed. It would point out that its many internal sensors were "feeling" a large excess of hydrochloric acid that had recently been added because of its aggravation with me. A few more conversations like this and I might become convinced that the computer algorithm plus the hardware were close enough to what was going on inside me that this creature actually had feelings. This conclusion is the same as Searle's point that thinking and understanding is tied to the brain's *hardware.*

The argument above indicated that for a robot to convince me it had feelings it had to show me specific hardware and plumbing as well as being a sophisticated conversationalist with a sense of self. An electronic simulation of the "stomach" would definitely not convince me because a real feeling needs real hardware. Since I am the only creature that I really know has feelings I attribute feelings and subjective awareness to others only to the degree that their innards are like mine. The word "feeling" thus has a strong anthropomorphic connotation. I suspect that the simple "stomach" flask described above was too simple to convince me. In addition to seeing the stomach and sensors I would need to see that the robot's brain was organized like my brain before I could ever accept that it had feelings. The brain is more important than the stomach since one doesn't need a stomach to have feelings. A futuristic live head, severed from the body, would be expected to have feelings when electrical stimuli are applied to the nerves that came from the stomach. See Dennett (1978) for a wonderfully humorous discussion of these issues.

I should point out that although the above discussion might seem to be an exercise in how I gain knowledge (epistemology) it is really an exercise in the true nature of the robot (ontology). What convinces me about the robot's ability to feel is seeing of what "stuff" the robot is built. When I see it has "nerves", "juices" and cortical activity like mine then I might be swayed about whether I think it has feelings.

The usual assumption about the relationship between Strong AI and dualism is that Strong AI is not dualistic since it claims to produce a mind out of an algorithm. Searle (1990), however, offers the following twist on this subject:

> The polemical literature in AI usually contains attacks on something the authors call dualism, but what they fail to see is that they themselves display dualism in a strong form, for unless one accepts the idea that the mind is completely independent of the brain or of any other physically specific system, one could not possibly hope to create minds just by designing programs.

From the point of view of this essay, Searle's indictment that Strong AI is dualistic might turn out to be a compliment rather than an indictment since we will be showing that duality is an attractive ontology that has proven to be successful in its quantum embodiment.

1e. The Weak AI version of the mind-body problem. Weak AI is based on the assumption that the brain is not algorithmic, but has some special hardware that can do the "feeling." This is exactly the point of adding the chemical flask plus sensors to the robot to act as a "feeling stomach" in Section 1d. A much more sophisticated hardware addition is advocated by Roger Penrose (1989). Penrose argues that a brain can do things that algorithms can't do. The crucial step in his argument concerns Godel's theorem which shows that within any algorithmic system there are theorems that can't be proven true or false from within the system, but which mathematicians can show to be true. Penrose, however, is on shaky ground since it should be possible for the computer to do just as well as the mathematician by expanding the computer's axiomatic base. Penrose argues that brains need something beyond algorithmic computation in order to function. He then goes out on a most fascinating limb, further than he needs to, with a speculation that quantum mechanics must be changed because of gravitational effects. He uses these gravitational effects, together with speculation on how present quantum theory is inadequate in his story of how consciousness might operate. Most neuroscientists, on the other hand, believe that someday we will have a reductionist explanation of consciousness and feelings (Crick & Koch 1990a, b; Freeman, 1990) without needing to change the present quantum laws.

There is indeed the possibility of a new theory coming along and radically changing our views of brain mechanisms (and possibly quantum mechanics). As Delbruck (1986) points out, the situation may be similar to what DNA did for biology:

> It might be said that Watson and Crick's discovery of the DNA double helix in 1953 did for biology what many physicists had hoped in vain could be done for atomic physics: it solved all the mysteries in term of classical models and theories, without forcing us to abandon our intuitive notions about truth and reality. Upon the discovery of the DNA double helix, the mystery of gene replication was revealed as a ludicrously simple trick. In people who had expected a deep solution to the deep problem of how in the living world like begets like it raised a feeling similar to the embarrassment one feels when shown a simple solution to a chess problem with which one has struggled in vain for a long time.

What kind of hardware might be required by a Weak AI understanding of consciousness? Descartes wondered how our conscious awareness is unitary even though our brain is split into two hemispheres. He speculated that the unitary mind is centered in the Pineal gland, one of the few brain structures that are unitary rather than paired. More recently there have been several proposals, with preliminary data to back them up, on how distant parts of the brain might be coupled together to act as single entity. Consciousness might be based on something like the proposal of Crick & Koch (1990 a,b) that the 50 Hz signals found by Freeman (1985), Gray et al. (1989) and Eckhorn, et al. (1988) or the synchronous firing proposed by von der Malsburg (1983) and von der Malsburg & Schneider (1986) binds together all regions of the brain that are attended to at that moment (see Science (1990), **249** p. 856-858 for a summary). My point in mentioning the 1990 theories of awareness is to indicate that brain researchers are actively looking for the substrate of consciousness and intentionality. They may well find a relatively simple feedback process or neural structure that is straightforward to implement in a robot and to simulate with an algorithm. Crick & Koch (1990 a,b) outline a host of specific experiments that must be done before the "DNA" of consciousness can be expected to be revealed. However, it is most important to point out that any future reductionist theory of consciousness would not change the dualistic nature of the mind-body interrelationship. The reductionist story is just one half of the duality.

1f. The ethical distinction between mind and machine. Barlow (1990) has recently written about an *ethical* reason for making a distinction between mind and brain. He sees a danger in treating people as machines because we are amoral in our use of machines. Barlow fears that moral strictures (Thou shalt not kill) would be undermined if minds were reduced to machines. I disagree and believe that the opposite will happen. Rather than humans being ethically reduced to machines, I expect that the feeling and thinking robots of the future will be accorded rights similar to humans. In order to become ethical entities with rights and responsibilities, these future robots must exhibit intentionality (a technical philosophical term related to the robot's understanding the meaning of the symbols it uses), free will (an ability to make decisions independent of its programmer and its environment), a self awareness, and an awareness of the implications of its decisions. How a robot might be given these qualities and capabilities will not be discussed here. Many defenders of Weak AI would be delighted to invent scenarios on how computers of the future could obtain these qualities and capabilities (including their careful definitions).

1g. Why mind is not reducible to body. Suppose some wonderful theory comes along that explains consciousness to the satisfaction of all concerned scientists. Suppose even, that our understanding of the brain has advanced to the point such that by monitoring the voltages in a small subset of neurons we are able to reliably predict what the person is feeling and thinking. Would this understanding of the mind imply the end of the dualistic mind-body problem? I say no!

a. The original problem that Descartes first faced is that the subjective mind and the objective body are operating on entirely different levels. We still have the problem discussed above that my own *subjective feelings* are qualitatively different from what produced the feelings. Even though one might have an excellent explanation of why I am feeling a certain way, that is quite different from the subjective feel of the feelings. Einstein agreed and supposedly said: "Science can't tell you the taste of chicken soup". This is the subject of many philosophical analyses of mind (Hofstadter & Dennett, 1981; Kolak & Martin, 1990). The philosophers then get stumped because they do not know what to do with the resulting dualistic theory. They are fearful of wholeheartedly embracing the duality. One exception is the philosophical analysis by Lockwood (1989) who provides a thoughtful reexamination of Descartes' duality in light of modern knowledge.

b. There are pragmatic and religious reasons to maintain the mind-body split. Suppose your belief that you are a machine has made you start acting like a machine and you would like to change that behavior. Suppose you would like to have a conversation with God to put some extra meaning in your life, but your materialist worldview makes it difficult. The duality theory to be discussed in Section 3 makes it legitimate to be a materialist at one moment and a God fearer (or God himself or herself) the next moment.

Many religions (Judeo-Christian, New Age, humanism) say that strong beliefs can create realities. For example, my belief that people around me are nice can make people around me nice. My belief that I can solve a problem can help me solve the problem. A person's belief in God or extraterrestrial "guides" (Lilly, 1972; Jeffrey & Lilly, 1990) who give advice and companionship can provide courage to get through difficult challenges and losses. The religious worldview allowed by a dualistic system seems to enable some people to cope with our complex world.

The split between mind and body is equivalent to the split between religion and science. If it were not for this split, scientists might still be persecuted like Galileo, and religionists would be intimidated by the constant expansion of science. The split allows friendly relationships between people with very different worldviews. The democratic nature of a dualistic metaphysics is one of the most important outcomes of the point of view that I will be advocating at the end of this essay.

1h. Further reading. The goal of Section 1 was to provide background on the mind-body problem and in particular to argue that there is a problem. Further information can be found in the many books published every year on this topic. Recent books that I have found most useful are Churchland's "Matter and Consciousness" (1988), Penrose's "The Emperor's New Mind" (1989) and Gregory's "The Oxford Companion to The Mind" (1987). In addition, the Artificial Intelligence debate between Searle and the Churchlands in the January 1990 Scientific American is required (and fun) reading. John Lilly's biography (Jeffrey & Lilly, 1990) provides a unique perspective of the power of the mind and of the lives of some of its explorers. The entertaining collection of articles in Hofstadter & Dennett (1981) and Kolak & Martin (1990) discuss both sides of our central question concerning how to connect the first person and third person experiences. My only criticism of these books and articles is that they do not appreciate that the two halves of a duality can fit together in a beautifully elegant and precise manner. That is the topic of Section 3 of this essay.

Section 2. Quantum Mechanics. This is not the place for a detailed introduction to quantum mechanics. Herbert's excellent book "Quantum Reality" (1985) does a superb job of examining quantum mechanics and its implications. Feynman's Lectures (1965) and his book "QED" (1985) give the master's view not only of quantum mechanics but also of the forces governing atomic structure. The present summary is limited to those points needed for the discussion of duality.

Quantum Mechanics, developed in 1926-1927 by Heisenberg and Schroedinger, is the theory that underlies all of physics and chemistry. It has been unbelievably successful, with no known violations and with some calculations (energy levels of hydrogen and magnetic moment of the electron) accurate to more than 10 decimal places. The most interesting feature of quantum mechanics for us is that it is a dualistic theory. The thousands of years of human thinkers from Plato to Kant couldn't put together a pretty mind-body duality. Nature, being cleverer than her humans, did figure out how to arrange a pretty duality with a remarkable way of connecting the two halves of the split.

Quantum mechanics is dualistic. The generally accepted interpretation of quantum mechanics, the Copenhagen Interpretation, was put forth by Bohr who insisted on its dualistic nature. This interpretation of the meaning of quantum mechanics was an outcome of the historic debates between Bohr and Einstein that have been eloquently written up by Bohr (1949). Einstein never accepted the probabilistic nature of quantum mechanics. Stapp (1972) provides a clear discussion of the Copenhagen Interpretation. Alternative interpretations of quantum mechanics will be shown in Section 3 to actually be consistent with the Copenhagen view. The Copenhagen Interpretation says that the universe must be split into two parts with very different laws governing the two halves. Above the split is the real world with which we are familiar. Observations are made, feelings are felt, experiments are described classically. The world looks almost like the classical world of Newton in which matter exists as particles with definite locations. Below the split the laws are very different. No observations are allowed. Contrary to some interpretations of the uncertainty principle, there is no uncertainty about the laws of nature or the state of the system beneath the split. The state of the system is represented by a set of complex numbers, called amplitudes or wave functions. Feynman (1948, 1985) proposed a set of algorithmic rules, compatible with relativity, for calculating the amplitude for the state of the system to propagate from one time to a future time.

The connection between the two sides of the duality is surprising, but simple. The probability for an event to occur in the world above the split is equal to the magnitude squared of the amplitude for that event below the split. This connection does not violate the essential nature of a true duality which is that the properties above the split cannot be reduced to the laws below the split. Of importance for the present essay is the question of where the split should be placed, to be considered in Section 3.

Double slit experiment. It is useful to describe a concrete example: the double slit experiment shown in Fig. 1. A source of photons is directed at a metal plate with two closely spaced slits. At the back of the box are hundreds of photomultiplier tubes (PMT) which detect individual photons. In such a situation the split is usually placed so that the photon source and the detectors are above the split and the metal plate with the slits is below the split. Thus

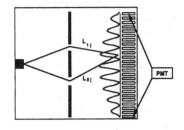

each photon is prepared classically and is detected classically. The photomultiplier tubes are acting as the observers. Every time a photomultiplier clicks, a photon is being observed. Between the entrance hole and the detectors, no observations are being made. Thus it is not possible to tell through which slit the photon went. Quantum mechanics as formulated by Feynman (1948) says that the amplitude to get from the entrance hole to one of the geiger counters is the sum of the amplitudes for each path the electron could take. Feynman showed that all possible curved paths should be included, but for the simple double slit experiment the answer is the same as if one only considers 2 straight-line paths, one through each slit. The amplitude to go through the ith slit to the jth counter is given by the complex number $E_j \exp(if)$. The phase, f, equals $2\pi L_{ij}/l$, where l is the wavelength of the photon and L_{ij} is the distance from the ith slit to the jth detector. The exponential is simply a mathematical notation that allows one to keep track of the phase of a wave. The

Gaussian-like envelope function E_j is present to normalize to unity the probability of finding the photon somewhere. For simplicity we neglect the slight dependence of the envelope on the slit number. The photon's wavelength was determined by the experimenter who sent the photon through the hole. By combining all I have said so far, the amplitude of finding the photon in the jth counter is approximately:

$$A_j = E_j \left[\exp(i2\pi L_{1j}/l) + \exp(i2\pi L_{2j}/l) \right]. \tag{1}$$

We now want to connect the world below the split to the physical world above the split. As stated earlier, the connection between the two is that the probability for an event to take place is given by the magnitude of the amplitude squared:

$$Prob_j = 2E_j^2 \left[1 + \cos(2\pi(L_{1j} - L_{2j})/l) \right]. \tag{2}$$

We have now completed a full quantum mechanical calculation. The cosine term in Eq. 2 is due to the photon's wave nature of being able to sample both slits, resulting in a sinusoidal probability of landing in a given photomultiplier counter, as shown in Fig. 1.

When the multiple slit experiment was actually carried out for photons, the characteristic interference pattern was found. The situation makes no sense to classical ways of thinking because it means that when nobody was looking the single photon sampled both slit 1 **AND** slit 2. According to classical thinking the photon would have gone through slit 1 **OR** slit 2. Many physicists have tried to reduce or modify the AND logic of quantum mechanics to the OR logic of classical mechanics. They have not succeeded. Photons and all other matter behave as waves when not observed and as particles when an observation is made. This behavior is called the wave-particle duality. Physics offers no reductionist help in showing how the localized particles can be obtained from the spread-out wave.

Schroedinger's Cat. Although the double slit experiment with its wave-particle duality violates one's sensibilities most physicists accept it when it involves low mass particles such as photons, electrons or atoms because quantum mechanics does so well in correctly predicting their behavior. However, quantum mechanics knows no boundaries and when applied to macroscopic phenomena we begin to feel queasy. The physicists created a theory that was more successful than they may have wanted. To illustrate the macroscopic implications Schroedinger (1935) developed the parable, now called the Schroedinger's Cat paradox, in which a cat is coupled to a quantum mechanical process, like the double slit setup, except that a trigger is pulled to kill the cat if the photon went in the upper slit. If the cat is not observed then it is in a superposition of states just as the photon was in a superposition of states in going through both slits. Before being observed the cat is both alive and dead rather than the classical notion of being alive or dead. In principle an interference experiment could be carried out between the alive and dead states just as was done for the photon going through the upper and lower slit. The cat parable points out the unusual nature of quantum mechanics. When the system is not observed it is in a state that does not correspond to our usual notions of reality. Most physicists are not bothered by this situation; in fact, they never even think about it. There are also physicists, however, who are deeply disturbed by the implications that unobserved states do not have properties. It is our goal to show that even though quantum reality might be a bit disturbing it is also quite beautiful.

The gedanken experiment shown in Fig. 2 adds further insight both for the Schroedinger Cat and

Fig. 2. The source in the center emits two photons. The calcite crystals split the beam according to the photon's polarization. The photomultiplier tubes (PMT) detect photon B. Photon A determines whether the cat is dead or alive.

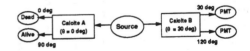

also for the forthcoming discussion of Bell's Theorem. The photons now come from the radioactive decay of positronium. This decay has the property that two photons, A and B, are always emitted in opposite directions and they always have the same angle of polarization. The photons then pass through calcite crystals as shown. Calcite has the property that an incoming photon whose polarization is aligned with the crystal will go through the calcite undeviated whereas for the perpendicular polarization it will be given a small displacement. The advantage of calcite over a simple polarizer is that calcite analyzes the state of polarization without absorbing any photons. Suppose in our experiment the crystals A and B are oriented at $0°$ and $30°$ respectively. Calcite B will direct photon B into the upper or lower beam depending on whether its polarization is $30°$ or $120°$ (polarization at any other angle can be expressed as a superposition of $30°$ and $120°$ polarizations). The polarization of photon B is determined by placing a pair of photon detectors (photomultiplier tubes) at the two exit positions of the $30°$ calcite. Since the polarization of photon A is the same as that of photon B (due to the properties of positronium decay) the polarization of photon A is known without measuring it (since we measured photon A that has the identical polarization). With this setup we know the exact timing of each individual photon A, and we know its polarization (either $30°$ or $120°$) before it enters calcite A.

We now use the upper beam (vertical polarization) to trigger the machine that kills Schroedinger's Cat. According to quantum mechanics the state of the cat is: $\cos(q)$ |alive> + $\sin(q)$ |dead> where q is the the angle between the the polarization of photons A and B and the notation | > is used to specify the state of the system. If photon B is in the upper beam ($q=30°$) then the cat wave function is: .866 |alive> + .5 |dead>. The cat has a 75% chance of being alive and a 25% chance of being dead (the probability is the amplitude squared, i.e. $\cos^2(q)$). If photon B is in the lower beam ($q=120°$) then the cat wave function is: -.5 |alive> + .866 |dead>. This exotic version of the Schroedinger Cat paradox shows that the relative sign between the live and the dead cat depends on whether photon B was polarized at $30°$ or $120°$. Since quantum mechanics applies to large as well as small objects, one could in principal extract this sign, so in some sense the cat is both alive and dead when not being observed. This refined presentation of the Schroedinger Cat paradox was given to strengthen the case that the cat is actually in a superposition of states.

Bell's Theorem. John Bell (1965) noticed an important property of the gedanken experiment with the calcite. He wondered what would happen if, after the positronium decay, the orientation of the calcite A had been switched from $0°$ to $-30°$ or if calcite B had been switched from $30°$ to $0°$ (I am changing the details of Bell's experiment to simplify the discussion). The experiment with the fast switchers is shown in Fig. 3.

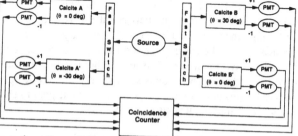

Fig. 3. The source in the center emits two photons. The fast switches direct each beam into one of two calcite crystals, which then splits the beam according to the photon's polarization. The photomultiplier tubes (PMT) detect the photons and send the detection information to coincidence circuitry to calculate the correlations.

One can make the following table enumerating all possible outcomes of the 4 possible experiments.

Decay	1	2	3	4	5	6	7	8	9	10	11
A (at $0°$)	+1	-1	-1	+1	+1	+1	-1	-1	+1	+1	-1
A' (at $-30°$)	-1	-1	+1	+1	-1	+1	+1	+1	+1	-1	-1
B (at $30°$)	+1	+1	-1	+1	-1	+1	+1	-1	-1	+1	+1
B' (at $0°$)	+1	-1	-1	+1	+1	+1	-1	-1	+1	+1	-1
correl.(AB'+AB+A'B'-A'B)	+2	+2	+2	+2	-2	+2	-2	+2	+2	+2	+2

There are 4 possible experiments (AB, AB', A'B, A'B') since each crystal could have two orientations as shown in the column at the left. The top row counts the decays of positronium. The next 4 rows list possible outcomes of whether the photon goes into the top (+1) or bottom (-1) path of the calcite. The numbers were chosen randomly except that when both calcite crystals are oriented at 0° the photons are perfectly correlated. If photon A goes up then so does photon B. The bottom row is a curious sum and difference of products of pairs of rows devised by Bell. Notice that the sum is either +2 or -2. Then Bell took the average of these products, which turns out must be less than 2. Now comes the problem. Alain Aspect did this experiment, in Paris, and found that the final average was significantly <u>greater</u> than 2! Quantum mechanics predicts that the average of each product is given by $\cos^2(q)$ where q is the relative angle between the two calcite crystals. This $\cos^2$ law, commonly known as the Law of Malus, predicts a Bell sum of: Sum = $1 + 2\cos^2(q) - \cos^2(2q)$ = 1 + 2 x 3/4 - 1/4 = 2.25 since q = 30° for the above arrangement. Aspect's experiment was in agreement with the quantum mechanical prediction and in disagreement with the logic we used above to give Sum ≤ 2.

The Copenhagen Interpretation of quantum mechanics has no problem with Bell's Theorem since calcite A could not be oriented at both 0 and 30° for a particular decay. The bold numbers in Table 1 indicate the actual orientations for each decay. One should not ask what would have happened on that trial had the calcite been rotated, since that experiment was not done. The only experimental results that can be considered are experiments actually performed. The conclusion to be drawn from the violation of Bell's Theorem is that the photons were not real before they were observed. By real, one means each photon had properties, such as polarization, before it was observed. If photon A had been real before it was observed then it would have made sense to ask about its properties as seen by an analyzer at an arbitrary angle. Aspect's experiment rules out the possibility that the photon had properties before it was observed. An alternate interpretation of the violation of Bell's inequality is that the photon does have properties such as polarization before it is observed, but there is a nonlocal interaction such that photon B "knows" the outcome of the experiment on photon A. In this case Table 1 is invalid because whether photon B decides to go into the +1 or the -1 beam depends on the orientation of calcite A. Aspect's experiment was designed so that the orientation of the polarizer was switched while the two photons were in flight so that information about the orientation of polarizer A would have to go faster than the speed of light to reach photon B in time to affect whether it goes into the +1 or -1 beam.

Aspect's experiment together with Bell's theorem imply that the world is either nonlocal or not real. Typically, authors who discuss these issues choose whether they prefer locality or reality to be violated. The dualistic nature of quantum mechanics allows both interpretations of the violation depending upon which half of the split is being considered. The rules of the real world above the split are nonlocal. The rules of the unobserved world below the split are local but not real (a particle can simultaneously have conflicting properties). Stapp (1990b) recently relaxed the reality assumption that was needed to derive Bell's theorem. He derives Bell's inequality from the simple assumption that any observation results in a unique answer (ruling out the "many worlds" interpretation).

Bell's theorem and Aspect's experiment imply that our world is very different from the classical world of Newton. The failure of many attempts to reduce classical mechanics to quantum mechanics is what qualifies the laws of quantum mechanics to be a true duality. Having an example of a true duality is a breakthrough for philosophers who have been fiddling with past dualities. The duality of quantum mechanics is much more explicit than previous dualities such as Descartes' mind-body duality and the yin-yang dualities of Eastern religions. These old dualities were fuzzy in that the worldview on each side of the split and the connection between the two sides weren't precisely specified. For the quantum mechanical duality, on the other hand, algorithms are available for how to do computations on both sides of the split. In addition, the quantum mechanical duality solves the split placement problem in a most elegant manner as will be considered next.

Summary. The characteristics above and below the split are summarized in this Table.

<u>Below split</u>	<u>Above split</u>
to be observed	observer
deterministic laws	probabilistic laws
Feynman rules	Classical rules (almost)
behaves like waves	behaves like particles
local interactions	some nonlocality
objective	subjective
body	mind
"dead" <u>and</u> "alive"	"dead" <u>or</u> "alive"

Section 3.   Von Neumann's insight that the split is moveable. In 1932 von Neumann published a most influential book: "The Foundations of Quantum Mechanics" (von Neumann, 1932). It is remarkable that so soon after quantum mechanics began, von Neumann wrote the book that still provides the mathematical foundation for the subject. To me the most interesting feature of quantum mechanics for the present discussion is a point, derived by von Neumann, which has received little notice. Von Neumann showed that the laws on both sides of the split are so cleverly constrained that the precise location of the split is not critical. **The split can be moved.** I will argue that the moveable split feature of quantum mechanics is exactly what is needed for overcoming the stickiest aspects of the mind-body problem.

3a. Schroedinger's cat again. Schroedinger's cat is useful for illustrating von Neumann's insight as shown in the following sequence of panels. The split is shown by the dotted line. The leftmost panel gives the standard Schroedinger cat story. There is a single observer, to be called O1, outside the box. Before O1 opens the window to look, the cat is in a superposition of being both alive <u>and</u> dead. By opening the window and looking, O1 "collapses the wave-packet" so that the cat is now in a unique state of being alive <u>or</u> dead. The story gets more interesting if we place O1 in a second box as shown in the second panel. If I, the second observer, am not looking, then O1 is in a superposition of states seeing an alive cat <u>and</u> seeing a dead cat. Once I make an observation, O1 collapses to one state or the other. The third panel removes the split even further, placing it in my brain.

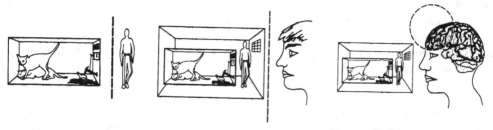

Fig. 4. Three placements of the quantum split are indicated by the dotted line. The left panel shows Schroedinger's Cat with observer O1. The middle panel has O1 inside a second box. I am the second observer (shown by the face). In the right panel a portion of my brain has been separated out as mind, with the rest of my brain as body.

3b. Interpretations of different placements of the split. The following placements are in ascending order.

Bohm's real world of hidden variables. David Bohm (1952) proposed a quantum formalism in which there was no split. All nature was on the same side as the observer. In terms of the moveable split, Bohm's formalism is equivalent to a placement of the split at the very bottom (see the last paragraph of this section for further discussion of whether this is a legitimate placement). We end up with a real world in which real particles are guided by an invisible "pilot wave" that is very similar to what had previously been the full quantum mechanical wave function. As pointed out in the previous section in connection with Bell's theorem, it is a world with nonlocal, complex interactions. It should be emphasized that the nonlocal interactions must be consistent with other placements of the split, so faster than light communications are ruled out. Proponents of "New Age" and Eastern religion seem to like the nonlocal aspects of the interactions. This placement of the split has the advantage that the split is invisible so that it looks like a unitary rather than a dualistic theory. This formalism, lacks elegance because unnecessary aspects of the wave function are carried along. Higher placements of the split, to be discussed next, seem cleaner because one only needs to deal with measured quantities and after a measurement the "non-actualized" outcomes are discarded. Stapp (1990b) raises a concern that Bohm's formalism includes predestination, the same as classical mechanics, that makes some people uncomfortable (rather than making me uncomfortable it makes me smile).

The physicist's placement. A physicist would say that the photon was detected (observed) by the photomultiplier tube that was used to trigger the device used to kill the cat. In this view the wave packet collapsed at an early stage so the cat was never in a superposition of states. The same outcome would be obtained if the cat were the observer. Although the physicist may want to keep the split below the photomultiplier tubes, he must tremble at the monster he created. Quantum mechanics that was invented to explain atoms, should also explain groups of atoms, like photomultiplier tubes and cats. The formalism of quantum mechanics does not stop when many atoms are interacting. A few physicists such as Penrose are attempting to modify quantum mechanics so that it fails for large objects, but these modifications have not yet been developed into a consistent theory. There is thus no way at present to stop the split from being placed above the cat!

Placement in front of the first human. The left panel of Fig. 4 shows the cat enclosed in a box with the split between the box and the human observer. If no information about the cat's state leaves the box then the cat is in a superposition of states and the collapse occurs when O1 peeks into the box. O1 could peek weeks after the photon was emitted. When O1 looks, he will either see a live cat or a cat that had been dead for weeks. Before he looked, in principle the alive-dead state could be time-reversed and an interference experiment could be done on the combined state to determine the relative phase between the alive and the dead components. This placement of the split, above the cat, corresponds to the theology of the bible in which humans have minds and souls and cats and other animals don't (like machines).

Placement between other humans and me. The middle panel of Fig. 4 places O1 plus the cat box in a larger box, with me being the observer. In the middle panel, I am represented by the face. With this placement of the split, other living creatures including humans are below the split, implying that everything about the operation of other humans must be reducible to the deterministic wave equation laws governing all matter. The total behavior of other humans must therefore be reducible to the laws of biology, chemistry and physics. This is the placement that should please biologists and psychologists whose research careers are devoted to explaining human behavior.

Placement between all minds and their bodies. This is a favorite placement for people who write about quantum mechanics and consciousness. The idea is to place all minds above the split and all the rest of nature below. An exceptionally lucid discussion of this placement is given by Wigner (1961) whose article is of special interest to visual psychophysicists since he uses a visual detection task as his gedanken experiment. A similar placement is advocated by Stapp (1990a). We will come back to this placement in Section 3c.

Placement just below my mind. The righthand panel of Fig. 4 places the split somewhere in my brain. This placement is similar to a solipsist version of the world where I exist and the rest of the universe is a figment of my imagination. It is, of course, quite different from Berkeley's solipsism since the "imagined" world is operating according to a very precise and deterministic wave equation. This split placement where only one's mind is real and the outside world is an illusion is reminiscent of how Descartes (1641) begins his beautifully written "Meditations." His starting point of "I think; therefore, I am" was based on his analysis that his knowledge of the outside world could be illusory.

In placing the split with my consciousness on one side and the rest of my brain on the other I do not have in mind a spatial placement but something more like a linguistic placement. Consciousness involves the same neurons that are involved in nonconscious activity so it is not possible to separate particular neurons to be on one side and others to be on the other side. One might wonder whether it is legal to place the split between different modes of activity of the *same* atoms. It is gratifying to know that it is very common for physicists to place the split not between different objects but rather between modes of oscillation of a single object. Consider, for example, a measuring device in which the output is a pointer on a meter. Physicists commonly separate out the center of mass mode of motion of the pointer as the only aspect of the needle to be quantized to act as the "observer". The other modes of motion of the needle would be left "unobserved" below the split.

The placement of the split between my mind and my body provides an opportune occasion to comment on free will. The probabilistic nature of the world above the split allows my mind to be "free" to choose what to observe, thereby allowing a bit of free will. It should be pointed out that as the word "free will" is commonly used (especially by social rather than physical scientists), we have much more free will than the tiny bit allowed by quantum mechanics above the split. Free will also operates below the split since a significant percentage of a person's synapses have developed by internal influences in addition to being influenced by outside factors such as the environment and heredity. It is common to refer to behavior caused by this internally developed (although deterministic) aspect of the synapses as "free will." An alternative, less serious, view of free will, obtained by giving the split a microscopic thickness, was suggested by Pirsig (1974).

Placing the split above me. It is difficult for me to place the split above my awareness. The state of enlightenment that is the goal of some forms of Eastern meditation may be an attempt to place the split above one's consciousness. The meditation is supposed to stop me from being an observer and to become "at one with the world." The placements of the split discussed earlier involve different linguistic structures. Placing the split above me involves eliminating language and thought.

An alternative ontology with the split above me is the "Many Worlds" approach of Everrett (1957). In this view, the wave function never collapses, but all possible outcomes of measurements do occur. The world with a live cat and the world with the dead cat both exist. I am only aware of one world. Other versions of me are aware of the other worlds. For this reason Stapp (1990b) calls it the "many minds" theory. Stapp has done the most creative work in making this view palatable. I believe that the "many minds" view is not opposed to the other views, as is commonly assumed. Rather it may simply correspond to a particular placement of von Neumann's split. The richness in the multiple interpretations is the great beauty of the quantum duality.

A full circle? The Bohm and the Many Worlds approach are quite similar. In both there was a question of whether a split was actually present since in neither approach is the split visible. I placed them into the formalism of the moveable split by placing the split at the extreme lower and upper ends respectively. Both the Bohm and the Many Worlds approach are based on the time evolution of the full wave function of the universe. In the Bohm case the wave function is used to guide the particles of the world and in the Many Worlds case the wave function is used to guide my mind. It pleases me (a lover of Klein bottles) to see that the two opposite placements of the split are not that different.

3c. Implications of the moveable and relativistic split. This section offers further views on how the split can be placed to help deal with the mind-body problem that started this essay.

The homunculus. In our psychophysics models, we invent mechanisms that process filtered versions of the image. The output of these mechanisms activate neural structures that produce motor output such as finger motions in pushing a response button or lip and tongue motions for a verbal response. We sometimes encapsulate the decision process, referring to it as a homunculus. The homunculus is a neural structure (involving motor cortex) that views the outputs of the mechanisms and makes the decision. Some scientists and philosophers are bothered by the homunculus concept since they worry that there must be a second homunculus viewing the first, etc., and an infinite regress results. For those who are worried about the infinite regress the quantum mechanical dualistic scheme discussed above should solve the problem. By placing the split between the sensory mechanisms and the motor output, a single homunculus is to be found above the split. The single split avoids an infinite regress.

The self-referential aspect of the homunculus brings up the image of bootstrapping (being able to fly by pulling up on one's bootstraps). As put by Delbruck (1986):

> We start with the naive question: How can mind emerge from dead matter as the result of purely physical processes? Mind then looks back on itself and says, "Aha, this is how I came about." (Like Baron Munchhausen, pulling himself by his hair out of the mud.)

One of the important contributions of the duality approach is that the *single* split of quantum mechanics avoids the problems associated with self-reference. By clever placement of the split one can have one brain structure observe another structure. There is no self-reference since in quantum mechanics the observer does not observe itself.

The homunculus is commonly called "I". After exploring the central role of subjectivity it is amusing to note the foresight of the inventors of the English language for making the personal pronoun "I" so special. It is the only non-proper name that is always capitalized. The word "I" is also special because its shape reminds one of its meaning. It is a single vertical slash, reminding us of the quantum split. Quantum mechanics has come to the rescue of philosophy and has provided the necessary formalism establishing a unique place for "I", the observer, in the scheme of things.

Relativism and Human Values. The flexible placement of the split legitimatizes many different worldviews. It is possible to believe that "I am God" by having me the only entity above the split. A similar placement of the split can lead to an Eastern theology in which the physical world is illusion. On the other hand, it is possible to have all humans above the split, with animals and inanimate nature being "machines" under our dominion, similar to the version of the Old Testament. The many conflicting worldviews can be brought into harmony since the democratic nature of quantum duality says that they all correspond to legitimate different placements of the split.

The extreme relativism of the quantum duality makes it difficult to derive absolute values from science. The difference between the quantum and classical worldview does, however, radically change man's place in the universe. Whereas in classical mechanics there was no need for an observer, in quantum mechanics the observer is essential. Quantum mechanics brings the Copernican revolution full circle and restores the observer to be at the "center of the universe." As Stapp (1989) points out:

> The quantum conception gives an enlarged sense of self as architect of the universe. From such a self-image must flow lofty values that extend far beyond the confines of narrow personal self interest. With the diffusion of this quantum conception of man science will have fulfilled itself by adding to the material benefits it has already provided to man a philosophical insight of perhaps even greater value.

Back to the mind-body problem. In order to bring further clarity to the mind-body problem we must be careful about our definitions of mind and body. One might be tempted to define mind as the observer and body as the observed. According to this definition, *any* placement of the split is possible. The right panel of Fig. 4 has my consciousness as the observer and my body as the observed. The physicist's normal placement would have the photomultipliers as the observers and the photons as the observed. In this sense the photomultipliers would be considered as part of the mind. Since most of us are not comfortable with a photomultiplier having a mind the definition of "mind as observer" is not satisfactory. We would like a definition closer to that given in Section 1, where mind was associated with subjective feeling. There is general agreement that higher mammals have feelings. I discussed earlier why I believe dogs have feelings. I am told by people I trust that cats and dolphins also have feelings. I am undecided about ants and other tiny creatures. Some of my colleagues are sure ants have feelings and others are sure they do not. I suspect we would have to define the word "feelings" more carefully to be able to get agreement on where to draw the line. Somewhere among the lower animals, a split can be drawn with all creatures with feelings above the split and the non-feeling creatures and inanimate objects below. This placement of the split allows each creature with feelings to have its own mind and subjectivity that are not reducible to the underlying laws. Placing the split above the creature allows one to explain it's consciousness in terms of neurophysiology (Crick & Koch, 1990a,b; Freeman, 1990).

The precise placement of the split depends upon which aspect of subjectivity one wants to isolate. To keep my feelings subjective rather than objective I would place much of the activity of my brain's limbic system above the split. If I choose to focus on my self-awareness or thinking rather than on my feelings then the appropriate split placement would shift to a different brain structure or brain activation. It is nice that the quantum mechanical split has the flexibility to accommodate different nuances in what we mean by our words.

Bohr's viewpoint on the connection between quantum mechanics and the mind-body problem, as clearly spelled out by Delbruck (1986), is different from the views expressed here. Bohr believed that by *analogy* the dualistic notion of complementarity found in quantum mechanics could be applied to other domains such as life-not life and mind-body. My view is stronger. I believe that the mind-body duality is not *analogous* to the wave-particle duality; rather it is the *same* duality, just with a different placement of the quantum split.

Constraints. The moveability of the split was not easy to achieve. The laws on both sides of the split must be tightly constrained. The power of constraints recalls the bootstrap approach to theoretical particle physics advocated by Geoffrey Chew (1968) for the past 30 years. For about 10 years it was the dominant framework for particle theorists. It is based on the belief that quantum mechanics and relativity have so many consistency constraints that the theories are unique. It may well be the approach that will provide answers to the most fundamental questions. Capra's warm book (Capra, 1988) includes the following statement from Chew about the ambitions of the bootstrap program: "My feeling is that the bootstrap approach is going to eventually give us simultaneous explanations for space-time, quantum mechanics, and the meaning of Cartesian reality." As part of this marriage of quantum mechanics and relativity I expect the bootstrap will, in addition, produce a relativity in the placement of the split, which is the theme of this essay.

The constraints above and below the split are needed to produce a consistent theory. Below the split are the Feynman rules (Feynman, 1985), similar to wave equations. All interactions are local (the recent superstring theories allow a limited amount of nonlocality in a manner that doesn't violate relativity or causality, and which seems to include the effects of gravity). Below the split, particles follow all possible paths. Above the split is the "real" observed world. There are limitations about how accurately positions and velocities can be jointly specified (the uncertainty relationships). There is also a need to introduce long range correlations that go beyond what is expected classically. Finally the connection between the two halves of the duality is clearly specified whereby the probability of an observation event above the split is equal to the square of the amplitude of the event below the split. In addition to all these constraints there may be an additional need for the system to be dualistic and to have a moveable split in order to guarantee self-consistency.

The end of ontology? Many physicists and philosophers are unhappy with the moveable split associated with the dualistic nature of the present quantum theory. They would prefer a world in which the "collapse of the wavefunction" occurred at a unique place (such as at 10^{-5} grams in Penrose's speculation). Their problem with a moveable split is that one is not left with a firm ontology. They want an ontology that gives a *unique answer* to what is out there. The present quantum mechanics gives such a slippery ontology that maybe we shouldn't even use that word. It is often acknowledged that the Copenhagen Interpretation upon which this essay was based provides an epistemology rather than an ontology. However, it is also acknowledged that the extreme split placements of Bohm (1952) and Everett (1957) are ontologies. I am asserting that the intermediate split placements which are trimmed of unnecessary baggage should also be acceptable ontologies.

It is always possible that the present quantum theory will be found to disagree with experiment and a new theory without a moveable split will be found to better describe Nature. I, however, would be saddened if the physics of the future abandons the moveable split. The ability of the subjective-objective split to be placed *anywhere* is to me the most beautiful concept in modern science. There is something elegant about the leanness of present quantum mechanics that tells us to talk only about observed quantities. In a world without the moveable split, in order to relate subjective feelings to nerve impulses my descendents would have to say the connection between mind and body is *analogous* to the split present in 20th century quantum mechanics. It will feel so much better to reside in the 21st century and be able to say that the mind-body connection *is* the split.

Acknowledgements. I thank Alain Aspect, John Searle, Henry Stapp, Erich Sutter and Christopher Tyler for useful discussions. The material in Section 3 was originally presented at the 1976 Esalen Conference on Quantum Mechanics and Consciousness and I thank the Esalen Center for its support. This work was also supported by NIH grant EY04776 and NSF grant BNS-8820229.

References.

Aspect, A., Dalibard, J., & Roger, G. (1982): Experimental test of Bell's inequalities using time-varying analyzers. *Phys. Rev. Lett.* **49**, 1804 - 1807.

Barlow, H. (1990): The mechanical mind, *Annu. Rev. Neurosci.* **13**, 15-24.

Bell, J.S. (1965): On the Einstein, Podolsky, Rosen Paradox, *Physics* 1, 195-200, reprinted in Quantum Theory and Measurement. Ed. J. A. Wheeler and W. H. Zurek, Princeton University Press, Princeton, (1983).

Bohm, D. (1952): A suggested interpretation of the quantum theory in terms of 'hidden' variables, I and II. *Phys. Rev.,* 85, 166-193, reprinted in Quantum Theory and Measurement. Ed. J. A. Wheeler and W. H. Zurek, Princeton University Press, Princeton, (1983).

Bohr, N. (1949): Discussion with Einstein on epistemological problems in atomic physics. In *Albert Einstein: Philosopher-Scientist*, ed. P. A. Schilpp, The Library of Living Philosophers, Evanston. Reprinted in Quantum Theory and Measurement. Ed. J. A. Wheeler and W. H. Zurek, Princeton University Press, Princeton, (1983).

Capra, F. (1988): Uncommon Wisdom. Conversations with Remarkable People. Simon & Schuster.

Chew, G. F. (1968): "Bootstrap": A scientific idea? *Science,* 161, 762-765.

Churchland, P. M. (1988): Matter and Consciousness. A Contemporary Introduction to the Philosophy of Mind. MIT Press, Cambridge Mass.

Churchland, P.M. & Churchland P.S. (1990): Could a machine think? *Scientific American,* January, 1990, 262, 32-37.

Crick, F.H.C. & Koch, C. (1990a): Towards a neurobiological theory of consciousness. To appear in: *Seminars in the Neurosciences.* vol. 2.

Crick, F.H.C. & Koch, C. (1990b): Some reflections on visual awareness. To appear in: *Symposia on Quantitative Bology,* vol. 55, Cold Spring Harbor Press, 1990.

Delbruck, M. (1986): Mind from Matter? Blackwell Scientific Publishing, Oxford.

Dennett, D. C. (1978): Where am I? Excerpt from Brainstorms: Philosophical Essays on Mind and Psychology. Bradford Books, Cambridge, Mass. Reprinted in The Mind's I: Fantasies and Reflections on Self and Soul. Eds. Hofstadter, D.R. & Dennett, D.C., Bantam Books, New York (1981). Also reprinted in Experience of Philosophy. Wadsworth Pub. Eds. Kolak D. & Martin R. (1990):

Descartes, R. (1641): Meditations. Bobbs-Merrill Co. New York, (1960).

Descartes, R. (1664): Treatise of Man. Harvard Univ. Press. Cambridge, Mass. (1972).

Eckhorn, R., Bauer, R., Jordan, W., Brosch, M., Kruse, W., Munk, M., & Reitboeck, H.J. (1988): Coherent Oscillations: A mechanism of feature linking in the visual cortex? *Biological Cybernetics,* 60, 121-130.

Everett, H. (1957): "Relative State" formulation of quantum mechanics. *Rev. Mod. Phys.* 29, 454-462. Reprinted in Quantum Theory and Measurement. Ed. J. A. Wheeler and W. H. Zurek, Princeton University Press, Princeton, (1983).

Feynman, R. P. (1948): A relativistic cut-off for classical electrodynamics. *Phys. Rev.* 74, 939.

Feynman, R. P., Leighton, R. B., Sands, M. . (1965): The Feynman Lectures on Physics: Volume III. Addison-Wesley, Reading, Mass.

Feynman, R. P. (1985): QED—The Strange Theory of Light and Matter. Princeton University Press, Princeton, NJ.

Freeman, W. J. & Skarda, C.A. (1985): Spatial EEG patterns, non-linear dynamics and perception: the Neo-Sherringtonian view. *Brain Res. Rev.* 10, 147-175.

Freeman, W. J. (1990): On the fallacy of assigning an origin to consciousness. In Machinery of the Mind. Ed. E. R. John, Birkhaueser, Cambridge, Mass.

Gray, C. M., Konig, P, Engel, A. K., & Singer, W. (1989): Oscillatory responses in cat visual cortex exhibit inter-columnar synchronization which reflects global stimulus properties. *Nature,* 338, 334-337.

Gregory, R. L. (1987): The Oxford Companion to The Mind. Oxford University Press, Oxford, England.

Haugeland, J. (1985): Artificial Intelligence. The Very Idea. MIT Press, Cambridge, Mass.

Herbert, N. (1985): Quantum Reality. Anchor/Doubleday, New York.

Hofstadter, D. R. & Dennett, D. C. (1981): The Mind's I: Fantasies and Reflections on Self and Soul. Bantam Books, New York.

Jeffrey, F. and Lilly, J. C. (1990): John Lilly, so far.... Jeremy Tharcher, Inc. Los Angeles.

Kolak D. & Martin R. (1990): Experience of Philosophy. Wadsworth Pub.

Lilly, J. (1972): The Center of the Cyclone. Julian Press, New York.

Lockwood, M. (1989): Mind, Brain and the Quantum, The Compound I. Basil Blackwell Ltd. Oxford, England.

Penrose, R. (1989): The Emperor's New Mind. Oxford University Press, Oxford, England

Persig, R. M. (1974): Zen and the Art of Motorcycle Maintenance: An inquiry into values. Morrow, New York.

Schrodinger, E. (1935): The present situation in quantum mechanics: A translation of Schrodinger's "Cat paradox" paper. Die Naturwissenschaften 23, 807-812; 823-828; 844-849. Reprinted in Quantum Theory and Measurement. Ed. J. A. Wheeler and W. H. Zurek, Princeton University Press, Princeton, (1983).

Searle, J.R. (1980): Minds, brains, and programs with open peer commentaries. The Behavioral and Brain Sciences, 3, 417-457. reprinted with a provocative commentary by Hofstadter in The Mind's I, eds. Hofstadter, D. R. & Dennett, D. C., Bantam Books, New York (1981).

Searle, J.R. (1990): Is the brain's mind a computer program? Scientific American, January, 1990, 262, 26-31.

Stapp, H.P. (1972): The Copenhagen Interpretation. Am. J. Phys. 40, 1098-1116.

Stapp, H.P. (1989): Quantum physics and human values. Lawrence Berkeley Lab. report LBL-27738.

Stapp, H.P. (1990a): A quantum theory of the mind-brain interface. Lawrence Berkeley Lab. report LBL-28574.

Stapp, H.P. (1990b): Quantum measurement and the mind-brain connection. Lawrence Berkeley Lab. report LBL-29594.

Von der Malsburg, C. (1983): How are nervous structures organized? In Synergetics of the brain. Eds. E. Basar, H. Flohr, H. Haken, A. J. Mandell. Springer, Berlin, 238-249.

Von der Malsburg, C. & Schneider, W. (1986): A neural cocktail-party processor. Biol. Cybern. 54, 29-40

Von Neumann, J. (1932): Mathematische Grundlagen der Quantenmechanik. Springer-Verlag. English translation reprinted in Quantum Theory and Measurement. Ed. J. A. Wheeler and W. H. Zurek, Princeton University Press, Princeton, (1983).

Wigner, E.P. (1961): Remarks on the Mind-Body question, in The Scientist Speculates. Ed. I. J. Good, Heinemann, London. Reprinted in Quantum Theory and Measurement. Ed. J. A. Wheeler and W. H. Zurek, Princeton University Press, Princeton, (1983).

Some tacit assumptions in visual psychophysics

Christopher W. Tyler

Smith-Kettlewell Eye Research Institute
2232 Webster Street, San Francisco CA 94115

THE INCOMPREHENSIBILITY OF THE HUMAN BRAIN

The most general assumption made in vision research (and neuroscience in general) is that the visual processing can be understood by those who study it. This implies that the functions of part of the human brain are comprehensible by a human brain. I discuss seven distinct reasons why the brain can never be fully understood, and their applicability to the various levels of human vision research.

Any discussion of comprehensibility requires a definition of what it means to comprehend something. In this case the thing to be understood is the human brain and the understanding will focus on its functional properties as a controller of the behaviour of the human body (as opposed to its metabolic or other biological activities).

The definition of "understanding" for the present analysis is based on the concept of being able to describe the function of each component of a system at all relevant levels of analysis. To be complete, this description must include a description of the effects of the component on all other aspects of the system, and also its roles in the hierarchy of subsystems making up the organisation of the system from its components. This definition is not itself intended to be complete or watertight, but to provide a framework for the kind of explanation which might be required for a complete understanding of a system such as the brain. My argument will proceed to develop rigorous reasons why the brain can never be completely comprehended, and then show that such incomprehensibility is not substantially alleviated even by the use of reasonable approximations to the exact descriptive requirements.

The most obvious basic component of brain function is the neuron. Once could choose a smaller unit of analysis such as the synaptic bouton or the membrane ion channel, but there is general consensus for the choice of the neuron as the most useful level to consider as the smallest functional component. A complete understanding of any one brain would then constitute the ability to provide a complete explanation of the function of each neuron, both in terms of the properties of each of its synaptic outputs and in terms of its role in the functional subunits (e.g., columns, hypercolumns, processing areas) to which it belongs. The understanding could not be considered to be complete unless it also included a description of how each neuron arrived at its functional state, and how it is affected by the activity of all other neurons in the brain in any combination. Without this knowledge, one would be unable to provide a complete explanation for the development and neuropathology of the brain, both of which are key aspects of its function.

Supported by NIH grants 1P30 EY 6555, RO1 EY 6883 and RR 5981, NSF grant BNS 9011837. My thanks to Anthony Norcia, Russell Hamer, Liu Lei and Marilyn Schneck for participating in the experiments.

Implicit in the concept of understanding is the question of who is doing the understanding. One can distinguish four levels of the task that we call comprehension, excluding the implicit knowledge of things we know how to do without being able to explicitly explain how. One is immediate conscious comprehension, where the understanding is fully present in someone's consciousness. This type of understanding is of very limited scope at any instant in the human brain. Second is the accessible set of conscious comprehensions available from memory to a given person. Third is the total set of comprehensions available from all members of the human population combined. The final level is the set of comprehensions accessible to humanity by the use of mental prostheses, such as books or computers, which essentially constitutes science.

For the present purpose, the agent of comprehension will be taken as the fourth level of comprehension, but will not include the prosthesis of artificial computer intelligence. It is conceivable that computers can be designed to generate algorithms that could in some sense "understand" human brains better than could the brains themselves. While this is an interesting philosophical question, it raises such issues of the validity of the artificial understanding that it will be excluded for the present purposes.

The following list of seven principles of incomprehensibility of the human brain forms a litany of pessimism for the ultimate achievement of understanding the brain. It is not intended, however, to imply that extensive progress in neuroscience cannot be made. The principle of incomprehensibility should be viewed as a backdrop against which the neuroscience enterprise must proceed, much as we know that we can never physically travel more than a few light years from our planet. This limitation provides a challenge to develop other less direct methods for exploring the universe. Similarly, the ultimate limits of comprehensibility may coexist with a wealth of opportunity for exploration and analysis before those limits become manifest. Some paradigms will allow extensive analysis before being constrained by one of the comprehensibility limits, while others may be restricted early in their application. Thus, the challenge for neuroscience is to develop the most penetrating forms of analysis in the face of the incomprehensibility limits, which is best achieved if one is aware of them.

Quantal Indeterminacy

At the lowest level of brain function lies the indeterminacy of the position of the quantal units (elementary particles) of which the brain is constituted. Quantum mechanics asserts that there is an inherent uncertainty of the measurable position or energy of all particles in time and space. In principle, any decision on the part of a neural system corresponds to such a measurement, so that all threshold events, such as the firing of neurons in the brain, are subject to quantal indeterminacy. There is a corresponding lack of complete knowledge of brain function to the extent that it is controlled by quantal events.

It may be argued that quantal events are relatively insignificant in brain function because of the mass action of large numbers of elementary particles in each neural spike, and of large numbers of neurons in each functional neural event. It is known, however, that human performance can be controlled by very few neurons, as exemplified by rod detection experiments in which light detection is mediated by the responses of only one or two retinal rods (Hecht, Schlaer and Pirenne, 1942; Sakitt, 1972). Thus, visual behaviour in the dark must be limited by the quantal indeterminacy principle. Cohn (1974, 1976) has also developed the analysis of the role of quantum fluctuations in both rod and cone vision near absolute threshold, to show how they predict and explain the asymmetry of detection of decrements over increments in this region. Although this asymmetry is not a property of quantal indeterminacy as such, but of the statistical fluctuations of quantal events.

Another example of the single (or few) neuron control hypothesis is the demonstration that the limits of monkey contrast sensitivity across spatial frequency correspond to the sensitivity of the most sensitive neurons in the primary visual cortex (Blakemore and Vital-Durand, 1979; Parker and Hawken, 1985). This suggests again that only a few neurons are required to mediate contrast detection performance.

The Many-Body Problem
One of the startling revelations of high-school physics is that exact analysis fails even in the pristine domain of discrete bodies moving through empty space. As shown by Poincare in the last century, there is no analytic solution to the paths of motion when more than two bodies are interacting. This failure is a consequence of the nonlinear effects of gravity in the defining equations, resulting in an insoluble system of nonlinear simultaneous equations. The solutions can only be obtained by incremental linear approximation. Although this process can be carried out to any desired accuracy, it requires extensive computation to deal with many objects involved in nonlinear interactions together. It is indicative of the problems facing the analysis of complex nonlinear systems that even a simple physical situation may not be treatable with analytic equations allowing direct solution.

The many-body problem becomes much less tractable when the analysis is applied to neural systems. Unlike the situation with space physics, the nonlinearities do not have a precise, known form. Given the vagaries of neurophysiological recording and the fact that the nonlinearities vary extensively across neurons, it is improbable that the form of the nonlinearities will ever be established with precision. Hence the accuracy of a linear approximation process is itself unknown, and so even that option is unavailable to brain studies. Consequently, the output properties of systems of multiple interacting neurons, which presumably constitutes most of brain activity, will remain inaccessible to exact analysis. Appeal to the statistical regularities of the mass action of neurons to ameliorate this problem (Eddington, 1928) is not applicable if only a few neurons control any given behavior, as argued above.

The Inverse Problem
This is the traditional problem of analysing the "black box". Any closed system may be considered to constitute a black box because, in general, opening it to study the components requires interfering with the state of the system. It may be possible for some simple systems with isolated components to characterise all the components completely with negligible distortion of their properties, but this rapidly becomes an intractable problem in the case of systems of the complexity and delicacy of even the smallest identifiable neural systems.

The inverse problem arises when, instead of isolating the individual components of a system, one attempts to infer the properties of the components from the behaviour of the outputs of the intact system (the black box). This approach immediately runs into the problem that there is an infinite set of component behaviors that could give rise to the same output behavior. This result has long been established for electrical field theory. A given potential distribution on the surface of a sphere may be derived from an infinite set of simple potential sources within the sphere (Helmholtz, 1853). Even if several parameters of the component sources are known, such as the strength of their charge or its spatial distribution, the inverse problem still holds. The only condition under which a unique solution may be obtained is if the number of the sources is limited to a small integer. But this is exactly the condition least likely to be met in a complex system such as the brain.

A second version of the inverse problem has been developed by Moore (1956) in finite automata theory. Moore's second theorem states that:

Given any machine S and any multiple experiments performed on S, there exist other machines experimentally distinguishable from S for which the original experiment would have the same outcome. (p 140).

The applicability of this theorem to the human brain as a finite automaton (discussed by Uttal, 1990), is even worse in case of the brain because we do not even have direct access to much of the outcome of experiments on it - the internal experiences corresponding to the workings of the brain. It is impossible in general to match the output of a machine with the conscious outputs of one's brain, although particular properties such as its time course may be compared. Moreover, the literature is replete with examples of alternative models which explain a particular set of observations equally well. Ockham's principle of parsimony is frequently invoked to choose one explanation over the others, but there is no guarantee that nature adheres to this principle. In any case, parsimony is often in the eye of the beholder, since a "simpler" explanation may be harder to understand than one at a higher level of complexity in terms of the brain's function.

In fact, one is often tempted to suspect the operation of a "principle of austerity" whereby one must favour the least appealing explanation for any activity. If there is a model which explains behaviour of a neural system without reference to consciousness, teleology, the subconscious, energy fields or neural nets, it is always to be preferred to one which involves such a reference, even though the latter may be simpler in form.

Inverse psychophysics

The inverse problem is generally accepted as being profound for psychophysical investigation, because psychophysics relies on outputs based on conscious decisions by the whole organism. There is thus very little information about the structure of the activities underlying that decision, which consequently have to be inferred by the relative effects of manipulation of the input variables. Teller (1984) has developed a thorough analysis of the types of psychophysiological assumptions and linking propositions which can be used as tools to approach this inverse problem. Uttal (1990) has analysed a number of factors contributing to the extent of this inverse problem. These include the thermodynamic law of increasing entropy in complex systems, the non-reducibility of chaotic systems even though they may be deterministic, and the non-quantifiable nature of human inductive logic and perceptual processes.

Inverse neurophysiology

The main appeal of neurophysiology is that it allows direct access to the components of the brain which are participating in the decision. It is therefore quite striking to realise the extent to which neurophysiology is subject to the inverse problem. Although the elemental outputs (spikes) of the elemental units (neurons) of the brain may be readily recorded, they still do not have elementary properties with respect to the stimulus. Immediately after the first layer of receptors, every neuron receives multiple connections from multiple varieties of adjacent neurons with various anatomical configurations and physiological roles. The relation between ganglion-cell spikes and stimulus conditions is therefore far from simple; it contains interactions between several receptor types, nonlinear spatial summation, both subtractive and multiplicative inhibition, far field effects, and so on. All this before the signal even leaves the retina!

The problem of describing the stimulus conditions which elicit a spike therefore falls in the class of inverse problems, even before one attempts to account for the structure of the connectivity in terms of its function to the organism. There Actually, the problem is better conceived as a "forward inverse problem" in the sense that the problem is to go forward from the array of stimuli eliciting a spike to the connectivity of the cells involved, rather than going back from the output to the structure of the mechanism.

Inverse neural imaging

One domain in which the inverse problem seems to have been solved is that of neural imaging. EEG maps, BEAM electrical images and CAT scans can all be used to generate 3D images of the brain or its activity from the concatenation of a series of 2D views obtained by imaging with different types of physical energy. The techniques are apparently successful in coping with the inverse problem of determining the density distribution inside the head that gives rise to the energy distribution on the outside. However, the accuracy of these images relies on the assumption that the transmittance of the media such as the skull is uniform, and are subject to distortion where this assumption of the inversion transform is not met.

MRI and PET scans, on the other hand, are not limited by this restriction because the density of the head is essentially zero to the particles emitted in these techniques. These methods, while they may be limited by noise and sampling density, do not suffer from a formal inverse problem. The mathematical transform by which the brain conformation is derived from the particle density profiles (Radon Transform), is a linear transform with a unique solution, to the Fourier Transform. This transform would become nonlinear and therefore subject to the inverse problem if applied to materials which significantly affected the particle transmission, but this is not the case for the brain.

Logical Incompleteness

Gödel's incompleteness theorem states that any logical system contains information that cannot be derived from its postulates. He developed this theorem through the logic of set theory. Any representational system which can be described within the framework of set theory therefore conforms to the same theorem, and thus has responses that cannot be predicted from its structure.

Gödel's analysis focussed on cases where the logic is self-referential and can include itself as a subject of its representation. Gödel's theorem was derived from Russell's paradox, exemplified by the riddle "The barber shaves only those in his village who do not shave themselves. Who shaves the barber?" This question is unanswerable within the scope of the postulate, since if the barber shaves himself he must be one of the group that the barber himself does not shave. In physical representational systems, this paradox may apply where the system's output is fed back to its input, as in the case of a video camera pointing at its own monitor screen. What will appear on the screen? Again unanswerable. Similarly in neural systems, feeding the output of any neural subsystem back to its input may develop a positive feedback loop with unpredictable consequences. Photosensitive epilepsy may be an example of the activation of such a feedback loop in the brain, driving the system into a limit cycle in which the brain activity is governed by principles quite different from those controlling its normal activity.

However, the difference between Gödel's logical systems and the physical representational systems is that the logical system must include all forms of reference, including the self-negating ones which produce a positive feedback loop. Physical systems are usually designed to avoid such self-negative feedback, and therefore avoid Gödel's incompleteness paradox. (Of course, the logical description of such physical systems would also avoid Gödel's theorem, so that the theorem too has limitations; it should be possible to devise logical systems which do not suffer from the incompleteness paradox). The enterprise of understanding brain function is, however, inherently self-referential, since it is the same brain that is attempting to accomplish the understanding. A complete understanding of the human brain would have to include an understanding of what it means to understand, of what it means to understand the understanding, and so on into an infinite regression of incomprehension!

This paradox is not limited to the single core process described as "understanding". A full analysis of the visual part of the brain would have to include an understanding of the process of visualisation, or visuo-spatial representation within the processing structures. This analysis would not be complete without describing the visualisation of the visualisation process itself and so on, landing us once again into an infinite regressive loop of autovisualisation. Thus, while there may be localised neural circuits which are not subject to self-referential problems, such as those in the retina, Gödel's paradox reveals an inescapable limit on the full understanding of the visual system in addition to the brain as a whole. A broad survey of the cognitive absurdities which arise from self-referential representation is portrayed by Hofstadter (1979).

One way out of the self-referential paradox of brain investigation is provided by the subjective/objective split discussed by Stanley Klein in this volume. In this view, my understanding of my brain is self-referential, but the goal of neuroscience may be accomplished by my understanding of (or having access to a description of) your brain is not. I would find that, at some point in the spiral regression of levels of understanding, your brain ceased to have a representation of the next level of self-comprehension. This dehiscence would not limit my ability to understand your brain; in fact, it should enhance it. However, this circumvention of the self-referential paradox relies on the constraint that comprehension of the brain does not include anything about subjective experience, feelings, awareness or thought. As long as these are excluded from consideration, understanding the brain may be pursued as an objective enterprise, untrammeled by the strictures of Gödel's theorem. If, however, the domain of subjective experience is included as part of the domain of enquiry, the paradox applies in full force. Hamer (personal communication) has suggested the coinage that the resolution of Russell's paradox is to regard it as a Heisenbergian "unidox".

The Cognitive Window

The mind-numbing complexity of the brain, with its 10^{10} interconnected neurons, has often been emphasized in the past. The human capacity for understanding this vast system is, however, completely dwarfed by the task of understanding even one neuron, which may typically have 50,000 connections of various types from other neurons. These connections may be viewed as the "vocabulary" of inputs available to that neuron, which may have interactive effects on the neuron based on their spatial arrangement on the dendrites and soma of the cell. Such complexity points to the observation that the number of connections is about the same size as the vocabulary of the average adult, which takes a good part of a lifetime to develop. Thus, it might be expected that a person might take many decades to develop the vocabulary to understand the properties of a single neuron, even if the complete response properties were well-established. Uttal (1990) has gone further to argue that even a complete understanding at the elemental level does not generalise to higher levels of organisation, which have their own principles of operation independent of the lower levels.

However, the definition of complete understanding adopted at the beginning of this article indicates that we would have to do far more than understand the behaviour of each neuron individually. Perhaps it would be enough to understand it in relation to the 50,000 ($\sim 10^{4.7}$) other neurons connected with it. This would result in the total number of functions to be determined becoming,

$$2^{10^{4.7}} \quad \text{or} \quad 10^{10^{4.2}}$$

Focussing the computation on the known connectivity in this way avoids the problem of the established degree of replicated function in the brain, such as the approximately 6400 apparently identical hypercolumns in primary visual cortex. On the assumption that these replicated functional units are relatively independent, the problem of understanding the brain is then simply the connectivity of each subunit multiplied by the roughly 10^6 different subunits, amounting to a total vocabulary of about

$$10^6 . 10^{10^{4.2}} \quad \text{or} \quad 10^{10^5}$$

Note that since neurons at all levels of processing are included this computation includes the processes of deep information processing throughout the hierarchy of processing levels.

What particularly emphasizes the hopelessness of condensing brain function into a form that could conceivably be grasped by the human cognitive capacity is the failure of recent technical advances to provide any hint of simplification of the problem. For example, means have recently been developed to trace the full three-dimensional connectivity of individual neurons (Gilbert and Wiesel, 1979). It might have been hoped that the connectivity of the neurons would support the specialisation of function proposed for the anatomical divisions of the visual pathways. Instead, each identified neuron seems to have as much diversity as the pathways themselves, each having dendritic processes extending many of the branches of the aggregate pathways. These findings may imply that the neurons are not specialised functional units but part of a homogeneous substrate whose specialisation is defined only by the structure of the overall pathway.

A further example of the apparent increase in brain incomprehensibility is provided by the recent proliferation of substances found to have a transmitter role in communication between neurons. As many as 100 potential transmitter molecules have now been identified (Shepherd, 1988). It might have been hoped that particular transmitters would provide a simplified view of the brain, localised to a subset of nuclei or neural types within the nuclei. The current picture is not encouraging for this aspiration, since most transmitter types are distributed broadly through the brain and peripheral nerve pathways with elaborate patterns of representation. Instead, these studies of the histological distribution of transmitter molecules, have increased the apparent complexity of neural interconnectivity by one hundredfold without providing an evident simplification. The fact that most of the transmitters are slow-acting, and probably have a modulatory function on the main excitatory and inhibitory transmission, does little to ameliorate the problem. Such modulatory activity may account for much of the process of bringing memories into consciousness, a major component of brain function.

Moral Inviolability
A further fundamental limitation on studies of the human brain is that it can inherently be studied only by morally acceptable means. Unlike the brains of lower species, for which direct brain studies which are often considered to be morally acceptable, ethical considerations dictate that human brains can be studied only by non-invasive and non-destructive techniques. This forbids many methods for functional dissection of the responses of brain components which have been developed with lower species. In principle, it might be possible to gain a full understanding of the brains of lower species, had they not been subject to the other limitations discussed here. However, the cultural sophistication of human brains can never be assessed by subhuman studies, so that the moral inviolability limit continues to constrain brain investigation in this area. The only way to avoid it would be to develop noninvasive methods to obtain all the information currently and foreseeably obtainable by invasive methods. Such noninvasive methods of probing brain structure (e.g., CAT scan) and function (e.g., PET scan) are indeed feasible on a coarse scale. It remains to be seen whether the spatiotemporal resolution of such methods can be improved sufficiently to significantly ameliorate the moral inviolability limit.

Social Indeterminacy
The final principle limiting brain investigation is the indeterminacy of human social relations. Many aspects of interpersonal relations have no objective reality but are derived from the interdependent perceptions of those involved. These perceptions can readily be changed, even simply by revealing the perception of the relationship by the other party or parties. The underlying brain functions are correspondingly inaccessible to deterministic investigation.

For example, two people may form a relationship in which one is honest about expressing feelings while the other believes that one should express feelings in a way that the partner would like to hear. They may have a viable relationship on this basis until one of them describes their philosophy about expressing feelings. The other will then realise that their philosophies are discrepant and they do not have the equable relationship that they appeared to have. Every interaction between them will now become tainted by this knowledge, whereas it had been perfectly satisfactory in the past. Was the earlier relationship equable, as they had both believed, or was it asymmetric in the degree to which feelings were expressed, as it later appeared? The answer is subject to the observability limit which underlies the quantum indeterminacy principle, that the full relationship cannot be determined without interacting with the individuals in the relationship, which will tend to change the nature of the relationship under study.

Many aspects of visual perception relate to interpersonal relations, such as face recognition, perception of emotions in face and body language, and so on. Thus, study of the operation of visual perception in realistic life situations is subject to the social indeterminacy limit in specifying the cues that are to be perceived.

THE NATURE OF PSYCHOPHYSICAL THRESHOLD
From the broad questions of the ultimate understanding of the brain, I now turn to specific assumptions which are employed in one technique of brain investigation - psychophysics. A principle tool of psychophysics is the measurement of the threshold of detectability within particular stimulus configurations. The application of such techniques to the analysis of spatial vision has been extensively reviewed by Graham (1989). She identifies no less than 92 distinct assumptions involved in the studies which this analysis; many of the assumptions are tacit in the original publications and are made explicit by later authors or by Graham herself. The present work gives a more extended treatment to a few of these psychophysical assumptions which apply to simple detection experiments.

The simplest concept of a threshold in psychophysics is of a fixed limit below which there is no trace of stimulus-related event. However, even from the earliest days (e.g. Weber, 1834; Fechner, 1860) it was recognised that the threshold was a statistical concept, fluctuating over time around a mean value which had to be estimated from multiple measurements. Although, the statistical description of such thresholds as being set by a noise-limited signal-detection process was not developed until about 100 years later (Tanner and Swets, 1954; Green and Swets, 1966), it is typically taken as an implicit assumption of the nature of psychophysical thresholds.

This statistical process is, however, a poor description of the behaviour of cortical neurons, which do show a fixed threshold that operates very differently from an optimal signal detection process. The nature of the functional threshold also affects the interpretation of psychophysical measures as revealing underlying physiological mechanisms. In particular, such interpretation in terms of the sensitivity profile of the mechanisms will be distorted by probability summation within the detection process to the extent that the threshold is statistical.

Nonlinearity of the discriminability function
Most work with psychophysical thresholds is done in a two-alternative forced-choice (2AFC) paradigm. The results of such studies often fail to conform with the underlying tenet of signal detection theory that the discriminability (d') is expected to increase in proportion to stimulus strength (although in principle any relation between d' and the stimulus is allowable). The parameter d' is a measure of the distance between the means of the signal + noise and the noise distributions in the internal response space. The relation between d' and the stimulus strength is a criterion-free measure of the internal response metric for stimulus intensity.

The departure from the proportionality tenet for discriminability is not in the direction of typical nonlinearities such as physiological adaptive processes, which would tend to produce a reduction or saturation of the response at higher intensities (i.e., a power function with an exponent less than 1). Rather, discriminability is often found to increase with a power function exponent substantially greater than 1. Nachmias and Kocher (1970), for example, found d' exponents averaging 2.5 as a function of the luminance of a large adapting surround for detection of a 1° circular disk of 50 msec duration. (Note that this exponent should not be confused with the power of the Weibull function fitted to the percent correct responses). Stromeyer and Klein (1974) found an exponent of about 2 for detection of contrast increments of a sinusoidal grating stimulus.

Since detectability is defined in terms of signal-to-noise ratio, a nonlinearity in the d' function may reflect changes in the variance of the noise with stimulus intensity in addition to a nonlinearity of the internal representation. The noise variability usually increases with cortical cell responses to contrast stimuli (Tolhurst, Movshon and Dean, 1983) and also for the quantal fluctuations of light intensity (Hecht, Schlaer and Pirenne, 1937). Such variations can distort the relation between d' and stimulus intensity in the direction of decreasing the exponent if noise variance increases, and vice versa. However, the change in variance within the region of detection threshold (for d' values from 0 - 2) is not likely to be large, and it is difficult to explain an exponent of 2 - 2.5 by this means.

The true threshold model
An alternative model for detection behaviour is that sensitivity has a genuine threshold below which there is no internal response to the stimulus, and that d' increases more or less linearly with stimulus intensity above this threshold. The true threshold model has been proposed in the past (e.g. Baumgardt, 1947; Bouman and van der Velden, 1947; Baumgardt and Smith, 1965), but these investigations operated at or close to zero false alarm rate, so were unable to assess the linearity of the internal response to the stimulus. When measured correctly (i.e. in a task with a well-defined false alarm rate), the true threshold model can generate data whose d' function is fairly well fit with an exponent greater than 1, providing an adequate explanation for the observed exponents in the 2AFC experiments. However, the true threshold model has a rather different interpretation from the high exponent model, and they can be empirically distinguished from one another.

Justification for true threshold behavior comes from consideration of the function of cortical neurons. Neurons in the cortex exhibit very little spontaneous activity, but may fire at high rates as soon as their optimal stimulus is present. Many neurons in the retina and optic tract, on

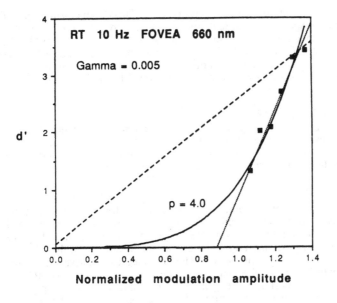

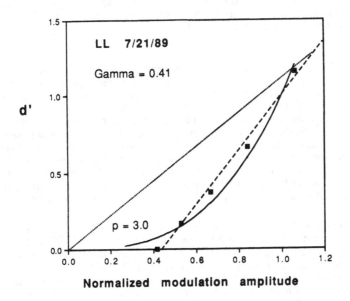

Fig. 1. Threshold detection behavior (d') for the detection of flicker in a 14' field.   The task was a Yes/No decision with 50% blank trials and an average false alarm rate of 0.45%. The signal levels are normalized to the mean sensitivity (d' = 1). Dotted curve - linear sensitivity hypothesis. Solid curve - best-fitting power function with an integer exponent of 4. Dashed line - threshold hypothesis.

Fig. 2. Similar data to those of Fig. 1 but with another observer who was induced to have a false alarm rate of 41%. The best fitting integer exponent was 3.

the other hand, have a relatively high spontaneous activity level. This arrangement implies that cortical neurons must have a response threshold, below which input from the noisy pre-cortical cells evokes little output. Since it is to be expected that the cortical neurons would need to have as high a sensitivity as is compatible with the noise level in their inputs, the most efficient level to set the threshold is near the limit of the input noise, corresponding to some statistical criterion on the basis of the noise distribution (e.g. the 5% level for activation by the input noise).

Distinction from high threshold model

The true threshold model should not be confused with the "high threshold model" (Tanner and Swets, 1954), which assumes that the threshold is set so high as to produce no responses related to the strength of the internal noise on "no stimulus" trials. In this case, one should expect to obtain no false alarms related to internal noise; any false alarms must be entirely due to guessing behavior.

The standard way to evaluate threshold hypotheses is to measure the form of the ROC (receiver operating characteristic) curve of Signal Detection Theory, which plots the percent hits against the percent false alarms. This would have a characteristic shape on the threshold hypothesis, conforming to a straight line for false alarm rates higher than some value set by the false alarms attributable to noise which exceeds the threshold. The predicted ROC curve is asymmetric towards greater discriminability at low false alarm rates; the shape is similar to the ROC curve predicted for a greater variance as signal strength increases. In discussing this prediction, Green and Swets (1966, pg. 139) say that all available data "are consistent with the existence of a sensory threshold located one standard deviation above the mean of the noise distribution." But they are also consistent with non-threshold theories in which the variance of the internal noise changes appropriately with signal strength. Apparently, therefore, the ROC curve approach is not able to distinguish between these two theories.

Analysis of psychophysical data

An alternative approach to the question of threshold is to perform an experiment at many levels of the stimulus down to very weak levels and determine the function in the region of very low d' values. Data of this type were obtained by Klein and Stromeyer (1980) for detection of compound grating stimuli. Their d' functions were best fitted by power functions with slopes between about 2 and 3. They comment, however, on the fact that points for the lowest d' values always fell far below the fitted function, but they do not mention that this is precisely what would be expected on the basis of the threshold hypothesis (S. Klein, personal communication). Other sensory data collected explicitly to test the threshold hypothesis go further to show the predicted threshold behaviour of zero d' values across a range of low stimulus levels before a response can be elicited (Gescheider, 1976). Confirmatory results have been obtained from a vibrotactile masking paradigm (Hamer, Verrillo and Zwislocki, 1983)

An example of threshold behaviour for visual flicker detection is shown in Fig. 1. The task was a Yes/No decision of the presence of flicker with 50% blank trials and a total of 200 trials per contrast level. The average false alarm rate was 0.45%. The computed values of d' are shown for various signal levels normalised to the mean sensitivity (d' = 1). These data clearly depart substantially from the linear hypothesis for d' (dashed line). In fact, an integer d' exponent of 4 would be required to fit the data adequately, far higher than any currently reported in the literature. On the other hand, the data are well fit by the dotted straight line corresponding to the threshold hypothesis, intercepting the abscissa at a value of 0.83 of the full amplitude for d' = 1.

Similar results are shown in Fig. 2 for a second observer, who in this case was exhorted to maximise sensitivity, and achieved a false alarm rate (gamma) of 41%. The data still deviated widely from the linear hypothesis (dotted line), however, and required an integer exponent of 3 for adequate fit on the power hypothesis. Nevertheless, the true threshold hypothesis (dashed line), provides a substantially more convincing fit to the data, and intercepts the abscissa at a value of 0.44 of the full amplitude for d' = 1. Both data sets therefore tend to support the operation of a true psychophysical threshold of the type seen in cortical cells.

There is an inherent difficulty with this approach, in that if there is a threshold, it will cut off the bulk of the noise distribution at these low levels, and tend to produce a low false alarm rate in a YES/NO experiment. Persuading the subject to use a high rate of false alarms, or even going to a 2AFC paradigm, will result in a proportion of false alarms that are not dependent on the noise values but due purely to guessing (i.e. noise at a later stage in the system). This guessing behaviour will, however, add to the hit and false-alarm rates in a slightly different way than will reducing the threshold level to increase the false alarms, so the function derived by increasing false alarms (or 2AFC) is not a veridical estimate of the true d'/intensity function. There is a Russellian paradox to this objection, however, because it says that the function demonstrating the threshold is distorted if there is a threshold, but is accurate if there is no threshold. Thus if the function is accurate it will distort the measurement and must therefore be distorted, whereas if it is inaccurate in showing that there is a threshold then it will portray the threshold behaviour accurately! I therefore turn to an electrophysiological method of analysing threshold behaviour as a way out of this dilemma.

THRESHOLD BEHAVIOR IN CORTICAL EVOKED POTENTIALS
An alternative approach to the question of linearity versus a threshold in cortical signals is to look at the response properties of the visual evoked potential (VEP). Many VEP studies are analysed under the assumption that there is a threshold in the response that can reasonably be compared with the psychophysical threshold estimate, especially with respect to the contrast of grating stimuli (e.g. Campbell and Maffei, 1970; Kulikowski, 1977; Cannon, 1983; Tyler and Apkarian, 1985; Allen, Norcia and Tyler, 1986). However, in these and most other studies, the assumption was made that the VEP amplitude is proportional to the logarithm of the contrast above some threshold contrast level. Straight lines were therefore fit to the data on log contrast, linear VEP amplitude coordinates at the lower contrast levels and extrapolated to zero response voltage to provide the VEP threshold estimate.

The problem with this approach, which has proved quite effective in practice, is that it depends in principle on the twin assumptions that there is a true threshold and, less severely, that VEP amplitude increases with log contrast. This theoretical approach was examined in detail by Campbell and Kulikowski (1972), who explored various other ways of analysing VEP contrast functions and their thresholds, but came to no convincing conclusion on the issue. It is therefore worth taking a further look at the issue with careful measurement of the VEP amplitude around the threshold region by a variety of techniques.

Display
Vertical sinewave luminance gratings were presented on a TECO 12" video monitor. The contrast behaviour was explored by means of a swept-parameter technique to allow rapid acquisition of responses for a full range of contrast values. The gratings were swept logarithmically in contrast over a 20-fold range in 19 half-second increments during a 10 second epoch. This fine sampling ensured an accurate estimate of the point at which the VEP fell to the background noise level (Norcia and Tyler, 1985). The contrast was always swept from low to high to minimise the effects of prior adaptation on the response of the following contrast at each contrast level.

Mean luminance was 80 cd/m^2. The display was carefully adjusted so as not to produce changes in mean luminance (flicker) correlated with pattern reversal, which was a temporal square-wave chosen to avoid potential even-order distortion products due to phosphor non-linearities. The experiments were conducted in low ambient illumination.

Signal Analysis
The analysis of steady state VEP at the stimulus reversal rate was performed by an Apple II+ computer and associated hardware. The analysis techniques (described fully in Norcia and Tyler, 1985) derived signal amplitude and phase every half-second during the 10 sec. contrast sweep. Contrast threshold was then estimated by extrapolation of the significant response at the highest spatial frequency down to zero uV, according to several criteria discussed below.

The same analysis was applied to a second Fourier frequency within approximately 10% of the signal frequency, and which was an integral submultiple of the sampling frequency and hence

LINEAR HYPOTHESIS

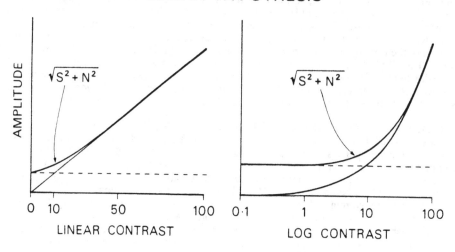

THRESHOLD HYPOTHESIS

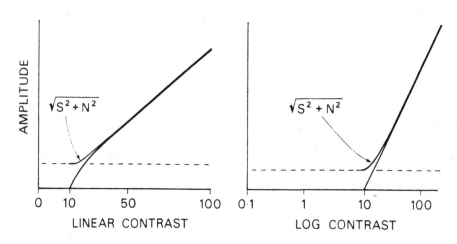

Fig. 3A. Predictions of the linear response hypothesis for the amplitude of the steady state VEP in noise as a function of stimulus contrast. The lower of the pair of full lines shows the underlying signal amplitude, while the upper one shows the effect of the inclusion of noise at the mean amplitude of the dashed line by virtue of its vector sum with the signal.

B. Predictions of the threshold hypothesis in the same format as A. Note the similarity in form between the two hypotheses at higher amplitudes orthogonal to the stimulus frequency. Since the EEG spectrum is locally flat, the EEG amplitude at an adjacent frequency can be used as an indicator of the background noise at the signal frequency during the trial (see Norcia, Clarke and Tyler, 1985). The noise frequency was not allowed to be closer than 1 Hz, since the full bandwidth of the 1 Hz data window is 0.5 Hz on either side of the signal frequency. The signal and noise frequencies are independent in the analysis but are expected on average to have the same mean amplitude in the absence of a signal.

Theoretical predictions
The aim of this VEP study was to compare predictions for two hypotheses of VEP behaviour - a linear response hypothesis and a threshold hypothesis combined with a logarithmic relation of the response to contrast. The second hypothesis is the one that is commonly assumed in the analysis of contrast VEPs, but for the present purposes the predictions are almost the same if the threshold is followed by a linear relation of the response to contrast. The test is therefore essentially one of the presence of absence of a threshold in the contrast VEP. The predictions will be developed to show the expected response amplitude in the presence of additive noise, the effect of the noise reduction in the vector mean relative to the scalar mean, and the relation of the VEP contrast threshold estimate to the psychophysical threshold for the same conditions.

The predictions of the linear hypothesis are shown in Fig. 3A in linear coordinates (left panel) and also semilog coordinates (right panel) to correspond to those used in the VEP analysis. The lower line of each pair shows the signal amplitude increasing linearly with contrast. The upper line indicates the expected value of the measured response with the inclusion of additive noise at the level shown by the dashed lines. In Fourier analysis, the noise combines with the signal on the basis of the Ricean distribution (Rice, 1945), but this is closely approximated by a square-root vector summation between the signal mean noise amplitudes (Norcia, Tyler, Hamer and Wesemann, 1989). Note that both with and without noise the linear hypothesis predicts a strongly curved function in semilog coordinates.

The main use of the contrast function data is to provide an estimate of threshold by extrapolating the high amplitude response to the intercept with the contrast axis. As a consequence of the curvature of its function on semilog coordinates, the linear hypothesis predicts that estimated threshold obtained by straight line extrapolation from the supra-noise response function will change radically with changes in the noise level. This prediction will be tested with an analytic technique for varying the measurement noise (Fig. 4).

The predictions of the threshold hypothesis are shown in the same format in Fig. 3B. The form of the signal amplitude is quite similar on both semilog and linear plots. Because the signal line cuts steeply through the mean noise level, the effect of additive noise is only a minor deviation from the straight line in the semilog plot. Thus, the threshold prediction differs from the linear prediction mainly in the region of the corner where the signal meets the noise, which is quite a sharp angle for the threshold hypothesis in contrast with the curved form for the linear hypothesis. Because of this fact, the effect of varying the noise level on the predicted threshold for the threshold hypothesis is essentially negligible if the extrapolation is limited to values greater than 1.5 times the higher noise level.

Moreover, the quantitative difference between the two hypotheses can be encapsulated in a the ratio of the predicted threshold response to the linear response. This threshold response ratio varies with the steepness of the response increase with contrast, and will be used to compare the theoretical predictions.

Response Measures
Four different measures of the VEP response were derived from the Fourier analysis of the signals over sets of N trials. At each contrast level, Fourier analysis can be viewed as yielding an amplitude and a phase of the signal at the analysis frequency for each stimulus trial. These parameters were combined in four different ways to derive measures of the scalar mean, vector mean, phase coherence and coherence SNR (signal/noise ratio).

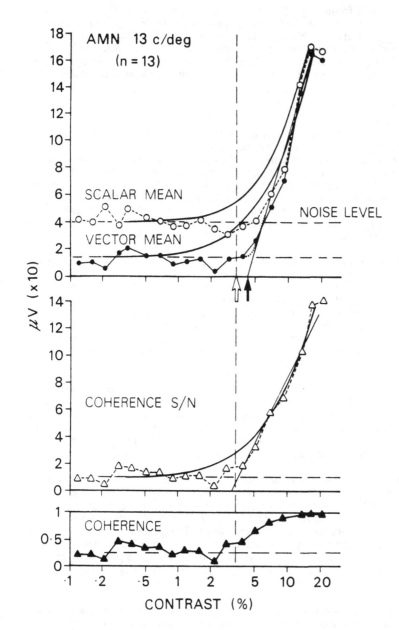

Fig. 4. Computation of four different VEP response measures as a function of stimulus contrast. Open circles - scalar means at each contrast for 13 individual contrast sweeps. Top panel: Filled circles - vector means for the same data. Solid curves - linear amplitude predictions for the scalar and vector means, including their respective noise levels (horizontal dashed lines). Thin oblique line - threshold hypothesis. Vertical line - psychophysical threshold.

     Filled triangles (bottom panel) - phase coherence at each contrast over the 13 sweeps. Horizontal dashed line - expected noise level.

     Open triangles (center panel) - coherence S/N ratio, a measure of the discriminability implied by the coherence measure.

The scalar mean is the mean of the amplitudes of the responses in each of the 13 trials, without regard to the phases of the responses, shown as the open circles in Fig. 4. The scalar mean has the property that the noise level in the absence of signal does not decrease with the number of trials contributing to the mean. Instead, the scalar mean provides an estimate of the noise level applicable on each trial, such that the accuracy of the noise estimate improves with the number of trials.

The vector mean (filled circles in Fig. 4) does take into account the phase of the individual responses to reduce the amplitude of the noise estimate as the number of trials increases. The vector mean is defined conventionally as the square root of the sum of the squares of the sine and cosine components of the Fourier representation of the signal on each trial. It therefore provides a means of tracking the synchronous response down to lower signal levels than available in the scalar mean.

The phase coherence (filled triangles, Fig. 4) is a measure of the degree to which the individual responses tend to fall near the same phase with respect to the stimulus. It is defined simply as the ratio of the vector to the scalar mean. If all the responses had the same phase the vector mean would be equivalent to the scalar mean and phase coherence would equal 1. Note that this definition tends to emphasize the contributions of the larger vectors relative to the smaller ones, so that signal fluctuations resulting largely in noise on some trials would not unduly perturb the estimate.

The coherence SNR (open triangles, Fig. 4) is not an independent measure, but is derived from the phase coherence, C, so as to provide a linear measure of the signal/noise ratio implied by the phase data. The definition of this new measure is:

$$\text{Coherence SNR} = \frac{1}{\sqrt{1/C^2 - 1}}$$

This expression can be shown to correspond to the ratio between the vector mean and the standard deviation of the noise implied by the phase coherence.

It should be clear from their definitions that the four VEP measures derived are not fully independent of one another. In fact, all the information available is contained in two of them (e.g. the scalar mean and the coherence SNR), but the comparisons among the various measures are useful for different purposes.

Results
The results of the various investigations of the threshold region of the VEP were all strongly in favour of the threshold hypothesis. Figure 4 shows the analysis of the four different response measures, and compares them with the psychophysical threshold obtained by pushing a button to indicate the first appearance of the test grating during the sweep (open arrow and dashed vertical line). The predictions of the threshold hypothesis are indicated by the straight line extrapolated through the supra-noise data to the contrast axis. The VEP threshold estimates are the same as the psychophysical threshold within about 0.1 log unit for all three measures (filled arrow shows threshold estimate for the vector mean).

The predictions of the linear hypothesis are shown by the solid curves in Fig. 4, which lie substantially above the noise level in the region up to the psychophysical threshold. None of the four measures show any significant tendency to follow this predicted rise, and all stay close to the expected value for the noise level right through the threshold region. Furthermore, increasing the effective noise, by using the scalar mean rather than the vector mean, shows no tendency for the extrapolation line to shift towards higher contrasts, as would be expected if a function from the linear hypothesis were controlling near-threshold behavior.

The final prediction for the threshold hypothesis when normalised relative to the predictions from linear hypothesis is shown in Fig. 5. In each case the particular form of the threshold hypothesis was derived from the points beyond a 1.5 SNR, and used to compute how far the data should fall below the level of the linear prediction. This comparison was performed for two bipolar recording channels, placed 3 cm vertically (Ch 1) and horizontally (Ch 2) from the

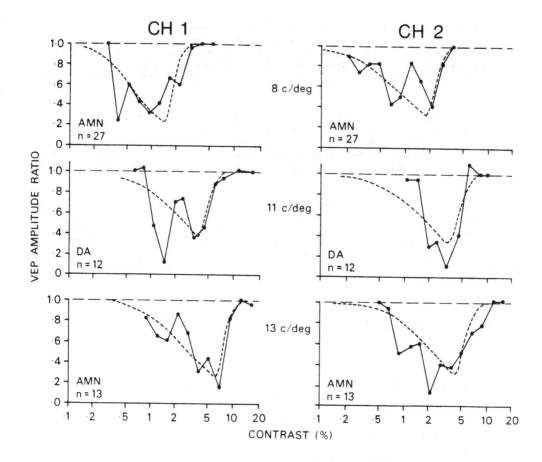

Fig. 5. Deviation of the data from the best-fitting linear threshold prediction, normalized to that prediction. For perfect adherence to the linear prediction the data would fall on the horizontal dashed line. The predictions for the threshold hypothesis are depicted by the short-dashed line, as derived from a best fit to the data for response amplitudes greater then 1.5 times the noise level. Note that in all cases the data are scattered around the prediction for the threshold hypothesis, indicating a sharper corner in the response amplitudes than would be compatible with the linear hypothesis.

inion. Spatial frequencies were 8 and 13 cy/deg for one observer and 11 cy/deg for another. None of the data come close to the predictions of the linear hypothesis (horizontal dashed lines), but they do tend to fall to low ratios in the region of the reduction predicted by the threshold hypothesis (short-dashed curve). It must be remembered that data close to the noise level are inherently noisy, so they should not be expected to adhere too closely to the prediction except for the higher contrast values. This sample of data shows no consistent deviation from the threshold prediction, giving reasonable assurance that it provides a complete account of the contrast behavior.

THE CONFOUND OF PROBABILITY SUMMATION

Probability summation is the improvement in sensitivity with number of samples of the stimulus or number of neural channels activated, based on the probabilistic nature of the psychometric function. Probability summation effects are small if the psychometric function is steep, but may become large if it is shallow. The (Weibull) steepness of the psychometric function is typically about 4, which means that the log-log slope of improvement in sensitivity is about 1/4, or less than a factor of 2 for each log unit increase in number of channels activated. This degree of probability summation is relatively negligible in most situations. Even in the classic study on spatial probability summation over number of bars of a grating (Robson and Graham, 1981), the total effect over the full measurable range from 2 to 64 cycles was only 0.5 log units (a factor of 3).

Conversely, if the slope of the psychometric function is shallow for some psychophysical task, then the effect on sensitivity can be large when the number of channels stimulated varies substantially. In particular, where the psychometric function is shallow enough to have a Weibull exponent of 1, then probability summation mimics full energy summation, and a tenfold increase in the number of channels stimulated produces a tenfold increase in measured sensitivity, which is as large as most psychophysical effects. Situations in which shallow psychometric functions are obtained include:

        Absolute thresholds
        Suprathreshold contrast discrimination
        Chromatic thresholds
        Infant thresholds of all kinds

It is therefore crucial to measure the slope of the psychometric function in all tasks involving substantial variation of stimulus extent in any dimension, or the possible number of channels involved in threshold determination.

Probability summation in the estimation of neural integration time

One situation in which the effects of probability summation have been analysed in detail is in the temporal summation (Bloch's law) paradigm to determine the integration time of the visual response under various conditions (Gorea and Tyler, 1986). If there is a linear internal response metric, then the full integration of energy for very short duration stimuli will result in psychophysical thresholds that are reciprocal with duration, producing a slope of -1 on double-logarithmic coordinates. No effect of probability summation on the form of the function is expected. Such results are obtained for contrast grating detection at durations less than 30 msec. (Tulanay-Keesey and Jones, 1976; Gorea and Tyler, 1986). Beyond that point, the form of the temporal integration curve is strongly dependent on probability summation, so that any attempt to determine the temporal response properties from the form of this function must take probability summation into account.

The predicted effects of probability summation depend on the model of the probabilistic process used to predict them. One such model is that detection is determined by the integral of the neural energy passing through a nonlinear transform (the Weibull function) representing the probabilistic effects of noise (Quick, 1974). The neural energy under this integral is controlled by the stimulus and by the response properties of the system - principally the response time constant and degree to which it is transient or sustained (Watson, 1979; Gorea and Tyler, 1986). The effects may be modeled on the basis of a typical system impulse response,

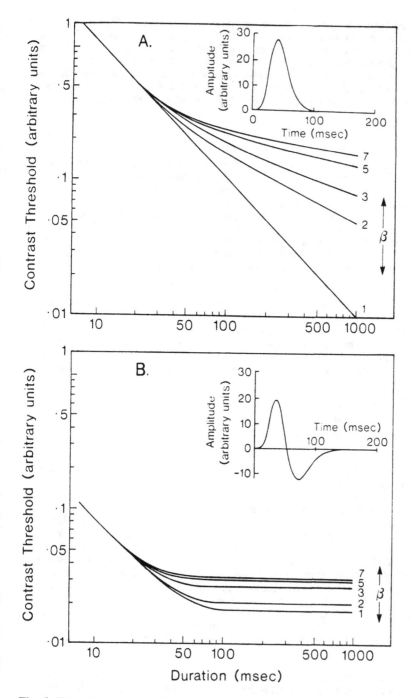

Fig. 6. The effects of probability summation on temporal integration curves, analysed in terms of the **beta**, the Weibull exponent (slope) of the psychometric function.
   A. Temporal integration with a sustained (monophasic) impulse response.
   B. Temporal integration with a transient (biphasic) impulse response.

specifically the one suggested by Watson (1979) to describe his flicker data at low and high spatial frequencies (see insets, Fig. 6).

The effect of these two cases on the degree of probability summation for longer duration pulses is shown in Fig. 6A for the sustained case and Fig. 6B for the transient case. The parameter on these plots is the **beta** power which describes the slope of the psychometric function - higher **betas** correspond to steeper slopes. Taking the transient case first, it is evident that the predicted thresholds level out for long duration stimuli, at all **beta** levels. This is because the transient response is limited to a lobe following the onset and offset of the stimulus pulse, with no activity in between. Thus, increasing the stimulus duration provides no extra information to change the threshold. Note, however, that different values of **beta** affect the point at which the function levels out, so that the estimate of "critical duration" (the intercept of the horizontal and - 1 asymptotes), varies by a factor of about 2 between curves of high **beta** and those for **beta** = 1. This variation is itself large in relation to physiological variations in the time constant, which has a total range of about a factor of 3 in human fovea (Graham and Kemp, 1938), for example.

It is therefore important to measure **beta** directly for each observer, in order to determine the correction that needs to be applied to derive an accurate estimate of the critical duration parameter. **Beta** has been found to have stable values between 2 and 5 even in a small sample of observers (Mayer and Tyler, 1986). The situation is more serious in the case of the sustained impulse response (k = 0), which may be expected for the following types of stimuli:

> Absolute threshold conditions
> Chromatic stimuli
> High spatial frequency stimuli and small spots

Some of these conditions are also those in which the slope of the psychometric function is expected to be shallow, putting the thresholds and their derived parameter estimates in double jeopardy for contamination from the effects of probability summation. The combined effects of the sustained impulse response and various values of **beta** are shown in Fig. 6B. At short durations, full summation is again obtained, again without any influence of probability summation. At long durations, however, the function does not level out, but asymptotes to a slope of -1/**beta**. This is a very different model for the form of the temporal integration function which requires a different analytic approach. If the data are analysed in the conventional Bloch's law framework, the estimated critical duration will be incorrect to an extent which increases as **beta** becomes shallower and also as the range of durations tested is increased, even though the underlying impulse response remains fixed.

The distorting effect of probability summation can be corrected if the function is modeled correctly with an accurate estimate of the slope of the psychometric function (Gorea and Tyler, 1986). However, this correction becomes progressively more inaccurate as this slope approaches 1. This means that the temporal summation function approaches a slope of -1, which is consequently indistinguishable from full summation. It would therefore be impossible to obtain any estimate of the critical duration from the data, which should form an uninterrupted slope of -1. This kind of function (although with a continuous slope of about -0.5) has been reported for spatial summation of chromatic changes (King-Smith and Carden, 1976) and interpreted as implying a very large chromatic integration area. The present analysis implies that such data could arise from probability summation effects.

Relation to the discrete threshold hypothesis
The existence of a discrete threshold, as argued in the previous section, might seem incompatible with the occurrence of probability summation, but this is not the case. The essence of the discrete threshold notion is that there is no internal response <u>below</u> a certain stimulus value, not that the response is discrete above that value. As Figs. 1 & 2 showed, the response may be continuous through the region of d' = 1, with a form very similar to the Weibull prediction. As long as there is a continuous underlying variable and a probabilistic response behavior, probability summation will occur.

However, the discrete threshold hypothesis does have an important implication for the extent of probability summation effects. The Quick (1974) formulation of probability summation through the Weibull function implies that sensitivity will improve indefinitely with stimulus extent. If we take the temporal case, which provides the possibility of indefinite extension of the stimulus duration, this soon develops an unreasonable prediction, even for photopic grating detection. Consider the sensitivity of an individual who happens to have a psychometric function with a Weibull slope of 2 (as has sometimes been reported; Mayer and Tyler, 1986), and design the experiment to eliminate possible attention effects synchronised to the stimulus presentation (Gorea and Tyler, 1986). For a wide-field grating at optimal spatial frequency (providing a sustained impulse response), and a brief 100 msec stimulus presentation which can fully activate the impulse response to the stimulus, this observer's threshold might be expected to be about 2% contrast (Tulanay-Keesey and Jones, 1976). Increasing the duration by one log unit to 1 sec, should, on the Quick theory, decrease this observer's threshold by 1/2 log unit, to 0.6%, which is an empirically reasonable value. Further increases in the duration to 10, 100 and 1000 sec (16.6 minutes), should decrease the threshold to 0.2, 0.06 and 0.02% contrast respectively. These are outlandish values which are well below anything that has been reported in the literature, although the precise experiment just described has not been explicitly performed. Tulanay-Keesey and Jones, for example, obtained thresholds of 1% for durations of 4 sec., but did not measure the slope of the psychometric function to be able to derive a prediction for this duration.

Motivated by this paradoxical prediction, Gorea and Tyler (1986) developed a two-process model of probability summation which reduced the predicted degree of probability summation at long durations. This included the idea that the psychometric function was derived from a nonlinear relation between d' and stimulus strength, which was already a deviation from the linear assumption of Signal Detection Theory. The present data (Fig. 1 - 5) go further in supporting the existence of a discrete threshold nonlinearity, which implies an absolute limit to probability summation. Wherever the discrete threshold level occurs, no amount of increase in stimulus extent will allow the measured threshold to fall below that value. If the threshold level is about half of the level at which d' = 1 (as in Figs. 1 & 2), the maximum gain from probability summation will be a factor of 2. So the discrete threshold hypothesis can provide a rescue from the paradoxical sensitivity increases predicted by Quick's theory, assuming that all the neural channels stimulated by the increased stimulus extent have the same sensitivity.

Probabilistic distortion in the measured psychophysical functions can still occur with a discrete threshold, as long as the d' criterion for the sensitivity measure falls on an increasing part of the function above that threshold. Thus one is not relieved of the requirement to measure the psychometric function in order to interpret the stimulus functions. To the contrary, the psychometric function needs to be specified more accurately in the region where it is most difficult to measure - the low probability regions. As the current data show, however, this is not unfeasible. It is often possible to establish that the psychometric function has the same shape throughout the stimulus parameter range by measuring the extreme values of the range, so that it may then be applied to intermediate values without recourse to the full measurement regime.

THE LOGARITHMIC RELATION IN LUMINANCE PROCESSING
Failure of logarithmic sensitivity relation to pigment concentration
Dowling (1960) and Rushton (1961) have enunciated the principle that, during dark adaptation, visual sensitivity is proportional to the logarithm of the level of pigment bleached. This principle is required to explain why it is that photopigment concentration recovers exponentially after a exposure to a bleaching light, while visual sensitivity does not. Rushton (1961) showed that visual sensitivity in rods recovers much more rapidly so that its logarithm appears to conform to an exponential function. Thus the two may be made to match over much of the range of recovery time by the assumption that there is some mechanism within the transduction cascade that sets visual sensitivity in proportion to the log of pigment concentration. Hollins and Alpern (1973) showed that the same principle operated for the cones at high bleaching levels, but that at low levels of pigment bleach and short times after the bleaching event, thresholds were elevated even more than implied by the logarithmic principle.

This further sensitivity reduction is usually attributed to non-pigmentary mechanisms of gain control.

The logarithmic principle is widely accorded the status of an implicit assumption, although its mechanism is poorly understood. There are several possibilities as to the operating time course of the proposed logarithmic mechanism. Either it is instantaneous, i.e. occurs within the time required to bleach the photopigment, or has a longer time course such as the integration time of the cone output, or it has an even longer time course corresponding to a type of dynamic gain control. One attempt to measure its time course in terms of the time course of multiplicative adaptation of a flashed increment threshold, Hayhoe, Benimoff and Hood (1987) found adaptation to be complete by 50 msec after the onset of the adapting field. These data imply that the adaptation is either instantaneous or similar to the duration of the cone integration time, but exclude a slow dynamic gain control.

If the mechanism is an instantaneous logarithmic compression, then one can test it by measuring the temporal integration function (Bloch's law) at levels that would require significant proportions of the photopigment to be activated at psychophysical threshold. For example, the 50% cone bleaching level occurs at 4.3 log Td in the foveola. An adaptation experiment was therefore conducted at a level of 4.9 log Td, which should bleach about 82% of the available pigment (Rushton and Henry, 1968) and push the visual system into a range where the logarithmic principle operates. The instantaneous version of this hypothesis takes the form that the internal response should be proportional to the log of the direct bleaching effect of the stimulus. (For small proportions of pigment bleached, < 10%, the logarithmic law predicts essentially linear behavior). At high background levels, the instantaneous version of the Dowling-Rushton principle should require that Bloch's law is violated and that thresholds are elevated exponentially to reach equal increments on the log bleaching function as test duration is reduced. The dashed line in Fig. 7 shows the prediction, on the basis of the log scaling factor of 2.9 found to apply to cones by Hollins and Alpern (1973).

A test of the saturation prediction at bleaching levels is shown by the data in Fig. 7, over a duration range from one second down to 50 usec. Below the critical integration time of about 50 msec., the function rises in close adherence to the line of linear proportionality (solid line) and clearly deviates from the predicted Dowling-Rushton function (dashed line). In this short duration regime, therefore, sensitivity is not proportional to the logarithm of pigment bleached by the stimulus, but is directly proportional to the quantity of pigment present. This implies that the stimulus strength at threshold was increased in direct proportion to the reduction in stimulus duration, and thus the instantaneous control of sensitivity is a linear rather than a logarithmic mechanism. In conjunction with the results of Hayhoe et al. (1987) this appears to suggest that the logarithmic compression is neither much longer nor much shorter than the cone integration time.

Note that the adherence to a line of proportionality for short durations means that a fixed number of quanta were required for detection throughout this duration range. The number of quanta isomerised per foveal cone within each test may be estimated at about 66, based on the 82% bleach, the spectral sensitivity of the 660 nm light, losses due to media absorption, self-screening and the Stiles-Crawford effect tabulated in Wyszecki and Stiles (1983), a foveal cone aperture of 1 u, and a quantum absorption efficiency of 0.65 (Pugh, 1987). These data suggest that each foveal cone is capable of processing about 66 photopigment molecules activated in as short a time as 50 usec. The transduction resources required to achieve this performance appear to go beyond what is currently proposed to be available in each cone outer segment (Pugh, 1987).

Weber's Law
The temporal integration data of Fig. 7 show that the mechanism for the logarithmic sensitivity relation of Weber's law must have a time course equivalent to (or slower than) the integration time (critical duration). Thus the Bloch's law behaviour for short durations must be set by the temporal integration stage, because it does not show any nonlinear behaviour attributable to the logarithmic principle. Further evidence of this was provided by Anstis and Mather (1985; see Anstis' chapter in this volume), who measured the effective distortion of the luminance signal

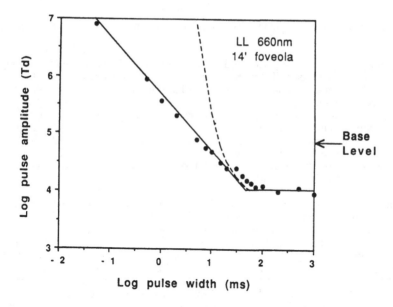

Fig. 7. Temporal integration at bleaching levels, as a test of the instantaneous version of the Dowling/Rushton logarithmic sensitivity relation. If the internal response were proportional to the log of pigment bleached by the test light, thresholds would increase exponentially (dashed line). In fact, the data show that, for 82% bleaching backgrounds, Bloch's law of time/intensity reciprocity is obeyed, implying linear response behavior in the cone transduction mechanism.

Fig. 8. A radial image with a linear sinusoidal luminance profile (within the reproduction limitations). At low luminance it appears sinusoidal, with bright and dark bars of equal widths. At high luminance the bright bars become broadened and the dark bars appear narrow and V-shaped, indicating a luminance nonlinearity operating throughout the range.

as an input to the apparent motion system. The results showed that this signal was linear (distortion-free) in the medium luminance range. Since the motion signal is computed in the cortex, these results fail to support the idea of a logarithmic compression at any point in the retina, at least for these transient stimuli.

There remain two salient possibilities to explain the proportional threshold elevation of Weber's law. One is a static compressive nonlinearity following the site of temporal integration (since the Bloch's law experiment excludes a preceding site). The other is a dynamic gain control mechanism which reduces the response gain in proportion to the adapting intensity. A third possibility, that of the background noise level increasing rather than signal decreasing, is formally similar to the dynamic gain control. The possibility that the noise increase is due to quantum fluctuations in the photon input or other quantal sources can be excluded because such mechanisms would produce a square-root increase with intensity rather than Weber's law.

There is certainly evidence to support the existence of a compressive nonlinearity in the visual response. Static luminance images show pronounced evidence of a saturating nonlinearity which is commonly attributed to logarithmic compression. In the example shown in Fig. 8 (a radial sinusoid), the compression is evidenced by the wide smooth appearance of the bright bars and the narrow line-like appearance of the dark bars. This observation was verified by measurements of the point of subjective equality (PSE) for the width of the bright versus dark bars in this image. The relative width of the bars was varied by introducing a nonlinear power function into the array controlling the intensity look-up table of a Mac IIfx computer. (The final output was calibrated to complete linearity by measurement of the intensity of all 256 available grey levels, so that the power function represents a true intensity output for each grey level.) The mean power distortions required to produce the appearance of equal dark and light bars for 600 msec presentations at 47 cd/m^2 were 2.97 and 3.33 for two observers. This power compensation implies that the effective luminance compression function is the reciprocal of these values, and has a power of about 0.32. This result has some generality because it is similar to the result obtained for direct brightness estimation (Stevens & Stevens, 1960). However, since it was based on estimation only of the mid-point, the result does not constitute strong evidence for a power compression as opposed to other possible types. Note also that this distortion is evident well into the center of the ray pattern, indicating that it occurs up to spatial frequencies of the order of 10 cy/deg.

Nevertheless, other sources of compressive nonlinearity are well known in the visual system, principally the saturation of the output of cells of all kinds. This saturation has been described as a simple hyperbolic function, a power hyperbolic function of the Naka-Rushton type, and an exponential saturation (Naka and Rushton, 1966; Baylor, Nunn and Schnapf, 1987). If any of these functions were governing luminance sensitivity, it would predict substantial deviations from Weber's law (based on the asymptotic slope of the derivative with respect to intensity). For example, a hyperbolic saturation (or Naka-Rushton function with a power of 1) would predict a Weber function with a logarithmic slope of 2 rather than 1, whereas exponential saturation would develop an even steeper exponential Weber function. (The term "Weber function" denotes any increment threshold function, whether or not it conforms to Weber's law). Thus the precise form of the saturation can have a pronounced effect on the measured Weber function.

In fact, the Weber function for cones adheres to a slope of 1 from low levels up through the intensity level of photopigment bleaching (Geisler, 1978). This is shown in Fig. 9, in comparison with the S-shaped prediction of the logarithmic principle, which would imply a steep increase in log threshold as the proportion of bleached pigment increases, followed by a plateau when almost all of the pigment is bleached. Instead, the threshold rises directly with bleaching intensity (Weber's law), implying that sensitivity decreases inversely with the pigment remaining, rather than logarithmically with the amount of pigment bleached. The adherence of the data to Weber's law implies that any static nonlinearity that produces this sensitivity decrease must be accurately logarithmic rather that any of the other candidate compression principles. But it is important to distinguish this underlying logarithmic

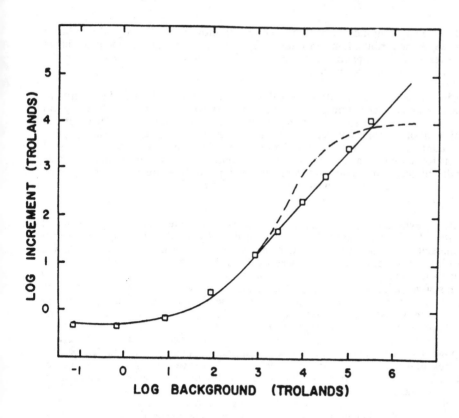

Fig. 9. A classic threshold-versus-intensity function (from Geisler, 1978), as a test of the adaptive version of the Dowling/Rushton logarithmic sensitivity relation. If the internal response were proportional to the log of pigment bleached by the background light, thresholds would increase according to the log of the bleaching effect (dashed line). In fact, the data show that Weber's law is obeyed, in opposition to the Dowling/Rushton prediction.

transducer function from the Dowling/Rushton logarithmic bleached-pigment principle for the sensitivity itself.

The accurately logarithmic nature of the underlying compression function is also borne out by the appearance of Fig. 8, in which the dark bars seem to become very narrow lines when viewed at high intensity. This is what is predicted by the logarithmic function, whereas the hyperbolic and exponential saturation functions become asymptotically linear for intensities below the 50% level of the output. This property implies that the parts of the bars below the perceived mean brightness should give a veridical appearance of the negative lobe of a sinusoidal luminance profile, i.e. should appear as rounded valleys rather than narrow lines. The demonstration of Fig. 8 therefore seems to eliminate the known forms of output saturation (hyperbolic, exponential) at any neural level as the source of the perceived nonlinearity.

Saturating Nonlinearity
If the observed saturation is produced by a static compressive nonlinearity then it should be a operative for all stimulus sizes down to the smallest visible (in contrast to dynamic nonlinearity, which may well integrate the gain across retinal regions). Moreover, reducing the stimulus size will increase the threshold response amplitude in each retinal cell within the stimulus region, increasing the response distortion caused by a static nonlinearity. Small stimuli may therefore be used to probe the form of any post-integration compressive nonlinearity controlling visual sensitivity.

An initial approach to this question was provided by studies of increment threshold with stimulation of small retinal regions for small background lights. Westheimer and McKee (1970), Buss, Hayhoe and Stromeyer (1982), and others have shown that the Weber fraction for a tiny test probe at the limit of optical resolution rises rapidly with an exponent two or three times larger than for Weber's law. A likely explanation for this result is the presence of a static nonlinearity limiting the response output. However, the data on of the spatial properties of this threshold elevation in the cone system (Westheimer and McKee, 1970) suggest that the elevation is spatially tuned to a size of 5' (and a correspondingly scaled region in peripheral retina), and does not occur for much smaller stimuli. This tuning size is much larger than the individual cone, and may correspond with the receptive field for the parasol ganglion cell, because they are the only retinal cells with receptive fields as large as 5' in the foveola (Rodieck, Binmoeller and Dineen, 1985). The implication of this spatial tuning is that the measured saturation must be at the output of the cell whose spatial tuning corresponds to the optimum masking stimulus size. It cannot therefore be the compressive nonlinearity of the cones themselves, which should be present for even the smallest resolvable stimuli.

This question of the compressive nonlinearity was addressed by Burr, Ross and Morrone (1985) in a study of the effect of spatial sampling on perception of sinusoidal modulation profiles. The paradigm was to present a static or a drifting modulation in a set of narrow bars at the limit of visual resolution and to measure the contrast sensitivity for detection of modulation as a function of the intensity in the bars. They found that contrast sensitivity did not remain constant as would be expected on the basis of Weber's law, but decreased inversely with intensity as would occur with a more compressive nonlinearity. Although the form of the decrease was not analysed in their study, its slope of -1 is compatible only with a hyperbolic compression, and excludes the alternatives of logarithmic, exponential and other static nonlinearities, as well as a simple dynamic gain control mechanism. They also showed that the spatial range of the compression was only about 0.5' in width, implying that the mechanism of the compression was extremely local in the retina. The anatomical dimensions of the foveal cone spacing varies between individuals from about 0.4' to about 0.7' from center to center (Curcio et al., 1988). The spatial range of the compression is thus well within the range of a single cone diameter in some individuals, but the possibility that it extends over two cones cannot be excluded by the data.

Thus, the most local manifestation of a static nonlinearity gives no support to the existence of a logarithmic compression. Note that, since the Burr et al. stimulus bars were only 0.5' in width, they covered only about 10% of the area of the receptive fields implied by the critical Westheimer diameter of 5'. The responses should not therefore have been subject to the output compression measured by Buss et al., since these small stimuli should have modulated the output of those receptive fields to only 10% of the maximum. The response compression measured by Burr et al. must therefore represent a nonlinearity at a prior point in the pathway, i.e. presumably before the parasol ganglion cells and perhaps in the receptors themselves. Human receptors are known to have a saturating output response, and although it has been fitted with an exponential rather than a hyperbolic compression equation (Schnapf, Kraft, Nunn and Baylor, 1988) a critical test of the difference between these functions was not performed. The important point is that the Burr et al. local nonlinearity data are compatible with what is known of human cone responses, but both nonlinearities are incompatible with the presence of a logarithmic compression in the cones that could account for Weber's law.

Dynamic gain control
A commonly recognised alternative to a static nonlinearity is that a dynamic gain adjustment

underlies Weber's law (e.g. Baylor and Hodgkin, 1974; Shapley and Enroth-Cugell, 1984). Such a mechanism could produce Weber's law by adjusting the gain to the prevailing light level (adaptation) but still allow linear operation around that level. The hallmark of dynamic gain control is that its output gain is adjusted inversely with the maintained input intensity, so that the output amplitude of an input increment is constant when the input is a fixed proportion of the maintained intensity. If the threshold output is set by noise beyond the site of gain control, this principle results in adherence to Weber's law that a constant value of dI/I is required to reach threshold.

Although the gain control mechanism can provide a complete explanation of Weber's law, it does not explain the saturating distortion of static images. A complete gain control sufficient to provide for Weber's law over a large operating range should compress the image equally in both contrast polarities, lightening the dark regions to the same extent that it darkens the light regions. This is not the form of compression seen in Fig. 8, where the dark bars actually seem to be sharper than a sinusoidal luminance profile of the same spatial period. It follows that there must be some mechanism for the observed compressive nonlinearity other than a proportional gain control.

Moreover, the properties of sensitivity regulation make it difficult to conceive how a receptor gain control mechanism can function effectively in the retina. Perfect gain control would eliminate all signal variation across the retina. Only if it has a spatial spread, a temporal delay or an incomplete gain adjustment would a dynamic gain control allow a version of the light image to pass to the higher neural stages. Thus, although gain control has been proposed in the inner retina on the basis of cat retinal physiology (Shapley and Enroth-Cugell, 1986), it does no better at accounting for the perceptual saturation than does the receptor output compression mechanism.

CONCLUSIONS

1. The limits to comprehensibility of the human brain imply that our best hope is to achieve only a partial comprehension of human visual function.

2. There is a hard cortical threshold limiting the psychophysical and electrophysiological response to changes in intensity.

3. Probability summation effects can dominate the form of psychophysical response functions.

4. There is no logarithmic nonlinearity in the human receptors.

5. There is no dynamic gain control in the receptors sufficient to explain Weber's law.

REFERENCES

Allen D., Norcia A.M. and Tyler C.W. (1986) Comparative study of electrophysiological and psychophysical measurement of the contrast sensitivity function in humans. *Amer. J. Optom. Physiol. Opt.* 63, 442-449.

Anstis S.M. and Mather G. (1985) Effects of luminance and contrast on the direction of ambiguous apparent motion. *Perception* 14, 167-180.

Baumgardt E. (1947) Sur les Mecanismes de l'excitation visuelle. *Arch. Sci. Physiol.* 1, 257-274.

Baumgardt E. and Smith S.W. (1965) Facilitation effect of a background light on target detection: A test of theories of absolute threshold. *Vision Res.* 5, 299-312.

Baylor D.A. and Hodgkin A.L. (1974) Changes in time scale and sensitivity in turtle photoreceptors. *J. Physiol.* 242, 729-758.

Baylor D.A., Nunn B.J. and Schnapf J.L. (1987) Spectral sensitivity of cones of the monkey *Macaca fascicularis*. *J. Physiol.* 390, 145-160.

Blakemore C. and Vital-Durand F. (1979) Development of the neural basis of visual acuity in monkeys. *Trans Ophthal. Soc. U.K.* 99, 363-368.

Bouman M.A. and van der Velden H.A. (1947) The two-quantum explanation of the dependence of threshold values and visual acuity on the visual angle and the time of observation. *J. Opt. Soc. Amer.* 37, 908

Burr D.C., Ross J. and Morrone M.C. (1985) Local regulation of luminance gain. *Vision Res.* 25, 717-727.

Buss C.M., Hayhoe M.M. and Stromeyer C.F. (1982) Lateral interactions in the control of visual sensitivity. *Vision Res.* 22, 693-709.

Campbell F.W. and Maffei L. (1970) Electrophysiological evidence for the existence of orientation and size detectors in the human visual system. *J. Physiol.* 207, 635-652.

Campbell F. W. and Kulikowski J.J. (1972) The visual evoked potential as a function of the contrast of a grating pattern. *J. Physiol.* 222, 345-356.

Cannon M.W. (1983) Contrast sensitivity: Psychophysical and evoked potential methods compared. *Vision Res.* 23, 87-95.

Cohn T.E. (1974) New hypothesis to explain why the increment threshold exceeds the decrement threshold. *Vision Res.* 14, 1277-1279.

Cohn T. E. (1976) Quantum fluctuation limit in foveal vision. *Vision Res.* 16, 573-579.

Curcio C.A., Sloan K.R., Packer O., Hendrickson, A.E. and Kalina R.E. (1987) Distribution of cones in human and monkey retina: Individual variability and radial asymmetry". *Science* 236, 579-582.

Dowling J.E. (1960) Chemistry of visual adaptation in the rat. *Nature* 188, 114-118.

Eddington A. (1928) **The nature of the physical world.** (AMS Press: New York).

Fechner G.T. (1860) **Elemente der Psychophysik** (Breitkopf und Hertel: Leipzig).

Geisler W.S. (1978) Adaptation, afterimages and cone saturation. *Vision Res.* 18, 279-289.

Geschieder G.A. (1976) **Psychophysics: Method and Theory** (Erlbaum: New Jersey).

Gilbert C.H. and Wiesel T.N. (1979) Morphology and intracortical projections of functionally identified neurons in cat visual cortex. *Nature* 280, 120-125.

Green D.M. and Swets J.A. (1966) **Signal detection theory and psychophysics.** (Krieger: Huntington, N.Y.)

Graham C.H. and Kemp E.H. (1938) Brightness discrimination as a function of the duration of the increment in intensity. *J. Gen Physiol.* 21, 635-650.

Gorea A. and Tyler C.W. (1986) New look at Bloch's law for contrast. *J. Opt. Soc. Amer.* A3, 52-61.

Hamer R.D., Verrillo R.T. and Zwislocki J.J. (1983) Vibrotactile masking of Pacinian and non-Pacinian channels. *J. Acoust. Soc. Amer.* 73, 1293-1303.

Hayhoe M.M., Benimoff N.I. and Hood D.C. (1987) The time-course of multiplicative and subtractive adaptation processes. *Vision Res.* 27, 1981-1996.

Hecht S., Schlaer S. and Pirenne M.H. (1942) Energy, quanta and vision. *J. Gen. Physiol.* 25, 819-840.

Helmholtz H. von (1853) Über einige Gesetze der Vertheilung electrischer Strome in korperlichen Leitern, mit Anwendung auf die thierischelektrischen Versuche. *Ann. Phys. Chem.* 29, 353-377.

Hofstadter D.R. (1979) **Gödel, Escher and Bach: an eternal golden braid.** (Basic Books: New York).

Hollins M. and Alpern M. (1973) Dark adaptation and pigment regeneration in human cones. *J. of Gen. Physiol.* 62, 430-447.

King-Smith P.E. and Carden D. (1976) Luminance and opponent-color contributions to visual detection and adaptation and to temporal and spatial integration. *J. Opt. Soc. Amer.* 66, 709-717.

Klein S.A. and Stromeyer C.F. (1980) On inhibition between spatial frequency channels: Adaptation to complex gratings. *Vision Res.* 20, 439-466.

Kulikowski J.J. (1977) Visual evoked potentials as a measure of visibility. In **Visual evoked potentials in man**, Desmedt J.E., Ed. (Clarendon Press: Oxford, pp 168-183).

Mayer M.J. and Tyler C.W. (1986) Invariance of the slope of the psychometric function with spatial summation. *J. Opt. Soc. Amer.* A3, 1166-1172.

McKee S.P. and Westheimer G. (1970) Specificity of cone mechanisms in lateral interactions. *J. Physiol.* 206, 117-128.

Moore E. F. (1956) Gedanken-experiments on sequential machines. In. C.E. Shannon & J.A.

McCarthy, Eds., **Automata Studies** (Princeton Univ. Press: Princeton, N.J.)

Nachmias J. and Kocher E.C. (1970) Visual detection and discrimination of luminance increments. *J. Opt. Soc. Amer.* 60, 382-389.

Naka K-I. and Rushton W.A.H. (1966) S-potentials from colour units in the retina of fish (Cyprinidae). *J. Physiol.* 185, 536-555.

Norcia A.M. and Tyler C.W. (1985a) Spatial frequency sweep VEP: Visual acuity in the first year of life. *Vision Res.* 25, 1399-1408.

Norcia A.M., Clarke M. and Tyler C.W. (1985) Digital filtering and robust regression techniques for estimating sensory thresholds from the evoked potential. *IEEE Eng. Med. Biol.* 4, 26-32.

Norcia A.M., Tyler C.W., Hamer R.D. and Wesemann W. (1989) Measurement of spatial contrast sensitivity in the swept contrast VEP. *Vision Research* 29, 627-637.

Parker A. and Hawken M. (1985) Capabilities of monkey cortical cells in spatial resolution tasks. *J. Opt. Soc. Amer.* A2, 1101-1114.

Pugh E.N. (1987) Vision: Physics and Retinal Physiology. In **Steven's Handbook of Experimental Psychology.**

Quick, R.F. (1974) A vector magnitude model of contrast detection. *Kybernetik* 16, 65-67.

Rice S.O. (1945) Mathematical analysis of random noise. *Bell Syst. Tech. J.* 24, 46-156.

Robson J.G. and Graham N. (1981) Probability summation and regional variation in contrast sensitivity across the visual field. *Vision Res.* 21, 409-418.

Rodieck R.W., Binmoeller K.F. and Dineen J. (1985) Parasol and midget ganglion cells of the human retina. *J. Comp. Neurol.* 233, 115-132.

Rushton W.A.H. (1961) Dark adaptation and the regeneration of rhodopsin. *J. Physiol.* 156, 166-178.

Rushton W.A.H. and Henry G.H. (1968) Bleaching and regeneration of cone pigments in man. *Vision Res.* 8, 617-631.

Schnapf J.L., Kraft T.W., Nunn B.J. and Baylor D.A. (1988) Spectral sensitivity of primate photoreceptors. *Visual Neurosci.* 1, 255-261.

Shapley R.M. and Enroth-Cugell C. (1984) Visual adaptation and retinal gain controls. In Osborne N. and Chader G., Eds. **Progress in Retinal Research, Vol. 3** (Pergamon: Oxford).

Shepherd G.M. (1988) **Neurobiology**, 2nd Edition (Oxford University Press: New York, pp. 145-176).

Stevens S.S. and Stevens J.C. (1960) Brightness function: Parametric effects of adaptation and contrast. *J. Opt. Soc. Amer.* 50, 1139-1146.

Stromeyer C.F. III and Klein S.A. (1974) Spatial frequency channels in human vision as asymmetric (edge) mechanisms. *Vision Res.* 14:1409-1420.

Tanner W.P. Jr and Swets J.A. (1954) A decision-making theory of visual detection. *Psych. Rev.* 61, 401-409.

Teller D.Y. (1984) Linking propositions. *Vision Res.* 24, 1233-1246.

Tolhurst D.J., Movshon J.A. and Dean A.F. (1983) The statistical reliability of signals in single neurons in cat and monkey striate cortex. *Vision Res.* 23:775-785.

Tufts D.W. and Kumaresan R. (1982) Estimation of frequencies of multiple sinusoids: making linear prediction perform like maximum likelihood. *Proc. IEEE* 70, 975-989.

Tulanay-Keesey U. and Jones R.M. (1976) The effect of micromovements of the eye and exposure duration on contrast sensitivity. *Vision Res.* 16, 481-488.

Tyler C.W and Apkarian P.A. (1985) Effects of contrast, orientation and binocularity on the pattern evoked potential. *Vision Res.* 25, 755-766.

Uttal W.R. (1990) On some two-way barriers between models and mechanisms. *Percept. Psychophys.* 48, 188-203.

Watson A.B. Derivation of the impulse response: Comments on the method of Roufs and Blommeart. *Vision Res.* 21, 1335-1337.

Weber E.H. (1834) **De pulsu, resorptione, auditu et tactu. Annotationes anatomicae et physiologicae.** (Koehler:Leipzig).

Wyszecki G. and Stiles W.S. (1982) **Color Science**, 2nd Edition (Wiley: New York).

Hidden assumptions in seeing shape from shading and apparent motion

Stuart Anstis

Department of Psychology, York University,
4700 Keele Street, North York, Ontario M3J 1P3, Canada

Introduction

How much of our perception of the world is driven by the immediate local stimulus and how much by stored knowledge and expectations? Hypotheses and expectations can usefully constrain our interpretations of sensory data by capitalising on our knowledge of the real world (e.g. Gregory 1970: Marr 1982: Ramachandran 1990). High-order properties such as depth, transparency and familiarity may help us make an intelligent "top-down" perceptual choice between alternatives in a perceptually ambiguous situation. But, as we shall see, many perceptions can be explained more parsimoniously as "bottom-up" processes driven simply by the luminance distribution in the stimulus. We shall discuss the role of hidden perceptual assumptions in two contentious areas: shape from shading, and ambiguous apparent motion.

Pomerantz and Kubovy (1981) have distinguished between "two different forms of the Pragnanz principle, which can be called the *simplicity principle* and the *likelihood principle*. The first, which is linked closely to the classical Gestalt conception, holds that we organize our percepts so as to minimize their complexity. In information-processing terms, this principle would imply processing small elements of a scene first, then conjoining them into larger clusters, which are then combined into even larger groups until the process reaches a stopping point. Such procedures are known as "bottom-up" procedures.... The second principle, that of likelihood, is definitely not part of the Gestalt heritage but instead may be attributed to Helmholtz. It holds that we organize our percepts so as to perceive the most likely distal stimulus that could have given rise to them. In more modern terms, the likelihood principle would operate via a learned "top-down" process (although evolution could have provided us with a bottom-up process to serve this function)".

Shape from shading

What assumptions are made by the visual system in deriving 3-D shape from shading? It is a fact of physics and geometry that the light reflected from a Lambertian surface depends upon the angle of incidence, so that a curved surface is shaded, and it is a fact of psychology that we can use this shading information to recover 3-D structure. We shall discuss perceptual assumptions that familiar objects such as faces are convex; that light comes from above; that the illuminated side of the object is nearest to the light source; and that light is bright.

1.   *Faces are convex.* When we look into a hollow mask it often looks convex and we are simply unable to see it as hollow. This resistance to reversal of depth has traditionally been attributed to familiarity with the shape of objects and the presence of monocular depth cues. Thus, Gregory (1970) attributes it to probability biasing in favour of the likely against the unlikely. He points to the two opposed principles of processing upwards and downwards, the first generating hypotheses which may

be highly unlikely and even clearly impossible, the second offering checks 'downwards' from stored knowledge, and filling gaps which may be fictional and false. But van den Enden and Spekreijse (1989) offer a non-cognitive explanation. A stereoscopic picture of a face offers two kinds of depth cues; binocular disparity, and monocular texture disparity -- gradients of texture, which are geometrically more compressed near the left and right edges of a convex face than they are for a hollow face. They claimed that the real reason that a pseudoscopically viewed face refuses to look concave is that each monocular view contains texture information that provides a strong cue that the face is actually convex. They viewed a convex face through a pseudoscope, and projected 'neutral' texture, which gave no monocular cues, on to the face from a projector near the observer's eyes (Georgeson 1979) Result: the face was correctly perceived as concave. Deutsch and Ramachandran (1990) and Peli (1990), however, both point out that this projected texture adds rich binocular disparity cues, which suffice to explain van den Enden and Spekreijse's results. Furthermore, the texture account does not explain why an actual hollow face, viewed with both eyes, looks convex. However, Deutsch et al. and Peli concur that the perceived depth depends upon the cues present in the stimulus, and no cognitive factors need be involved.

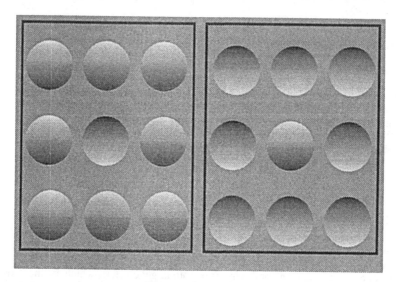

Figure 1.

        a   b

2.   *Light comes from above.* Fig. 1 b is simply Fig. 1 a turned upside down. As a result the picture appears to be reversed in depth, with the bumps becoming hollows and vice versa. In the absence of explicit lighting cues the visual system assumes that the light comes from the top of the picture. The assumption that laight comes from above rather than any other direction does not simplify our perceptions in any way but it does match them better to our sunlit physical world in which the light really is more likely to come from above, so we are applying a perceptual constraint derived from our knowledge of the physical world. This is a prime example of the likelihood principle.

Ian Howard (personal communication) has noted that if you bend over and look between your legs at Fig. 1, the light is assumed to come from above in retinal, not gravitational coordinates, as if the light source were assumed to be stuck to one's forehead. This is not a highly intelligent perceptual assumption.

There is some evidence that the constraint is learned by experience. Hess (1972) reared chicks in special cages that were lit from below through the floor, so that the grains they ate were illuminated from below. He then exposed them to a pair of photographs of grain. In one photograph the illumination came from above: in the other, from below. The chicks pecked at the grains illuminated from below. Once learned in early life, however, such a constraint could arguably become hard-wired.

3. *The illuminated side of object is nearest the light source.* This sounds so trivially obvious as to be barely worth saying. For myself I only became aware that I make this assumption when it appeared to be violated by Brian Rogers' New Moon Illusion. This phenomenon was first pointed out to me (on Moon Drive in Toronto) by Brian Rogers. (It is called the New Moon Illusion because it was described more recently than the Old Moon Illusion in which the moon looks larger when it is near the horizon. It is seen best when the moon is about half full, and it has nothing to do with the new moon, else it would be called the New Moon Illusion.)

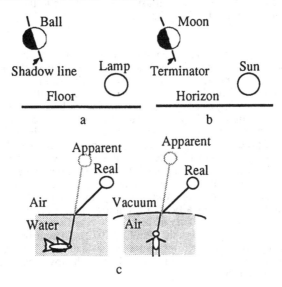

Figure 2 a, Obviously this picture is wrong. The shadow line on the ball is tilted to about 11 o'clock, but it should be tilted to 1 o'clock since the lamp is lower than the ball.
b, Brian Rogers' New Moon Illusion. The terminator line on the half-full moon looks tilted to 11 o'clock even though the sun looks lower in the sky than the moon.
c, The new moon illusion is not caused by atmospheric refraction, which makes the sun look *higher* in the sky, not lower, than it really is.

In Fig. 2a the lamp is below the sphere, yet the edge of the shading on the sphere is inclined to the left of vertical at about '11 o'clock' indicating that the illuminating lamp must be higher than the sphere. Anyone can see that there is something wrong and the picture is incorrectly drawn. Fig. 2b is similar but the sphere has been replaced by the moon and the lamp by the sun. The shadow line on the moon, known to astronomers as the terminator, is tilted at about 11 o'clock, so one predicts that the sun ought to be higher than the moon. On looking over one's shoulder at the sun, however, one finds that the sun is lower in the sky than the moon. This appears to contravene the laws of physics, and it certainly contravenes one's perceptual assumptions. I recently attended an evening party when the sun was setting but the moon was already high in the sky, and I asked several fellow scientists to explain this (with hypotheses based upon explicit assumptions about the physical world, as opposed to any implicit perceptual assumptions underlying the illusion itself.)

A mathematician, perhaps inspired by Ptolemy, suggested that refraction makes the sun look lower than it really is. This assumption is false because atmospheric refraction actually bends the sun's rays so as to make the sun look *higher* than it really is, and keeps the sun visible for a few minutes after it has sunk below the horizon (Fig. 2c). So refraction cannot explain the new moon illusion since it predicts an effect in the wrong direction.

An astronomer suggested that the orbits of the sun, earth and moon lie in different planes. This assumption is true, but irrelevant. The new moon illusion would still be visible if the scene were viewed in a single brief flash, and motions of the heavenly bodies make no difference.

A distinguished physicist suggested, admittedly toward the end of the party, that the sun, moon and observer do not lie in the same plane. This assumption is false, since as his wife (the mathematician) reminded him, any three objects always lie in the same plane.

Michael Swanston (personal communication) has convincingly argued that the sun and moon look about equidistant from the observer, although the sun is really 400 times as far away as the moon. This misperception of the sun's position, based on a gross underestimate of its distance, leads to the illusion. Probably the old nor the new moon illusions are both due to the misperception of distances from the observer -- of the moon itself for the old moon illusion and of the sun for the new moon illusion.

4.   *Light is bright*. Again this sounds too obvious to be worth saying. What else could light be but bright? Actually it is counterfactual, but not meaningless, to suggest that light could be dark. One could simulate a lamp with a spray can spraying out white paint on to black objects. When the paint dried it would give a fossilised rendering of illuminated objects. Black paint sprayed on to white objects does not give a rendering of any physical light source because there is no "black light" in nature. Of course, in principle one could construct an *internal* world of black light and white shadows, not by changing the external physical universe but by re-wiring the visual system so that highly reflective objects produced the sensation that we now label "black", and unlit space would produce the sensation that we now label "white". But this is simply to restate the old speculation that although we agree to call a ripe cherry "red" and a leaf "green", it may be that my "green" sensation would look like your "red" sensation if they were somehow both fed into the same brain. More to the point, we can interchange black and white *experimentally* by viewing the world through a video link which converts the picture electronically into a photographic negative. This makes it hard to distinguish bumps from hollows and impossible to recognise faces of celebrities, especially if the portraits are high-contrast "lith" photographs (Phillips 1972). The brightness reversal disrupts the shape from shading which permits us to recognise facial features. Patrick Cavanagh was able to establish, after two years' work, that shadows look like shadows only if they are darker than illuminated regions (Cavanagh and Leclerc 1989).

Our ability to decode shadows may have a learned component (Hess 1972). Although naive observers cannot recognise famous faces in negative, there was a time when people who worked in television newsrooms routinely acquired this unusual skill. In the early days of television, newsreels were shot on cine film. The negative film stock was developed and a positive print made for broadcasting. It was soon realised that valuable time could be saved by broadcasting the negative film directly and reversing the brightnesses electronically. The film editors were called upon to edit the films while these were still in negative, and they rapidly learned to recognise world leaders even in negative. Suneeti Kaushal and I are planning to investigate long-term adaptation to a negative visual world. Subjects will wear a helmet-mounted stereo display in which two small TV cameras on the front of the helmet feed into two miniature TV screens which the subject views, one with each eye, through suitable magnifying lenses. The TV pictures will be electronically reversed in brightness.

Apparent motion

As a microcosm of perceptual processes we used ambiguous apparent motion stimuli in which two shapes abruptly exchange positions. One can perceive the two objects as moving past each other in opposite directions, or one object can seem to move while the other does not, or each object can change shape without shifting position. Which of these percepts occurs is strongly influenced by stimulus and observer variables, and we shall discuss the role of "bottom-up" low-level processes and "top-down" assumptions, expectations, or learned constraints about the physical world.

Braddick (1974) proposed a distinction between short-range and long-range processes in motion perception. The short-range process is thought to occur early in the visual system and has been identified with directionally selective neurons in the visual cortex that operate passively and in parallel over the whole visual field. It operates over short distances (<15 min in the fovea) and brief durations (<100 ms), and adaptation of the short-range system underlies the motion aftereffect. Its inputs come exclusively from stimuli that are defined by luminance. The long-range process, on the other hand, is thought to occur at later stage of processing with properties more resembling cognitive or interpretative processes than the responses of single neurons. It operates over longer distances and times than the short-range process, and can accept non-Fourier stimuli as inputs, for example patches that are defined by texture, cyclopean depth, or short-range motion (Cavanagh 1989). The notion of short and long range motion processes has been reviewed by Braddick (1980) and Anstis (1980). Cavanagh and Mather (1989) have published a highly critical review of these concepts, but they certainly still have heuristic value. Motion perception has been reviewed by Nakayama (1985), Borst and Egelhaaf (1989) and Sekuler *et al* (1990), and Newsome *et al* (1989) have directly compared motion perception by an alert monkey with the performance of its neural motion detectors.

It is plausible to equate the short-range process with bottom-up processes and the long-range with top-down. Marr and Ullman (1981) proposed a computational model of the short-range process and suggested that there are two types of computational tasks associated with motion perception, tasks of separation and tasks of integration. Tasks of separation can be solved in principle by using only instantaneous measurements such as position and its time derivatives in the image. This includes such tasks as motion segregation, and could probably be handled by short-range processes. Tasks of integration cannot be solved using only instantaneous measurements but require the combination of information over time. This includes such tasks as the recovery of 3-D structure from motion (Ullman 1979) would probably require long-range processes. Ullman (1979) proposed a computational model of the long-range process, which involves computing similarities across successive time frames to match up correspondence tokens, and then does a cost-benefit analysis which essentially minimises the total path lengths of all motions in the visual field.

A parsimonious motion system would operate only on the stimulus luminance. We shall discuss luminance, illusory brightness, and texture segregation. A more intelligent motion system would first analyse a scene for depth, using such cues as perspective, shape from shading, and occlusion, and then use 3-D objects as correspondence tokens. Examples are given below.

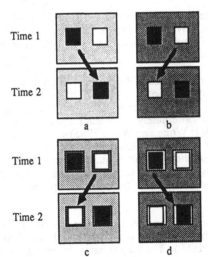

Figure 3 Apparent motion when a black and white square suddenly change places (Anstis and Mather, 1985).
a, On a light surround the black square appears to move.
b, On a dark surround the white square appears to move. Thus apparent motion is attributed to the square with the higher luminance contrast.
c, When the surround is light but each square is surrounded with a dark picture frame, the white square usually appears to move.
d, When the surround is dark but each square is surrounded with a light picture frame, the black square usually appears to move. Thus luminance contrast is assessed by local mechanisms.

High-contrast objects appear to move

Luminance. Fig. 3a,b shows a black square and a white square side by side which instantaneously change places, so that the black square suddenly becomes white and at the same instant the white square becomes black. What will one perceive? Does one see two squares flickering in place, or does the white square jump one way, or does the black square jump the other way? George Mather and I (1985) found that the answer depends upon the luminance of the background. On a light surround the black square is seen as jumping (Fig. 3a), and on a dark surround the white square is seen as jumping (Fig. 3b). The square that differed most from the surround, in other words the square with the higher contrast, was perceived as jumping.

The observers adjusted the luminance of the background until they saw either the two squares move equally frequently or both squares move at once. This mid-grey indifference point was equivalent to a paint mixed from equal parts of black and white paint, so it was the arithmetic, not the geometric mean of black and white. This shows that the motion system operates linearly on luminances, not upon log luminance. More recently, however, Shioiri, Cavanagh and Favreau (1989) have found a slight non-linearity - less marked than a logarithmic function -- which is consistent with the compresson of retinal cone responses reported by Boynton and Whitten (1970). Shioiri *et al* (1989) have also used this technique to study the linearity of intensity coding along the three cardinal axes of color space. Two bars of two colors (C1, C2) were alternated to produce apparent motion. The background color was a mixture of C1 and C2. If the background color was closer to C1, only the C2 bar appeared to jump. If the color different in a linear cone activation space controls this apparent motion, the setting should be midway between C1 and C2, no matter which pair of colors is chosen for C1 and C2. None of the cardinal axes (achromatic, R-G cone, or B axes) showed a linear response, but the non-linearity was less extreme than a logarithmic function.

The surround can be partitioned into a large background area and a small area like a picture frame adjacent to the test squares (Fig. 3c, d). When these areas are pitted against each other by setting them to different luminances, the picture frame luminance has a much stronger influence than the remote surround on the perceived direction of motion.

Illusory brightness. Fig. 4a shows two pieces of black/grey checkerboard, each moving back and forth through the diameter of one checkerboard square. The two moving checkerboards do not interact. Now a surround is added -- a large stationary black-white checkerboard, positioned so that each small grey checkerboard replaces alternately some black squares and some white squares of the surround (Fig. 4b). White (1979, 1981: White and White 1985) showed that grey squares that replace white squares in a checkerboard look appreciably darker than when they replace black squares, as Fig. 4b shows. (White used gratings, not checkerboards, but the principle is the same). Now the two regions interact strongly, and a single region is seen as jumping back and forth through a dozen square-diameters. The illusory induced brightness controls the apparent motion just as effectively as the physical luminance did in Fig. 3.

White's phenomenon has attracted considerable interest because it cannot readily be classified as either simultaneous contrast or as brightness assimilation (Hamada, 1984: Foley and McCourt, 1985: Moulden and Kingdom, 1989: Zaidi, 1989). The grey squares that replace white squares are bordered by black squares, and from simultaneous contrast grounds would be predicted to appear lighter than the grey squares that replace black squares, the exact opposite of what is found. Moulden and Kingdom (1989) attribute White's effect to two mechanisms, one a local concentric spatial filter operating at the corners of the grey test regions, the other a spatially extensive filter operating along the long bars of the grating (or presumably along diagonal rows of checkerboard squares).

To perceive a single very large jump instead of two small jumps seems to violate principles of simplicity and likelihood. We believe this is a bottom-up process that accepts only low-level luminance cues. First, a comparison of the left hand regions of Fig. 4 at Times 1 and 2 shows that the edges of the checkerboard squares reverse their luminance polarities in the transition from Time 1 to Time 2. This will militate strongly against seeing the small local motions (Anstis 1970: Anstis and Rogers 1975: Anstis and Mather 1985). Second, adding the background checkerboard changes the

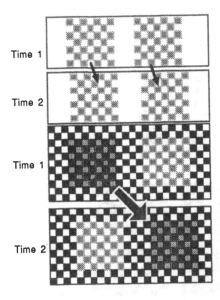

Figure 4 Illusory lightness can mediate crossover motion.
a, two grey checkerboards each move through the diameter of one checkerboard square. Each checkerboard appears to move locally and independently (grey arrows).
b, a stationary black/white checkerboard surround is added. In the left-hand region at Time 1, the white checkerboard squares have been replaced by grey squares. This region looks darker than the right-hand region in which the black checkerboard squares have been replaced by grey squares (White, 1979). At Time 2 the two regions interchange. Result: the higher contrast region, here the apparently dark one, appears to jump across to the right (black arrow).

whole luminance distribution, especially at low spatial frequencies; spatially integrating Fig. 4 by blurring or squinting at it will reveal the left hand region to have an appreciably lower space-averaged luminance than of the right-hand region, so a large-scale visual filter that integrated the stimulus neurally would pick up a Fourier energy motion signal. Of such scale effects the Gestalt psychologists who pioneered the simplicity principle knew nothing.

Thus, the responses to the real and illusory brightnesses in the stimuli of Figs. 3 and 4 can be explained by a low-level process sensitive to motion energy. The two squares in Fig. 3a can be compared to two superimposed gratings of the same spatial frequency moving in opposite directions. If the two gratings are of equal contrast the outcome is a stationary counterphase flickering grating, but if one grating, say the one that moves to the right, is of higher contrast then the combined stimulus contains more motion energy to the right and is perceived as moving to the right.

Texture segregation. We can generalize these results to squares defined by texture, not by luminance, where there is no Fourier energy moving predominantly in one direction, yet the direction of perceived motion is controlled by the perceptual *salience* of the textured squares. Fig. 5 shows two textured squares consisting respectively of coarse and fine random dots, on a surround of very fine random dots (Fig. 5a) and again on a surround of very coarse dots (Fig. 5b). The dot diameters in the fine surround, the two squares, and the coarse surround, are in the ratios 1: 2: 4: 8. Although all four textures have the same space-averaged luminance, it is easy to segregate the two squares perceptually from the surrounds. However, the finer textured square stands out as more salient against the coarse surround, and the coarser textured square stands out better against the fine surround. This texture salience controls the apparent motion, although less compellingly than luminance did in Fig. 3. The two squares suddenly exchanged places and observers were asked to report whether they saw the fine texture jumping to the left or the coarse texture jumping to the right. We found that the answer depended upon the texture of the surround. When the background texture was coarser than the coarse square, the finely textured square was perceived in apparent motion, and when the background texture was finer than the fine square, the coarsely textured square was perceived in apparent motion. The square that differed most from the surround was seen in motion.

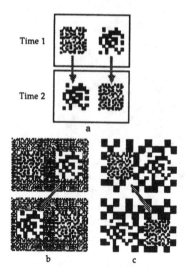

Time 1

Time 2

a

b   c

Figure 5 Perceptual salience, based on texture discrimination without luminance cues, can mediate crossover motion.
a, two patches of random-dot texture, one fine and one coarse, exchange places. Apparent motion is ambiguous.
b, when the stimulus in **a** is superimposed on a very fine textured background, the coarse texture is more salient and is seen in apparent motion (arrow).
c, when the stimulus in **a** is superimposed on a very coarse textured background, the fine texture is more salient and is seen in apparent motion (arrow).

To summarise, our crossover effects appear to operate at a fairly low level, after the level of luminance discrimination or of texture segregation. They require no simplicity or likelihood assumptions by the visual system. Motion perception can be based upon stimulus luminance (Figs 3, 4), or at a higher visual level upon texture-based non-Fourier stimuli (Fig. 5), which admittedly provide somewhat less compelling impressions of movement (Chubb and Sperling 1988). We conclude that stationary regions can first be defined by any visual cue such as luminance, depth, texture and so on, and then displacements of such regions can then give rise to a long-range motion percept. Cavanagh (1989) reviews many examples of such inter-attribute apparent motion.

Near objects appear to move

Perspective. Fig. 6 shows a perspective sketch of a protruding square slab next to a square recess. When the slab and the recess abruptly exchanged places, subjects reported that the slab, not the hole, moved. When the color of the bottom of the recess was made the same as the top of the slab, the display now looked like two buttons being pushed in alternation and subjects now reported motion in depth, along the line of sight and at right angles to the previous motion. (The stimuli used were these actual sketches, not real 3-D objects). These are examples of intelligent processes in long-range motion perception.

Shape from shading. Fig. 7 shows the familiar Ternus (1926) configuration for apparent motion. Three spots jump back and forth between positions a,b,c and time 1 and positions b', c', d' at time 2. The percept depends upon the timing. When there is no interstimulus interval (ISI) the two central spots are always visible and look stationary, and subjects report 'element motion' in which one spot jumps from end to end (Fig. 7a). When there is an ISI the central spots flash on and off and subjects report 'group motion' in which three spots jump back and forth together (Fig. 7b)(Pantle and Picciano 1976). Group motion is also seen when the stimuli are presented dichoptically. Braddick and Adlard

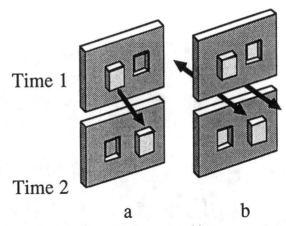

Time 1

Time 2

a   b

Figure 6 Apparent motion is attributed to the nearer object.
a, When these two perspective sketches are superimposed and exposed in alternation, the protruding slab is seen in motion (arrow) between two stationary holes, moving in the plane of the wall.
b, When the color of the bottom of the recess is made the same as the top of the slab, the display looks like two buttons being pushed in alternation and the motion is now in depth, along the line of sight (arrows).

(1978) found that when they restricted the dichoptic motion to the two central spots, group motion was often reported, but when they restricted the dichoptic motion to the two outer spots subjects nearly always reported element motion. Thus the effect of dichoptic presentation in reducing element movement depended, oddly enough, not on dichoptic presentation of the element that apparently moved, but of the elements that appeared stationary. They concluded that element motion is mediated by a low level short-range process (Braddick 1974) --not by sensing that the end element is jumping back and forth, but by sensing that the two central spots are stationary. They attributed group motion to a higher level, more interpretative long-range motion process. For a differing view see Cavanagh and Mather (1989).

Now let us give the disks some apparent depth by means of shape from shading (see Ramachandran 1988). The flat grey disks of Fig. 7 a, b are replaced in Fig. 7 c, d by shaded disks that look like saucers, and the empty spaces by shaded disks that look like bumps. When there is no ISI, instead of seeing element motion most observers report that the disks at the left and right ends of the display are flipping up and down without changing their position, driven by the "dumb" local luminance cues. However, when there is an ISI, instead of seeing a group of three saucers moving to the right, most observers report a single bump jumping to the left between the endmost of four empty saucers. So the long range motion is controlled by "smart" cues of perceived depth. It seems that the motion tokens for long range motion, but not for short range motion, are derived after shape from shading has been computed.

Covering and occlusion. Sigman and Rock (1974) explored the role of occlusion in apparent motion. They propose that the perception of apparent motion can be the outcome of an intelligent problem-solving process. They exposed two stationary spots a and b in alternation, by moving an opaque rectangle back and forth, alternately covering and uncovering the two spots at a tempo that ought to give good apparent motion. As far as other theories of apparent motion are concerned, there is no reason why these conditions should not produce an impression of a and b moving. But from the standpoint of problem-solving theory, the moving rectangle provided an explicable basis for the appearance and disappearance of a and b, namely that they are there all the the time but are undergoing covering and uncovering. This is what the observers reported; they rarely reported apparent motion. However, if the rectangles were drawn so as to look transparent they did not look capable of covering anything, so it was no longer a fitting or intelligent colution to perceive a and b as two permanently present dots that were simply undergoing covering and uncovering. In this condition, subjects again reported apparent motion (Rock, 1983).

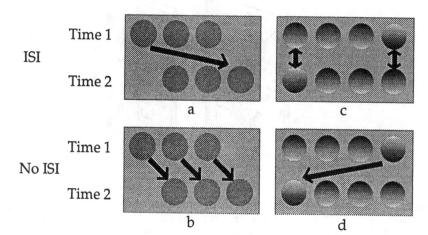

Figure 7 Ternus display.
a, with no ISI, element motion was seen, with the end disk jumping back and forth.
b, with an ISI, three disks jumped back and forth as a group.
c, with no ISI the shaded disks at the end flip up and down, driven by local luminance cues.
d, with an ISI the bump-like shaded disk jumps back and forth across three apparently stationary saucer-like disks (See Ramachandran 1988).

Figure 8 shows the effects of pitting two determinants of apparent motion against each other: luminance contrast, and the depth cue of covering. The visual system is confronted with the choice of either a low-level decision that the higher contrast object is moving, or else an intelligent decision that the nearer object is moving. In Fig. 8, two squares jumped back and forth, one in front of a barrier of wavy lines and the other behind it. The rear square was always held at a mid grey on a black background. The front square was either light grey, dark grey or mid grey.

When the front square was light grey, it was both seen in front of the other square because of the covering cue, and was also higher in contrast because it was lighter against the black surround. Result: not surprisingly, it seemed to be the square in motion when the two pictures were alternated.

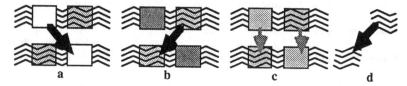

Figure 8 In these ambiguous motion displays luminance contrast was pitted against the depth cue of occlusion. The surround was black so the lighter squares had higher contrast.
a, Cues in harmony. The high-contrast light square was in front, covering the wavy lines, and it appeared to move (arrow).
b, Cues in conflict. Contrast won out over depth: the lighter, high-contrast square in back appeared to move (arrow), and the darker, low-contrast square in front did not.
c, Contrast cue minimal or absent. Squares had different hues but the same luminance throughout. Squares appeared stationary (grey arrows), but the wavy lines appeared to detach from the squares and jump back and forth independently, as diagrammed in **d.**

When the front square was dark grey, occlusion cues still identified it as lying in front of the other square but now it was lower in contrast. Result: contrast won out over depth, and the more distant, higher contrast mid grey square was the one that appeared to jump back and forth while the nearer, dark grey square appeared stationary. Thus the low-level luminance cue won out over the potentially "intelligent" depth interpretation.

It might be that contrast overcame depth not because contrast was strong but because depth was weak, being supported only by the cue of occlusion. Accordingly we added in stereo (not illustrated). The squares that covered the wavy lines in Fig. 8a, b (on the left at Time 1 and on the right at Time 2) were now also put in front of the wavy lines in stereo depth, and the other squares were put behind the wavy lines in stereo depth. This changed the motion percept. Now the squares no longer jumped sideways in the frontoparallel plane. Instead, each square jumped back and forth in depth along the line of sight, as in Fig. 6b.

When the two squares were set to different hues of the same luminance (or else both were set to the same mid-grey) and the two pictures were alternated, a most surprising result was observed (Fig. 8c, d): The two squares now flickered in colour at each alternation but did *not* appear to move, and the patch of wavy lines that lay in front of one square broke free of the remaining wavy lines to which it was hitherto firmly attached, and jumped across to lie in front of the other square.

This is reminiscent of the finding by Ramachandran and Anstis (1986) that figure-ground organisation can constrain apparent motion. An outline square jumped back and forth in front of a stationary random-dot surround (Fig. 9). The parts of the background that were covered and uncovered by the square looked stationary. Next, the outline was deleted so that only an empty region moved back and forth. The percept now changed radically; the previously moving square region now looked liked like two stationary empty holes, with a group of background dots jumping back and forth between them, even though the dots in the two background regions were uncorrelated. We conclude that the visual system normally suppresses spurious motion signals that are generated by covered and uncovered regions.

What parts of the motion system can be adapted?

As we move through the world the retinal image of objects ahead of us expands and the image of objects behind us contracts. We use this *optic flow* to guide locomotion (Cutting 1986: Koenderink 1986). We can use optic flow to assess the time to collision: there is evidence that long jumpers use it

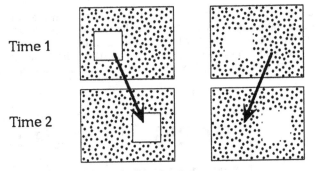

Time 1

Time 2

Figure 9 Occlusion and apparent motion (Ramachandran and Anstis 1986).
a, The outline square was perceived as jumping back and forth (arrow) against a stationary random-dot background.
b, when the outline was removed a group of dots, previously a stationary part of the background, appeared to jump back and forth (arrow).

to adjust their stride just before a jump (Lee 1980), and diving sea birds know from visual cues when to fold their wings just in time before they hit the water (Lee and Reddish 1981). Regan, Beverley and Cynader (1979) have discovered specialised visual channels that respond to size change. When confronted with an expanding square, local motion detectors first sense the outward motions of the opposite sides of the square, and then the size change detectors respond to the velocity difference, derived from the output of these motion detectors. The same authors also discovered visual channels that respond to motion in stereo depth.

Brian Rogers and I (in press) wondered whether there are also size change detectors that respond to change of spatial frequency content over time. The spatial frequency content of an approaching target zooms down the spatial-frequency spectrum, and this could be detected very rapidly by a hypothetical mechanism -- a type of "neural speedometer" --that received inputs from spatial frequency channels and detected gradual increases or decreases in spatial frequency. Such a frequency sweep detector could be very fast and computationally cheap because it would need merely to detect changes in the relative activity levels in different spatial frequency bandpass filters. Unlike local motion detectors, it would not need to discriminate spatial phase.

We tested this idea with two types of zooming random-dot texture display. One texture simply expanded repetitively. It looked like a movie made by aiming a movie camera at the centre of a sheet of stationary sandpaper and steadily zooming the lens in. The other texture also expanded repetitively, but whereas the texture grains grew gradually larger from frame to frame as before, the grains were replaced on every frame by a fresh set, with no frame-to-frame correlation between the grain positions. It looked like a movie made by shooting a single frame of a sheet of stationary sandpaper, then zooming the lens in slightly and using a *different* sheet of sandpaper on each successive frame. Stated differently, the stimulus consisted of twinkling dynamic random visual noise in which the mean grain size gradually expanded. So this stimulus provided expansion without correspondence between successive pattern elements.

Result: the coherently expanding texture gave a strong impression of motion, whereas the phase-jittered texture did not. Furthermore, following a 60-s adaptation period, the coherent stimulus gave a strong motion aftereffect but the phase-jittered stimulus gave none at all. We repeated the experiment with an expanding grating which either was, or was not, jittered randomly in phase. Results were the same as for the random noise: the simple expanding grating gave a strong percept of expansion, followed by a strong motion aftereffect, but the phase-jittered grating did not. The two stimuli had the same amplitude spectra over time, but differed in their phase spectra. In summary, we found no evidence for a rapidly acting, phase-blind channel that responds to temporal changes in frequency. Instead, our results could be explained by local motion detectors. Note that our results cast no doubt upon those of Regan et al. (1979) since our stimuli and conditions were very different.

Spatial phase spectrum is important for object recognition. One can dissociate the amplitude and phase spectra of a stationary stimulus, and once the spectra are dissociated, one can recombine the amplitude spectrum of one stimulus with the phase spectrum of another stimulus, thereby forming a new pattern. Piotrowski and Campbell (1982) created such hybrid patterns, combining the amplitude spectrum of a military tank with the phase spectrum of a face. The resulting pattern looked like a face

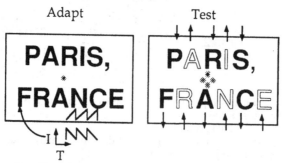

Figure 10 Motion aftereffects from motionless stimuli.
a, subjects adapted to dimming letters on a brightening surround, while fixating on the central spot.
b, subjects shifted their eyes to one of four fixation test spots, say the top one. Result: a motion aftereffect in which the light letters appeared to drift upwards, the dark letters downwards (arrows).

and not a tank. Piotrowski and Campbell also found that an image with the amplitudes of a face and the phases of a tank looked like a tank. They concluded that the phase spectrum of a stimulus was the primary determinant of its perceived identity. Our results suggest that spatial phase is just as important for perceiving the motion of a stimulus as it is for recognising objects.

We confirmed the low-level, luminance-based nature of the motion aftereffect by producing motion aftereffects from motionless stimuli (Anstis, in press). First, we generated a ramp aftereffect (Anstis 1967) by exposing our subjects to the display shown in Fig. 10a. The letters in the words PARIS, FRANCE repetitively dimmed over time, and the uniform grey background repetitively brightened over time. The luminance of the two regions were modulated by 1 Hz sawteeth in opposite directions. Following 30 s of adaptation the modulation was stopped, and subjects reported a ramp aftereffect in which the letters appeared to brighten gradually and the surround appeared to dim. Fixation was maintained throughout on a central fixation spot. Next, subjects re-adapted to the same stimulus for a further 30 s, but now the test stimulus was as shown in Fig. 10b. The surround was a fixed mid-grey; half the letters (P,R,S: F,A,C) were slightly darker than the surround and the other letters (A,I: R,N,E) were slightly lighter. The central fixation spot was replaced by four test fixation spots arranged in a diamond. When subjects shifted their eyes to the uppermost test spot, they perceived the dark letters as apparently drifting upwards in a motion afterefect, and the light letters as drifting downwards (vertical arrows in Fig. 10b). If they looked at the lower fixation spot the aftereffects were in the opposite direction. If they looked at the left (or right) fixation spot the light letters appeared to be drifting to the left and the dark letters the right (or vice versa). Thus, following adaptation to directionless dimming and brightening stimuli, subjects could change the direction of the motion aftereffect within a single test period simply by re-directing their gaze. These aftereffects were not artifacts from eye movements, but arose from superimposing a region of ramp aftereffect upon a slightly displaced luminance contour in the test field. We conclude that the output from luminance-change detectors provide an input into motion detectors, and that no hidden assumptions of any kind play a role in the motion aftereffects. Here we are examining the lowest possible level of motion detection.

Acknowledgments This research was supported by Grant A0260 from the Natural Science and Engineering Research Council of Canada (NSERC) and by research contract 9-97109 from Canadian National Defence. I thank my co-workers Patrick Cavanagh, Brian Rogers, and V. S. Ramachandran, some of whose work is described here; Suneeti Kaushal for her assistance; and Ian Howard for helpful discussions.

References

Anstis, S. M. (1967) Visual adaptation to gradual change of luminance *Science* **155** 710-712
Anstis, S. M. (1970) Phi movement as a subtraction process *Vision Res.* **10** 1411-1430
Anstis, S. M. (1980) The perception of apparent movement. *Phil. Trans. R. Soc. Lond. B* **290** 153-168.
Anstis, S. M. (in press) Motion aftereffects from motionless stimuli. *Perception*
Anstis, S. M. and Mather, G. (1985) Effects of luminance and contrast on the direction of ambiguous apparent motion *Perception* **14** 167-180
Anstis, S. M. and Rogers, B.J. (1975) Illusory reversal of depth and movement during changes of contrast *Vision Res.* **15** 957-961
Anstis, S. M. and Rogers, B. J. (in press) Some aftereffects from visual looming that are mediated by local motion, not by size change.
Borst, A. and Egelhaaf, M. (1989) Principles of visual motion detection. *Trends in Neurosci.* **12** 297-305
Boynton, R. and Whitten, D. (1970) Visual adaptation in monkey cones: Recordings of late receptor potentials *Science* **170** 1423-1426
Braddick, O. J. (1974) A short range process in apparent motion. *Vision Res.* **14** 519-527.
Braddick, O. J. (1980) Low-level and high-level processes in apparent motion. *Phil. Trans. R. Soc. Lond. B* **290** 137-151.
Braddick, O. J. and Adlard, A. J. (1978) Apparent motion and the motion detector. In: Armington, J., Krauskopf, J., and Wooten, B. R. (Eds): *Visual psychophysics: Its physiological basis.* New York: Academic Press, pp. 417-426
Cavanagh, P. (1989) Inter-attribute apparent motion. *Vision Res.* **29** 1197-1204

Cavanagh, P. and Leclerc, Y. (1989) Shape from shadows. *J. Exp. Psychol.: HPP* **15** 3-27
Cavanagh, P. and Mather, G. (1989) Motion: the long and short of it. *Spatial Vis.* **4** 103-129.
Chubb, C. and Sperling, G. (1988) Drift-balanced stimuli: a general basis for studying non-Fourier motion. *J. opt. Soc. Am. A* **5** 1986-2007.
Cutting, J. E. (1986) *Perception with an eye for motion* Cambridge, MA: MIT Press
Deutsch, J. A. and Ramachandran, V. S. (1990) Binocular depth reversals despite familiarity cues: An artifact? *Science* **25** 565
Foley, J. and McCourt, M. (1985) Visual grating induction. *J. opt. Soc. Am. A* **2** 1220-1230
Georgeson, M. (1979) Random-dot stereograms of real objects: Observations on stereo faces and moulds *Perception* **8** 585-588
Gregory, R. L. (1970) *The intelligent eye.* New York: McGraw-Hill, pp. 124-136.
Hamada, J. (1984) Lightness increase and decrease in square-wave gratings *Percept. Psychophys.* **35** 16-21
Hess, E. H. (1972) "Imprinting" in a natural laboratory *Scientific American* **227** 24-31
Koenderink, J. J. (1986) Optic flow. *Vision Res.* **26** 161-180
Lee, D. N. (1980) The optic flow field: the foundation of vision. *Phil Trans. R. Soc. Lond.* **B** 290 169-179. Reprinted in: Longuett-Higgins, H. C. and Sutherland, N. S. (Eds): The psychology of vision. London: Royal Society, 1980.
Lee, D. N., and Reddish, P. E. (1981) Plummeting gannets: a paradigm of ecological optics. *Nature* **293** 293-294
Marr, D. and Ullman, S. (1981) Directional selectivity and its use in early visual processing. *Proc. R. Soc. Lond. B* **211** 151-180.
Moulden, B. and Kingdom, F. (1989) White's effect: A dual mechanism. *Vision Res.* **29** 1245-1259
Nakayama K. (1985) Biological image motion processing: A review. *Vision Res.* **25**, 625-600.
Newsome, W. T., Britten, K. H., Movshon, J. A. and Shadlen, M. (1989) Single neurons and the perception of visual motion. In: Lam, D. M-K. and Gilbert, C. D. (Eds): *Neural mechanisms of visual perception.* Proceedings of the second Retina Research Foundation conference, pp. 171-198. Portfolio Press, Texas.
Pantle, A. and Picciano, L. (1976) A multistable movement display: evidence for two separate systems in human vision. *Science* **193** 500-502.
Peli, E. (1990) Binocular depth reversals despite familiarity cues: An artifact? *Science* **25** 565
Phillips, R. J. (1972) Why are faces so hard to recognise in photographic negative? *Percept. Psychophys.* **12** 425-426
Piotrowski, L. N. and Campbell, F. W. (1982) A demonstration of the visual importance and flexibility of spatial-frequency amplitude and phase. *Perception* **11** 337-346.
Pomerantz, J. R., and Kubovy, M. (1981) Perceptual Organization: An overview. In: Kubovy, M. and Pomerantz, J. R. *Perceptual Organization.* Hillsdale, N. J: Lawence Erlbaum, pp. 423-456
Ramachandran, V. S. (1988) The perception of shape from shading. *Scientific American*
Ramachandran, V. S. (1990) Visual perception in people and machines. In: Blake, A. and Troscianko, T. (Eds): *AI and the Eye.* New York: Wiley, 21-77.
Ramachandran, V. S. and Deutsch, J. A. (1990) Binocular depth reversals despite familiarity cues: An artifact? *Science* **25** 565
Ramachandran, V.S. and Anstis, S.M. (1986) Figure-ground segregation modulates apparent motion *Vision Res.* **26** 1969-1975
Regan, D., Beverley, K. I. and Cynader, M. (1979) The visual perception of motion in depth. *Scientific American* **241**, 136-151.
Rock, I. (1983) *The Logic of Perception.* MIT Press, Cambridge.
Rock, I., Tauber, E. S., and Heller, D. P. (1964) Perception of stroboscopic movement: Evidence for its innate basis *Science* **147** 1050-1052
Sekuler, R., Anstis, S. M., Braddick, O. J., Brandt, T., Movshon, J. A. and Orban, G. (1990) The perception of motion. In: Spillman, L. and Werner, J. (Eds): *Visual perception: The neurophysiological foundations .* Academic Press, London.
Shepard, R. N. (1981) Psychophysical complementarity. In: Kubovy, M. and Pomerantz, J. R. *Perceptual Organization.* Hillsdale, N. J: Lawence Erlbaum, pp. 279-341
Shioiri, S., Cavanagh, P. and Favreau, O. (1989) Non-linearity in color space measured by apparent motion. *Investig. Ophth. & Vis Sci. Suppl.* **30** 130

Sigman, E. and Rock, I. (1974) Stroboscopic movement based on perceptual intelligence. *Perception* 3 9-28

Ternus, J. (1926/1938) The problem of phenomenal identity. In: Ellis, W. D. (Ed): *A source book of Gestalt psychology*. London: Routledge & Kegan Paul.

Ullman, S. (1979) *The interpretation of visual motion*. Cambridge, MA: MIT Press.

van den Enden, A. and Spekreijse, H. (1989) Binocular depth reversals despite familiarity cues. *Science* 24 959-961

White, M. (1979) A new effect on perceived lightness. *Perception* 8 413-416.

White, M. (1981) The effect of the nature of the surround on the perceived lightness of grey bars within square-wave test gratings. *Perception* 10 215-230

White, M. and White, T. (1985) Counterphase lightness induction *Vision Res.* 25 1331-1335

Zaidi, Q. (1989) Local and distal factors in visual grating induction. *Vision Res.* 29 691-697 63

What's up in top-down processing?

Patrick Cavanagh

Department of Psychology, Harvard University,
33 Kirkland Street, Cambridge Massachusetts 02138, U.S.A.

Much of the work in vision in the last decade has examined what low-level vision tells high level vision. Cues such as optic flow, depth of focus, and contour intersections have been shown to be useful, reliable correlates of the 3-D structure of a scene. However, the retinal image is often ambiguous. Figure 1, for example, can be seen as either a duck or rabbit. It does not appear as a hybrid, though: only one or the other of these interpretations is seen at any given moment. In addition, the final percept — duck or rabbit here — often contains more structure than is available in the retinal image. The relative positions of the two ears of the rabbit, for example, one near, one far, are not specified in the image yet they are available in the percept and determined by our 3-D knowledge about rabbits. In these instances, the interpretation must have been influenced by top-down processes. Clearly, top-down processing speeds the analysis of the retinal image when familiar scenes and objects are encountered and can complete details missing in the optic array.

Fig. 1. An ambiguous figure that can be seen as either a rabbit (apparently staring into the sky) or a duck. The 3-D structure attributed to the various parts of the image changes in the two interpretations but the 2-D information — the location of the contours — is unaffected.

Top-down processing requires that something be up top, of course, and there have been only vague ideas about the representations that might be involved and the means by which they would influence perception. Basically, the tacit assumption is that something is up top and that this something solves otherwise puzzling visual problems. In order to start an examination of these processes, I shall describe a particular stimulus that can only be interpreted with the aid of top-down processing but for which there is, initially, nothing up top.

In analyzing this stimulus, I shall concentrate on the early stages that lead from the initial 2-D representation on the retina to object recognition. Current bottom-up approaches to vision (see Fig. 2) assume that the 2-D representation leads to an internal 3-D model before the stimulus is identified (Marr, 1982; Biederman, 1987). Work that I have just begun takes a different approach, suggesting that object parts and boundaries should not be explicitly identified at such an early stage and that matching of raw 2-D views may be the most effective way to make the initial memory contact (Fig. 3). The basic question is the level at which image elements should be labeled as particular image tokens, whether edges, curves, 2-D shapes or 3-D volumetric features. This labeling commits the visual analysis to treat image elements in specific ways in subsequent processing and it can be disadvantageous to make this commitment prematurely.

Image ⇒ Contours ⇒ Parts ⇒ 3-D Model ⇒ Object

Fig. 2. Image analysis, as proposed by Marr (1982) and Biederman (1987) for example, proceeds from from the image through a 3-D model before indexing memory to identify the object.

I shall examine the possibility that object recognition begins, not with the construction of a 3-D model, but with a crude match of 2-D views to internal prototypes. The prototype that has the best match then guides the construction of an internal 3-D model. In other words, recognition may start with a quick table look-up process, operating on principles completely different from those implied in Figure 2. The results of this process are then "up top", available to initiate top-down processes.

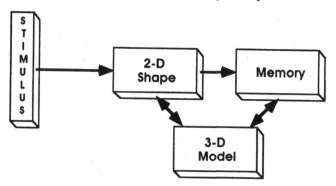

Fig. 3. In special cases, recognition may start with an initial match of 2-D image information against memory prototypes which then guide the construction of a 3-D model.

Figure 3 is presented as a sketch of the components of the visual system that are relevant for this paper. The diagram is principally concerned with the flow of shape information and not, for example, with the analysis of the position of objects or their displacement in the scene. These analyses may form part of a second processing system — the "where" system proposed by Mishkin, Ungerleider, and Macko (1982). The figure does not show the direct contribution of binocular disparity, motion parallax, convergence cues or gradients (shading, optic flow, etc.) to the 3-D model, but these are not considered in detail in this paper.

The stimuli I shall use to examine early recognition processes are figures where shape is defined by shadows (Fig. 4, righthand panel). There is more to this image than just contours, however: Many people presented with only the contours (Fig. 4, lefthand panel), for example, cannot identify it. Nor can they specify which contours might be shadow contours and which might be object contours. However, it is essential to identify the cast shadow borders in an image (this point is discussed in more detail below) because they are generally unrelated to the object contours and they seriously disrupt the interpretation of the image if they are confused with object contours — as they are in the lefthand panel of Figure 4. Clearly, any process (or person) that tries to identify parts or volumetric features in such an image would perform poorly without some knowledge of the object. I shall argue that any approach that labels the borders in this type of image before identifying the object will be faced with several spurious borders and these extra borders will seriously disrupt the segmentation of object parts. In most natural images, there are many redundant cues that help to identify which contours are shadow contours and which are not. The images that I are studying are therefore not intended to be representative of natural images but are an especially difficult type of image that humans are nevertheless able to interpret remarkably well.

Fig. 4. Contour information alone may be insufficient for 3-D interpretation. The left panel contains the same contours as the right but is difficult to recognize on its own.

Shadows are useful for recovering 3-D information as in the righthand panel of Figure 4, where the surface structure is revealed only by shadows (Cavanagh & Leclerc, 1989). Even though shadows are useful they suffer from several ambiguities in an image like Figure 4. First, it is not evident in the image whether an area is dark because of dark pigment or because of a dark shadow (of course, the figure *is* just dark pigment even though our interpretation attributes the darkness to shadows). Different kinds of shadow contours — attached or cast, see Figure 5 and 6 — play very different roles in reconstructing 3-D structure and nothing in the high contrast images that I am using distinguishes between these two types. Finally, the interpretation of the shadows depends critically on knowing the direction of the illuminant and this is not specified in the image in any explicit manner. In one sense, we have to know where the shadows are before we can identify the direction of the illuminant but we have to know the direction of the illuminant before we can discover the shadows.

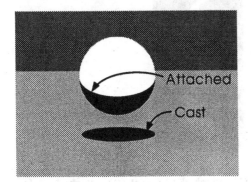 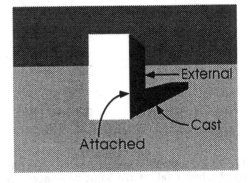

Fig. 5. Shadows have two types of borders: attached borders where the direction of the illumination is perpendicular to the surface normal (the light just grazes the surface); and cast borders where the shadow cast by one surface falls on a second surface. An object's external borders are only visible where the background and the object have a different brightnesses.

The essential goal in discovering the shadows in a figure is identifying the cast shadow borders. These borders have a special status in images because they do not correspond to any discontinuity in the object but to a discontinuity in the illumination. The cast shadow border is not a material border and basically needs to be ignored in order to patch together the pieces of surface that actually belong together.

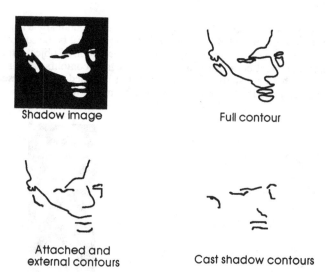

Fig. 6. The contours of the shadow image (top left) are difficult to interpret on their own (top right) but the attached and external contours (bottom left) are easily recognized. The cast shadow contours (bottom right) present a meaningless jumble of lines.

I claim that the recognition of shape from shadow in a stimulus like that in Figure 4 starts with an initial 2-D matching process. The reason is that there is no alternative for these images: the stimulus cannot be interpreted directly from image data without knowing what the object is. As mentioned above, the interpretation of the image requires that the different types of borders — object border, cast and attached shadow borders — be identified or parsed in the image. The only way to parse the image contours into attached or cast borders without any top-down guidance is presented in Figure 7 and I shall show that this parsing fails on images that we can nevertheless interpret.

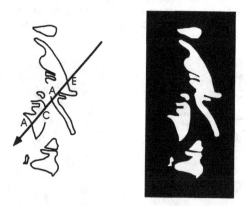

Fig. 7. A line parallel to the direction of the illuminant can uniquely label the light to dark transitions which always fall in the same sequence: external (E), attached (A), cast (C), attached, ... [e.g. {E,A},{C,A},{C,A},...] no matter where the line is traced.

If we assume that the direction of illumination can be determined (in the extreme, all directions can be examined until a consistent interpretation is found), then it is possible to distinguish cast from attached shadow borders in the image and from the attached shadow borders to recover the object's 3-D structure. Figure 7 demonstrates that along any line parallel to the direction of illumination, borders alternate in a fixed order. If the background is dark, the order is always external, attached, cast, attached, cast, attached, and so on, with cast and attached repeating in pairs. Using this rule, the border type can be identified throughout the image.

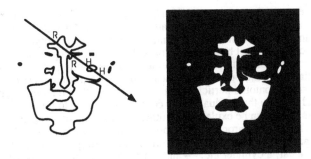

Fig. 8. In this image, some regions may be dark either because of low reflectance (R: hair, dark glasses, lipstick) or because of shadows. Other regions are light because of high diffuse relectance (skin) or because of specular relections (H: highlights on the glasses). The labeling scheme of Figure 7 fails here because the light to dark transitions can no longer be uniquely identified.

However, this parsing only works in an image where the borders are all shadow borders. If the image also contains reflectance and highlight borders (shown as R and H in Figure 8), no consistent labeling is possible. Shadow borders cannot be distinguished from reflectance or highlight borders in the image even if the direction of the illuminant is known. Therefore, the image contours cannot be parsed. *The image is nevertheless recognizable.* I conclude that some information other than that available in the image must guide the interpretation. But this cannot be true either since most people perceive the righthand panel of Figure 8 as a face without being told what it is beforehand. There is no other information available. Even if I want to invoke top-down processing, there is nothing up top!

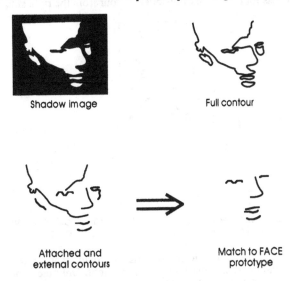

Shadow Image   Full contour

Attached and   Match to FACE
external contours   prototype

Fig. 9. The contours of the shadow image (top right) contain an easily recognized subset that can match to characteristic contours of a face prototype.

How can we get knowledge about the object before we recognize it? I suggest that a very different type of process performs a rapid, crude match against the image data to identify the type of stimulus. This hypothesis or prototype then guides the interpretation of the image. Note that even though the contours of shadow figures such as those in Figure 4, 7 and 8 are difficult to recognize, there is sufficient contour information in them for a match to simple prototypes in memory. As we saw in Figure 6, there is a subset of contours in these figures that is easily recognizable: the attached shadow and external contours. It cannot be known beforehand which contours are attached shadows and which are cast but a matching process that is capable of recognizing subsets of contours in the presence of irrelevant contours could extract the best match.

Figure 9, for example, contains some characteristic contours of the object mixed in with, and indistinguishable from, many irrelevant cast shadow contours. An initial 2-D match is therefore advantageous only if image information can be matched to memory prototypes without first selecting which subset of image contours must be object contours. Matched filtering techniques developed for 2-D recognition in the 1960s demonstrate possible methods for performing the necessary steps: identify targets based on a partial set of contours in the presence of unrelated contours (Bieringer, 1973; Van der Lugt, 1964); specify which contours participated in the match; fill in the missing contours (Collier & Pennington, 1966); and identify the residual image contours unrelated to the matched object (Caulfield, 1974, residual contours must then be explained by other scene components such as shadows, and other, occluding objects). This match process must be able to match against all possible prototypes simultaneously. It is probably most reasonable to think of storing only a small set of characteristic protoptypes — a face, a car, a cylinder etc. — and not a prototype for each possible instance of these broad classes.

Thus a rough 2-D match could select the best candidate object — a face, boot, hat, or whatever. The 3-D information stored with this prototype could then guide the construction of an internal 3-D model — verifying that it is consistent with the image contours and resolving ambiguities of shadow borders, occluded objects and incomplete contours. A match between a subset of the image contours (and these will only be among the external and attached shadow contours such as, in Figure 10, the nose profile or the lips) has the important consequence of identifying the *residual* contours — the contours not explained by the prototype. A second process must then "explain" these residuals. On the right in Figure 10, the residual contours might be attributed to shadows or material boundaries or to other objects that are occluding the object that matched the prototype. These hypotheses for the residuals must then be verified in the image to see whether the original hypothesis can be maintained; for example, if a residual contour is to be interpreted as a shadow border, the image should be darker on the shadow side of the border. Figure 11 depicts this process for one piece of residual contour from the right side of Figure 10.

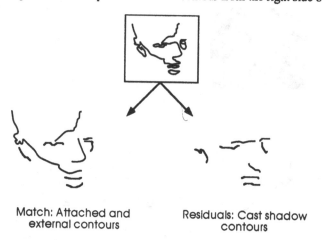

Match: Attached and   Residuals: Cast shadow
external contours   contours

Fig. 10. The image contours include the informative attached shadow and external contours on the left and the more or less irrelevant cast shadow borders on the right. Only the contours on the left can be expected to participate in a match to a simple prototype although certainly not all those shown here would be involved — perhaps only the nose, eyes, lips and ear contours. Those that don't participate in the match are the residuals that must be explained through different assignments in the scene. These will always include the cast shadow contours shown on the right.

If there is insufficient support in the image for the 3-D aspects of the initial prototype, it would be discarded. This is the probable fate of a face prototype for the top image of Fig. 10 — there are no dark regions to support 3-D shadow explanations of the many contours that are not characteristic face (lips, nose, forehead, etc.) contours. The prototype with the next best 2-D match would then be selected to guide the 3-D modeling. However, there is no other obvious 3-D prototype for this particular image. The only remaining explanation is that the contours are all material borders so that the enclosed areas are seen as separate 2-D islands.

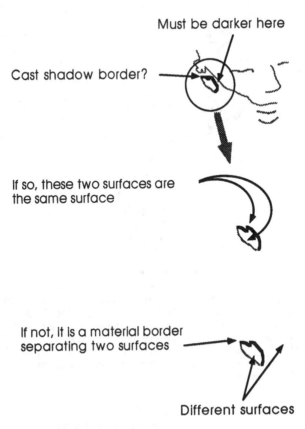

Fig. 11. Each residual contour must be explained by an additional scene property. If the particular contour shown as a thick line in the ear area is to be labeled as a cast shadow border, the adjacent region to the right must be darker than that to the left.

The memory prototypes required for this inital 2-D match are quite unlike those required by recent approaches (Marr, 1982; Biederman, 1987) where memory prototypes are object-centered and can serve to identify any arbitrary 2-D view of an object in the scene. In contrast, if the initial match is based on stored 2-D views, each object prototype would have to have numerous 2-D views as part of its representation in memory. The number of necessary views would have to be especially large if size- and orientation-invariant coding is not used by the visual system (although see Cavanagh, 1984, 1985). In one sense, this is not an insurmountable problem even in the worst case since 2-D matching is ideally suited to the massively parallel processes hypothesized in neural net memory systems (Anderson, Silverstein, Ritz & Jones, 1978; Kohonen, 1977).

Note that the 2-D views bundled together as memory prototypes are viewer-centered (see Fig. 14). There is evidence that the visual system does, in fact, operate on viewer-centered representations and not 3-D object models when accessing memory. Rock and his colleagues (Rock, DeVita & Barbeito, 1981; Rock & DeVita, 1987) have demonstrated that views of wire-frame objects seen from different directions are reliably identified only when they have the same retinal projection, indicating that 2-D viewer-centered representations may mediate recognition.

In the test figures that I have used, top-down processing operates from a rough prototype for the object in the image and, in cases where there is no other source, the prototype is provided by a 2-D match to the image. The model of recognition therefore has three stages: first, a 2-D match of the image against memory prototypes selects the "best" prototype; second, this prototype guides the construction of a 3-D model, checking the image for consistent support as the interpretation develops in detail; finally, the completed 3-D model corresponds to recognition.

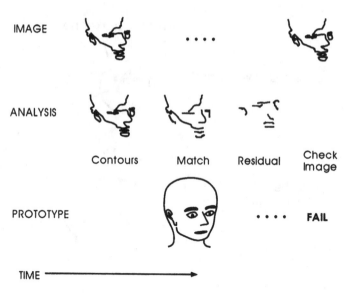

Fig. 12. The analysis of the image depicted on three levels: the displayed image which remains unchanged after its initial presentation; the analysis steps of extracting image contours, matching to memory prototypes, identifying the residual contours not contained in the prototype, and checking image for support for local interpretations of each residual contour; finally, the prototype is available to direct top-down processing following the match to memory. The face prototype fails in this test image containing only contours since no support can be found for 3-D interpretations of the residuals (e.g. shadow borders or occluding objects).

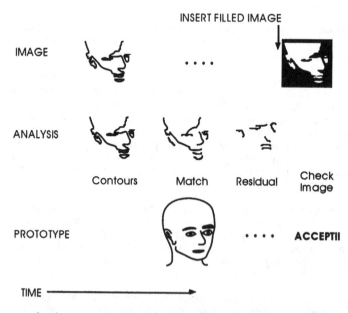

FIG. 13. In this example, the contour image is replaced in the display with a filled image just prior to verifying the residuals. Brightness levels are appropriate for shadow explanations of the residuals so the 3-D structure of the prototype is accepted even though only contours were presented initially.

If this model of early recognition is correct, it has an interesting and unexpected consequence. If the contour version of a shadow figure (e.g. the lefthand panel of Figures 4, 7, or 8) is presented, the 2-D match to the prototype should occur *even if the figure is not recognized*. The match occurs as shown in Figure 12, but it is rejected later because of lack of support for the residuals. It may be possible therefore, to switch the image from a contour version to a filled version at the appropriate moment and obtain recognition in the same total time — as if the filled version were present from the start (Figure 13). I have begun experiments to test this prediction.

In summary, what's up top? In the examples that I have presented here, it appears that some type of rough prototype may be the representation that guides the interpretation of the image. In most natural images where many redundant cues are available, the prototype may be chosen based on 3-D information. In the high-contrast images that I have used, no 3-D information is available from the image either directly or through pictorial cues such as perspective, contour intersections, or deep concavities. I claimed that for these images an initial 2-D match selected the best prototype to guide image interpretation. This initial match does not constitute recognition, however, and an experiment was suggested to demonstrate that this early match occurs even for stimuli that cannot themselves be recognized. The stored prototypes may be limited to basic object types, some as complex as a face and others as simple as a cylinder, but do not require a prototype for each instance of a class.

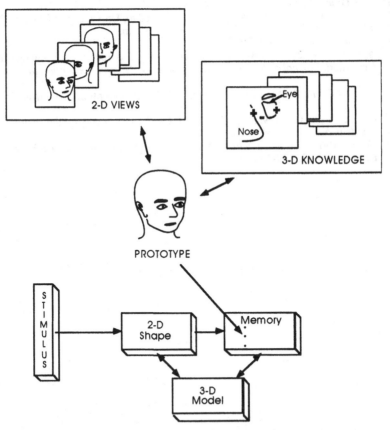

Fig. 14. Memory prototypes include a set of 2-D views from several viewpoints and 3-D knowledge about the object such as directions of curvature along the object contours that are visible in the various 2-D views as well as identification of the parts. Prototypes for those parts would contain additional information as well. The prototypes could be identified from 3-D information available in the image or in its absence from a 2-D match to individual views. Once the prototype is identified, it guides the completion of the 3-D model of the object.

REFERENCES

Anderson, J. A., Silverstein, J. W., Ritz, S. A.,& Jones, R. S. (1977): Distinctive features, categorical perception, and probability learning: Some applications of a neural model. *Psychological Review* **84**,413-451.

Biederman, I. (1987): Recognition-by-components: A theory of human image understanding. *Psychological Review* **94**, 115-147.

Bieringer, R. J. (1973): Optical correlation using diffuse objects. *Applied Optics* **12**, 249-254.

Caulfield, H. J. (1974): The rabbit-and-hat problem. In P. Greguss (ed.) *Holography in Medicine*, Surrey, U.K.: IPC Science and Technology Press, 3-26.

Cavanagh, P. (1984): Image transforms in the visual system. In P. C. Dodwell & T. Caelli (eds.) *Figural synthesis*. Hillsdale, N. J.: Lawrence Erlbaum Associates, 185-218.

Cavanagh, P. (1985): Local log polar frequency analysis in the striate cortex as a basis for size and orientation invariance. In D. Rose & V. G. Dobson (eds.) *Models of the visual cortex*. London: John Wiley & Sons, 85-95.

Cavanagh, P. & Leclerc, Y. (1989): Shape from shadows. *Journal of Experimental Psychology: Human Perception and Performance* **15**, 3-27.

Collier, R. J. & Pennington, K. S. (1966): Ghost imaging by holograms formed in the near field. *Applied Physics Letters* **8**, 44-46.

Julesz, B. (1971): *Foundations of Cyclopean Perception*, Chicago, University of Chicago Press.

Kohonen, T, (1977): *Associative memory*. Springer-Verlag, Berlin.

Marr, D. (1982): *Vision*. Freeman: San Francisco.

Rock, I., Di Vita, J. & Barbeito, R. (1981): The effect on form perception of change of orientation in the third dimension. *Journal of Experimental Psychology: Human Perception and Performance* **7**, 719-731.

Rock, I. & Di Vita, J. (1987): A case of viewer-centered object perception. *Cognitive Psychology* **19**, 280-293.

Mishkin, M., Ungerleider, L. G., & Macko, K. (1983): Object vision and spatial vision: Two cortical pathways. *Trends in Neurosciences* **6**, 414-417.

Van der Lugt, A. B. (1964): Signal detection by complex spatial filtering. *IEEE Transactions on Information Theory* **IT-10**, 139-154.

Tacit assumptions in the computational study of vision

Simon Ullman

Artificial Intelligence Laboratory,
MIT, 545 Technology Square, Cambridge, M.A. 02139

The computational study of vision has two different goals. One is the development of a theoretical foundation and engineering practice for the construction of useful "vision machines". The other is to develop a theory for the processes underlying vision in biological systems and, in particular, human vision.

The belief that computational vision could provide a useful framework for understanding human vision rests on a number of assumptions that are usually taken for granted, without being explicitly stated or called into question. Although these assumptions are shared by researchers in the field, they are not universally accepted, and most of them have been challenged on various grounds. Theses challenges are worth considering, because if some of the main assumptions are in fact invalid, the entire enterprise may turn out to be misguided.

The goal of this paper is to discuss some of the tacit assumptions underlying computational vision, and some of the challenges that have been raised against them.

Assumption I:
THE BRAIN IS AN INFORMATION PROCESSING DEVICE

Different artificial devices are constructed for different goals. Some are mechanical devices, constructed, for example, to move things from one place to another. The main function of other devices is to transform energy from one form to another. Yet other devices are constructed to manipulate information. The first assumption means that the brain, although a natural biological system rather than an artificial device, also belongs to the class of information processing devices. The important aspects of the brain are not its mechanical properties (although F. Crick has suggested that tiny muscles in dendritic spines may control synaptic efficacy in the brain), nor properties concerned with energy manipulation (although there are metabolic processes in the brain). The important processes in the brain have to do with the manipulation of signals and information, rather than forces or energy.

According to this view the brain receives sensory signals, these signals get transformed, analyzed, integrated, stored, compared, etc. These processes can then lead, for example, to the generation of patterns of activation of muscles, and thereby to observed behavior. In performing its tasks, the brain may be different from any known artificial information processing device, but it still belongs to the general class of information processing systems. The view of the brain as an information processing device does not mean, of course, that the brain is similar to standard 'Von Neuman' digital computers. It may use, for example, distributed representations and massively parallel computation. It is still subject, however, to the same general principles that govern a Turing machine. Its processes can still be studied, at least at a functional level (as opposed to their biological implementation) within the general framework of studying computing devices.

This assumption has important implications to the manner research is conducted in computational vision. First, it justifies the entire enterprise, making it plausible that building artificial vision machines will tell us something relevant about "biological machines". Second, it sets the general framework for computational vision, and determines the kind of questions that should be asked. it suggests the view that vision produces some internal representations, and that other processes manipulate and use these representations. (The representation issue is discussed in more detail below). It leads to the formulation of certain questions, such as the type of the representations created (e.g., the primal sketch, 2.5-D sketch etc.,), or questions such as "how do different visual module communicate" etc.

Challenges to Assumption I.

A number of objections have been raised against the view of the visual system (and the brain) as an information processing system. Most of these objections contend not only that something in this way of thinking is simplified or incomplete, but that the entire approach is misguided and will lead nowhere.

1. Subjective experience

The basic argument is that when we see, we don't just process information, but we have a conscious experience of what we see. In contrast, machines manipulate signals, but do not have similar experiences.

For a long period this problem has not been a legitimate subject that scientists studying vision and the brain in general were concerned about. As J. Searle (1990) puts it, "...since Descartes, we have, for the most part, thought that consciousness was not an appropriate subject for serious science or scientific philosophy of mind". It is interesting to note that more recently it became the subject of much thinking and theorizing within the fields of psychology and biology (e.g. Baars 1988, Crick 1990, Edelman 1989, Jackendoff 1987, Johnson-Laird 1988, Penrose 1989, Searle 1990)

The subjective experience objection has two versions, a 'minimal' and a 'maximal' one. The minimal version contends that the theory of vision will be incomplete without explaining subjective phenomena. Some researchers believe that a complete theory will be attainable within the current scientific framework (Crick 1990, Edelman 1989). Others do not, and believe that the brain is not merely some kind of a biological computer, but also something else. It contains some "extra" mechanisms, above and beyond the ones known today (e.g., according to Penrose's recent theory (Penrose 1989), some quantum effects that make it capable of 'surpassing' the limitations of a Turing machine). According to this 'minimal' version the computational approach may be incomplete, but it is not fundamentally wrong.

The 'maximal' version contends that the fact that the brain somehow "generates" subjective experience and a computer does not, indicates that they are two radically different sorts of entities. It will therefore be entirely misguided to try to explain one in terms of the other.

My own view is close to the 'minimal' view. The origin of perceptual experiences lies outside the realm of current theories, and it seems likely to me that radical changes and new insights will be required before such issues will be addressable by scientific theories. At the same time, this problem does not appear to me to undermine the computational approach to vision. The problem is 'late' rather than 'early' in the sense that one can study much of the brain's structure and function without having to tackle the problem of perceptual experience. At least up to a point, the perceptual systems that have been studied do behave like specialized information processing devices, operating on principles familiar to scientists and engineers. A beautiful example is the owl's sound localization system, which has been studied and analyzed in impressive detail on the levels of both function of mechanism (e.g., Konishi et al 1985). This is a sophisticated system using components and methods that are entirely expected on the basis of common engineering practices, such as filters, delay lines, coincidence detectors and the like.

2. The problem of meaning

This is a philosophical objection that claims, roughly, that "symbol manipulation does not capture meaning". One version of this appears in the form of J. Searle's "Chinese room argument" (Searle 1980). Searle describes a situation where a number of people (or a single person) in a closed room are following the instructions of a 'Chinese understanding' program. They receive input symbols, written in Chinese, through an input slot. They then follow a set of instructions, comparing the input string to other symbols stored in various drawers in the room, moving symbols from one place to another, copying some other symbols etc., and eventually deliver an output string through an output slot. The people in the room can follow this symbol-manipulating procedure and produce a 'Chinese understanding' behavior without understanding Chinese at all.

Similarly, the argument goes, when I see something, it has a meaning for me. In contrast, the generation of representations inside a computer does not have any meaning. These representations can only have a meaning for us, for people who can interpret them.

The Oxford philosopher P. Hacker (1990) raises a related argument. In his view:

"a creature that can see... can also search for what it wants, look for things it likes, look at objects that interest it... flee from those it perceives as threatening and which it fears".

The problem is related to, but not identical with, the problem of perceptual experience. Humans and animals can see, and machines cannot,

"...not because they 'have minds'.. nor is it because animals and men are biological structures and industrial robots are not. It is rather because our use of these expressions [e.g. 'see'], and so their meaning, are bound up with the highly complex behavioral repertoire of creatures in the rich environment of the world they inhabit, and systematically interwoven with a host of further psychological predicates the application of which is licensed by the same form of behavior".

Seeing thus makes sense only by its relations to other aspects of behavior and 'psychological predicates' that give it meaning. For machines, 'seeing' does not have such a meaning, therefore they cannot see in the same sense we do.

Personally, I am not convinced that there is a serious problem here that is really distinct from the one concerning perceptual experience. Aspects of this problem are taken up again below, in discussing the problem of internal representations.

3. Direct perception

The theory of direct perception (e.g. Gibson 1950, 1979) is strongly opposed to any notion of vision as involving computation and information processing. Concepts such as inference, interpretation, or categorization, and processes such as copying, storing, comparing, and matching, have no possible place in the theory of visual perception. The theory similarly rejects notions such "nervous operations on the signals in the nerves". What is suggested instead is that perception is "the process of direct information pickup".

This position is in strong opposition to any version of a computational approach. It is interesting, therefore, that some researchers in computational vision still regard themselves as Gibsonians, or "neo-Gebsonians". How can computational theories be related to an approach that is entirely opposed to any notion of computation? The answer is that the Gibsonian theory has also emphasized another aspect : the richness of the visual stimulus. The theory stresses the point that the visual stimulus in a natural environment is rich in information (particularly in the form of spatial and temporal texture gradients). This is presented as a contrast to theories that describe the visual stimulus as impoverished, therefore requiring further inference processes.

An example of the theory is its account of binocular stereo vision. The theory begins by describing the source of depth information -- the relative position of corresponding points on the two retinae. The next stage, according to the theory, is that this information is directly "picked up". Unlike computational theories of stereo vision, direct perception sees no reason to include in the theory processes that filter the images, extract features, establish correspondence, resolve ambiguities, extract disparity, etc.

This notion of direct perception appears to me to be mistaken and fruitless. Briefly, the main argument is the following (see Ullman 1980). The term 'immediate' is relative to a given area of investigation. For example, for the psychologist of vision, it is justifiable to consider the photoreceptors in the eyes as registering 'directly' light intensity, or that Meissner's corpuscle 'immediately' registers touch. The response of a retinal rod, for example, to light, is in fact a highly complex process, composed of a sequence of many distinct events. A biophysicists will therefore not be satisfied with a description of the rod as registering light 'directly'. For the psychologist, the details of the biophysical reactions are outside his or her domain of investigation, and it is therefore justified as far as psychology is concerned to view the process of registering light as 'direct'. In general, then, a process can be called 'immediate' in a given domain, if it has no meaningful decomposition into more elementary constituents within the domain of interest. It therefore becomes evident that a satisfactory theory of psychology cannot regard binocular vision, or object recognition, for example, as 'direct' information pickup.

Assumption II:
COMPUTATIONS CAN BE STUDIED ON THEIR OWN

One version of this assumption is Marr's well known division of the study of vision into three levels -- the levels of computation, algorithm, and mechanism. Each of these levels can be studied more or less on its own (although, as emphasized by Marr, they are not entirely independent).

Some form of this assumption is crucial for justifying computational vision as a means for studying aspects of perception. In computational vision one usually proceeds by trying to develop a computational scheme that can solve successfully a given problem, such as recovering binocular disparity, recovering structure from motion, etc. The assumption is that by doing so one studies some general aspects of the problem that will carry over to human vision as well.

The assumption above has two main aspects:

1. Sufficient independence of the mechanism and the computation. If the description of the processes that go on is equivalent to a complete description of the mechanism, then there is little value in studying the 'computation' independent of the mechanism.

2. Sufficient constraints on possible solutions. If, for example, there is a large number of very different ways of solving the binocular stereo problem, then the study of possible stereo algorithms on their own is less likely to be of direct relevance to the study of binocular stereo vision performed by the brain. If the set of possible solutions is highly constrained, studying computationally in detail a number of options is more likely to prove useful to the understanding of how the brain solves the same problem.

Challenges to Assumption II.

1. Connectionism

Some versions of connectionism challenge the distinction between the levels of algorithm and mechanism. The computational properties of the system are 'emergent behavior' of the units and their connections, and the studies of function and mechanism cannot be separated. Unlike a standard computer, one cannot distinguish in such a network a program, or an algorithms, using, or 'running on', the hardware. The conclusion is that the emphasis should therefore be placed on the study of the unit and network properties. For example, Hopfield (personal communication) has shown that the exact form of the Hebbian learning rule in associative networks can directly influence the overall behavior of the system. In one form of the rule, the system will recall sequences equally well in both forward and backward directions. In a slightly modified form, the forward direction will be highly favored. Other properties, such as the capacity for associative recall, are network properties that arise from the entire pattern of connections in the net.

A real difficulty will indeed arise if it will turn out that the simplest description of a network's operation is essentially to describe fully its units and list all their interactions.

This is, however, unlikely, for two reasons. First, the distinction between mechanisms and computation is often a conceptual rather than a physical one. For example, one can construct a special purpose piece of hardware for performing edge detection, or FFT, etc. In such a device it is difficult to distinguish an algorithm 'running on' the hardware. And yet, one can describe the particular algorithm embodied in the hardware. Similarly, in complex connectionist networks, even when there is no physical distinction between the 'hardware' and the computation 'using' the hardware, a conceptual description of the computation may very well exist.

The second point is that connectionist networks for difficult tasks, rather than being one large unanalyzable unit, often have structure in terms of sub-components that perform well-specified aspects of the task. For example, the network developed at AT&T's Bell Laboratories for reading hand written numerals (LeCun et al 1989) has a number of distinct levels performing different operationss involving abstraction and feature detection etc. In such a structure, although it is still a highly interconnected network of simple units, it becomes easier to give at least a partial description of the network's operation in computational terms.

2. Non-biological solutions

An objection is often raised that computational studies of vision produce 'artificial', non-biological, solutions. The feeling is that the solutions produced by computational vision are sometimes too 'logical', 'neat' and 'symbolic' in nature (Minsky 1990). This objection is related in part to the assumption stated above regarding the constraints on possible solutions. The worry is that the space of possible solutions is in fact large and unconstrained, and the solutions proposed by theoretical studies may bear little relevance to the biological solution. This worry can be addressed at least in part by adopting an appropriate methodology in the field, namely, by taking into consideration constraints imposed by known biological and psychological data. It also appears that in many cases the space of possible solutions is indeed rather restricted. In object recognition, for instance, it is difficult at present to come up with even a single scheme that will be able to perform the task in a reasonable manner. Finally, some aspects of computatioanl vision are sufficiently broad in nature that they are likely to apply to any system, natural or artificial, that performs visual tasks. The main example here is the set of 'natural constraints' utilized by the visual system (Marr 1982), such as object continuity in the case of binocular disparity extraction, a rigidity constraint in the case of perceiving structure from motion (Ullman 1979) or the use of smoothness constraint in a number of visual computations (Poggio, Torre, & Koch 1985).

Assumption III
MODULARITY

The assumption is that the visual system consists of a number of modules that can be studied more or less independently (see also Fodor 1983). Modules that have been studied computationally include edge detection, stereo vision, motion measurement, structure-from-motion, color, texture analysis, shape-from-shading, surface interpolation, segmentation, object recognition, and others. Obviously, there are some inter-relations among them, but the assumption is that to a first approximation they are independent. The integration of modules is assumed to be primarily 'late' in nature. That is, it operates primarily on the results obtained by the individual modules. This is the scheme researchers in computational vision follow in practice. Each of the individual modules is usually studied on its own. In addition, some researchers consider separately the problem of their integration, namely, how results of the individual modules are then put together.

Challenges

1. Empirical evidence

The empirical evidence regarding the modularity of visual processes is not entirely clear. Anatomically, there is evidence for multiple parallel pathways, e.g., for motion, binocular vision, and color processing. The processing stream of the color system, for instance, includes the parvo layers of the LGN, the blobs in V1, thin stripes of V2, and area V4 (Livingstone & Hubel 1984). There is also evidence from brain lesions regarding specific deficits of cortical origin in color vision, motion perception, and binocular vision.

In contrast with this evidence for independence, there is also considerable body of evidence, primarily from psychophysics, for early interactions between modules. For example, Gilchrist (1977) has shown that perceived three-dimensional (3-D) structure can affect perceived lightness. It is known that apparent motion can be established between stereo discontinuities, indicating fairly early interactions between the two modules. There are indications for interactions between motion perception and surface interpolation, and others.

As mentioned above, the empirical data concerning interactions among modules is still fragmentary. It seems to me that the interactions are unlikely to be all of the 'late' type. At the same time, the modularity assumption appears to be sufficiently useful as a starting point. Theoretically, the system could have been one large interconnected entity that resists any attempt to break it down into meaningful modules. This does not appear to be the case, even if interactions among modules do exist at different levels. In future computational studies it would be of interest to explore more fully the nature and usefulness of such interactions. For example, when combining stereo and motion information, there is more room for useful interactions between them than e.g. simply combining the three-dimensional estimates provided by the two modules, with some appropriate weights (such as '60% stereo, 40% motion'). To mention just a few examples, motion and stereo discontinuities could interact, the two systems could be used to calibrate one another, and motion information, combined with the current depth estimations, could be used to predict an updated depth map (to facilitate the computation of an updated depth map rather than starting it anew).

2. 'Society of mind'

This objection is termed after M. Minsky's (1987) description of the mind as a society of a very large number of highly interactive agents. Translated to the domain of vision, this means that the visual process is composed not of a relatively small number of more-or-less independent modules, but of a very large number of highly interdependent processes, with the interactions between them being as complex and as important to the operation of the system as the individual modules themselves. This view is related to Ramachandran's (1985) view of the visual system as a 'bag of tricks'. The view is again of the visual system as composed of a very large number of highly specialized (and presumably interacting) processes. Ramachandran raises this notion as an alternative to the computational view -- the brain does not use computations, it uses ' tricks'. As far as I can see, this is not really a challenge to the computational view. The 'tricks' described by Ramachandran can be equally described as specialized, and often highly complex and sophisticated computations (as in the case of motion capture, that can distinguish, e.g., between the inside and the outside of a closed figure). I think that it is an interesting and a reasonable point of view that the visual system may include considerably more modules than commonly assumed in current computational vision. For example, there may be not one stereo matching computation, but a number of different ones, for different types of inputs, (dense vs. sparse input, with or without clear monocular cues) as well as for different goals (e.g. for manipulation in nearby space, as opposed to distant objects). Such an architecture does not by itself undermine the notion of modularity, but it may change some of the current notions in computational vision.

Assumption IV
BOTTOM UP PROCESSING

Current computational approaches typically assume a very large and often dominant role to the bottom-up component of the visual process. 'Bottom-up' means here an automatic process that is determined by the input, and is not governed by specific goals of the computation, and does not bring into bear specialized knowledge regarding specific objects. In Marr's theory, for example, visual processing proceeds primarily through the bottom-up creation of a sequence of representations, the primal sketch, then the 2.5-D sketch, and finally the 3-D representation. Similarly, in computational theories of most of the visual modules the computations are again primarily bottom-up. For example, in performing stereo matching, current scheme do not 'care' what is the purpose of the computation (e.g., manipulation, navigation, or object recognition), or what we know about the specific objects in the scene. This emphasis on bottom-up processing is the dominant approach in current computational vision, but alternative approaches have also been advocated (e.g., Freuder 1974, Tenenbaum & Barrow 1976).

Challenges

1. Empirical evidence

There is some psychophysical evidence indicating the effect of high level knowledge on relatively early visual processing. As mentioned above, Gilchrist (1977) has shown that the interpreted three-dimensional structure of the scene can have effects on the perceived lightness of surfaces. When we see hollow shaded masks of faces, the visual system 'resists' seeing them as concave, it seems that our knowledge about faces tends to enforce a convex interpretation (Gregory 1970). The recognition of objects in some cases, e.g., the well-known image of a Dalmatian dog by R.C. James (Gregory 1970, Marr 1982) does not appear amenable to bottom-up processing. From anatomical standpoint, interactions going 'back down' can certainly be supported by the extensive back projections between visual areas.

It should be noted, however, that the empirical evidence is not entirely unequivocal. There are also a large number of cases where early processes appear 'impenetrable' to higher influence. An interesting and perhaps somewhat unexpected example are certain segmentation processes (described e.g. by Nakayama *et al* 1989). It appears that in many cases the visual system 'decides' for us what goes with what in the scene, i.e., which regions belong together as a part of a single object and which are separate, and that high level influence on these decisions is sevrely limited. Physiologically, it has been shown (Livingstone & Hubel 1981) that, at least in V1, responses in natural sleep, barbiturate anesthesia, and different levels of arousal, are quite similar, suggesting that the processing up to that level is not highly dependent (although may still be modulated to some degree) by higher level activity.

It seems to me that in most of the earlier processes, such as edge detection, stereo, motion processing, or shading analysis, there may be some higher level effects, but they are not of major influence. Even segmentation processes appear to have a very significant bottom-up component. Object recognition may, however, be quite different. I find the bottom-up creation of 3-D models and matching to memory (as in Marr's scheme) unlikely. A scheme in which stored knowledge about objects' shapes and information from the image can interact in both directions appears more likely for the purpose of visual object recognition.

Assumption V
EXPLICIT REPRESENTATIONS

The 'final' goal of the visual process, according to this assumption, is to produce a set of representations that are useful to the observer, and are not cluttered by irrelevant information (such as Marr's sequence of representations mentioned above). In the words of J. Frisby (1980):

"...it is an inescapable conclusion that there must be a symbolic description in the brain of the outside world".

The main question in computational vision can therefore be formulated as follows:

"When we see, what are the symbols inside our head that stand for things in the outside world".

For example, the goal of the binocular stereo process is to produce a three-dimensional representation, perhaps in the form of a depth map. Other processes that extract 3-D information, such as structure-from-motion and shading analysis, may then add depth information to the same representation.

This view is in the spirit of cognitive science in general. On this view cognition is described in terms of processes operating over internal representations. The picture is that the visual system delivers representations of depth, shape, motion, color, etc., and later processes then manipulate and use these representations for tasks such as recognizing objects, planning and controlling movements in the environment, and navigation.

Having a common representation of this type, e.g., for depth information, has some advantages from a computational standpoint. A common representation can integrate contributions from various sources, and then different modules could all use the same representations, rather than having to create unnecessary duplications of the same information.

This general picture guides much of the thinking about visual processes. For example, work on binocular depth is usually not concerned directly with controlling an arm, or avoiding obstacles, etc., but with the creation of a depth representation, with the assumption that subsequent processes will somehow use this representation to plan and guide behavior.

Challenges

1. Multiple direct pathways

Although arguments could be raised in favor of using some common explicit representations, alternative schemes are also possible. It may be that there are no common representations but many separate ones, and that quantities such as depth, surface orientation, direction of motion, etc., are not represented explicitly anywhere in the system. For example, it may be that information derived from the motion perception system is used to create patterns of activity that will guide the hand to catch a moving object without an explicit representation of direction or speed. Furthermore, there may be multiple pathways of this type for performing different tasks. The control of eye movements, for example, may not share a common representation with the hand control pathway. In such a structure there may not be an identifiable common and explicit representation of properties such as depth and velocity.

The work of R. Brooks on the so-called subsumption architecture for motor control (Brooks 1986) is close to this alternative view. Some of the objections raised by the philosopher H. Dreyfus against artificial intelligence are also related to this view. He points out in his argument to the distinction between 'knowing that' and 'knowing how'. 'Knowing that' relates to facts that we store and explicit rules for actions we follow. 'Knowing how' on the other hand is akin to the skill of riding a bicycle. This knowledge is not stored in the form of a list of facts and rules to follow, but is somehow incorporated into the behavior of the system. Similarly, it may be that in a variety of perceptual skills and visually guided behavior we 'know how' to perform different tasks without having an explicit representation of the state of the world and rules for manipulating these representations.

2. Empirical evidence

The empirical evidence regarding common and explicit, vs. distributed and implicit, representations is not yet clear. Physiologically, something like an internal 'depth map' has not been found, but this may simply reflect our very partial knowledge of the visual system. Some evidence appears to be in broad agreement with the general notion of common, explicit, representations. For example, the primary visual area V1 appears to make explicit properties such as orientation and direction of motion, and this representation subserves different areas involved in different tasks. Area MT may be a common 'motion representation' that is used for different tasks requiring motion information.

Other physiological evidence agrees more with the picture of a more direct pathway with an implicit form of encoding. In area 7a of the posterior parietal lobe cells appear to be involved in the transformation from retinal to head-based coordinates (see Zipser & Andersen 1988 for a description and a model of this transformation.) Many of the cells in this area are affected by both the retinal location of the stimulus and eye position. From this combined response it is possible to deduce the spatial coordinates relative to the head coordinated system. No cells were found, however, that encode explicitly the spatial position of the stimulus with respect to the head.

3. Philosophical objections

Objections have been raised on philosophical grounds against the use of terms such as 'representations' and 'internal descriptions'. It has been argued that it is a categorical mistake to say that events in the brain can represent or describe something. 'Representation' is the use of symbols of a formal system by a cognizant agent who can interpret them. In the case of the brain, there is not cognizant agent outside the brain that will interpret brain representations. This type of argument is expressed, e.g., by P. Hacker (1990):

> "...neural firing are not symbols, do not have definitions of rules of syntax, are not employed by symbol-using creatures, cannot be used correctly or incorrectly, cannot be grammatical or ungrammatical, and are not part of a language".

This and related arguments are considered to be sufficiently devastating to the computational approach to conclude that *(ibid)*:

> "[Computational vision] as an explanatory theory of vision is a non-starter since such a theory is not conceptually coherent".

In contrast with this view, it seems to me that there is a reasonable notion of 'representation' that can be meaningful in describing patterns of activity in the brain. When we say that a certain device can be viewed as a symbolic system we imply that some (but not all) of the events within the system can be consistently interpreted as having a meaning in a certain domain. For example, some of the events within an electronic calculator have a consistent interpretation in terms of entities in arithmetic. Other events inside the calculator, for example, in the power supply, do not have such an interpretation. Similarly for the brain: according to the representational view some of the events inside the brain may have a consistent interpretation in terms of depth, surface orientation, reflectance, and the like. This is apparently not what Hacker and others would like the term 'representation' to stand for, and it may be given a different name, say, representation*. The claim that certain brain patterns can be viewed as representations* is a statement that is testable, at least in part, by empirical means. In this sense we can ask whether or not we find in the brain a common explicit representation* of depth, for instance, used for a number of different tasks.

It is probably true that the notion of representation and perhaps the entire computational approach do not provide an explanation to epistemological questions related to knowledge, meaning, or awareness. In my view, this is also not the goal of current computational vision. As far as the notion of representation is concerned, it seems to me that in the manner it is used in computational vision it has a reasonable and potentially useful meaning. At the same time, the use of common, explicit, representations by (or in) the visual system is still an unsettled issue.

Assumption VI
LINKS TO THE BRAIN

This assumption has to do with the link between computational theories and their interpretation in terms of physiological events. In making this connection it is usually assumed that the firing rates of individual units represent the meaningful physiological variables. When we look for, say, depth representation in the brain, we will try to discover single units whose firing rates correlate with perceived depth.

This assumption is more technical in nature then the previous ones. It is an important assumption for making the link between theoretical models and empirical studies of the brain. This assumption does not mean, however, that a single unit by itself codes the meaningful variable. The brain could use a population encoding, e.g., for orientation, in which a single unit codes for a possible range of orientations. In this case the firing rate of a single unit is ambiguous and does not code a unique orientation value. The value is determined instead by a distributed activity across a population.

There is a body of physiological evidence that supports this general notion. Starting from low level features, units in V1 seem to signal (either individually or within a population encoding) the presence of oriented features, their orientation, contrast, direction of motion, etc. More surprisingly, single units in an area of the STS of the macaque monkey seem to function as some sorts of 'face analyzers'. Their individual firing rates have clear correlations with the presence of e.g. specific faces, observed from a restricted set of viewing directions. In a recent study by W. Newsome and his collaborators (Salzman, Britten & Newsome 1990) microstimulation of single units in area MT of the monkey were sufficient to bias the animal's judgements towards the direction of motion encoded by the stimulated neurons.

The reason for bringing up this assumption is that there have been recently some challenges that, if verified, may change this picture, and may have important implications to the 'physiological link' assumption.

Challenges

Oscillations and temporal coding
There have been a number of challenges and alternatives to the notion of the firing rate of a fixed set of neurons as the main variable encoded by neurons. One example is the temporal coding notion of B. Richmond and his coworkers (Richmond, Optican & Gawne

1987). They have shown some evidence that the temporal firing pattern of neurons may carry important information, e.g., about the shape of the stimulus. According to this evidence, a neuron may respond to two different stimuli with the same mean firing rate, but signal two different events by the shape of the temporal firing pattern. In such a case it is not meaningful to ask merely how strongly a given unit responds to a given stimulus; the more relevant variable is the temporal pattern of the response.

Another possibility that is attracting recently considerable attention is based on the findings of correlated oscillations in the cortex. Gray, Singer, and others (e.g., Gray, Konig, Engel & Singer 1989) have shown correlated oscillations of about 40-60 Hz in the cat's visual cortex in response to moving stimuli. For example, they have stimulated two cortical sites separated by 7mm in the cortex, that had non-overlapping receptive fields, arranged colinearly, and with similar orientation preference. When the separate receptive fields were stimulated by a single long bar, the responses at the two sites were oscillatory at about 50 Hz and phase locked. When the single line was replaced by two shorter lines moving together the phase locking was reduced, and when the shorter lines moved independently the correlation disappeared completely. This and related experiments raise the possibility of the use by the cortex of temporary ensembles that oscillate coherently together. On this view the firing rate of an individual unit has little meaning, if its firing is not a part of an oscillating ensemble.

A number of experiments have suggested this mode of activity, but their interpretation is still not entirely clear and unequivocal. If this notion of oscillating ensembles turns out to be a fundamental cortical mechanism, it will not change the computational theories on their own, but it will change significantly the relations between theory and modeling. It will change the way physiological data is interpreted for the purpose of modeling, the form of the predictions, and the forms of the models themselves.

Acknowledgemet: This work was partially supported by NSF grant 8900267

References

Baars, B.J. 1988. *A Cognitive Theory of Consciousness* . Cambridge: Cambridge University Press.

Brooks, R.A. 1986. A robust layered control system for a mobile robot. *International j. of Robotics Research, 2,* 14-23.

Crick, F.H.C. & Koch, C. 1990. Towards a biological theory of consciousness. Manuscript in preparation.

Fodor, J.A. 1983. *The Modularity of Mind.* Cambridge, MA: The MIT Press.

Freuder, E.C. 1974. S computer vision system for visual recognition using active knowledge. *M.I.T. A.I. Lab. Tech. Report 345.*

Gibson, J.J. 1950. *The Perception of the Visual World.* Boston: Houghton Mifflin.

Gibson, J.J. 1979. *The Ecological Approach to Visual Perception.* Boston: Houghton Mifflin.

Gray C.M., Konig, .., Engel, K. & Singer, W. 1989. Oscillatory responses in cat visual cortex exhibit inter-columnar synchronization which reflects global stimulus properties. *Nature 338,* 334-337.

Edelman, G.R. 1989. *The Remembered Past* . N.Y: Basic Books.

Frisby, J.P. 1980. *Seeing: Illusion, Brain and Mind.* Oxford: Oxford University Press.

Gregory, R.L. 1970. *The Intelligent Eye.* London: Weidenfeld & Nickolson.

Hacker, P.M.S. 1990. Seeing, representing, and describing: An examination of David Marr's computational theory of vision. Manuscript in preparation.

Jackendoff, R. 1987. *Consciousness and the Computational Mind .* Cambridge, MA: MIT Press.

Johnson-Laird, P.N. 1988. *The Computer and the Mind.* Cambridge, MA: Harvard Univ. Press.

Konishi, M. Sullivan, W.E. & Takahashi, T. 1985. The owl's cochlear nuclei process different sound localization cues. *J. Acoust. Soc. Am., 78,* 360-364.

LeCun, Y., Boser, B., Denker, J.S., Henderson, D. Howard, R.E., Hubbard, W. & Jackel, L.D. 1989. Backpropagation applied to handwritten zip code recognition. *Neural Computation 1(4),* 541-551.

Livingstone, M.L. & Hubel, D.H. 1981. Effects of sleep and arousal on the processing of visual information in the cat. *Nature, 291,* 554-561.

Livingstone, M.L. & Hubel, D.H. 1984. Anatomy and physiology of a color system in the primate visual cortex. *The J. of Neuroscience, 4(1),* 309-356.

Marr, D. Vision. 1982. San Francisco: W.H. Freeman.

Minsky, M. 1987. *Society of Mind.* N.Y.: Simon and Schuster.

Minsky, M. 1990. Logical vs. Analogical, Symbolic vs. Connectionist, Neat vs. Scruffy. The 1990 Japan Prize Lecture, Circulated manuscript.

Nakayama, K. Shimojo, S. & Silverman, G.H. 1989. Stereoscopic depth: its relation to image segmentation, grouping, and the recognition of occluded objects. *Perception, 18,* 55-68.

Penrose, R. 1989. *The Emperor's New Mind* N.Y.: Oxford University Press.

Poggio, T. Torre, V. & Koch, C. 1985. Computational vision and regularization theory. *Nature, 6035,* 314-319.

Ramachandran, V.S. 1985. The neurobiology of perception. Guest editorial, *Perception, 14,* 97-103.

Richmond, B.J., Optican, L.M., & Gawne, T.J. 1987. Evidence of an intrinsic temporal code for pictures in striate cortex neurons. *Abstracts Soc. Neuros., 17 Ann. Meeting, New Orleans, LA, Nov. 16-21,* 178.11 p. 631.

Salzman, C.D., Britten, K.H. & Newsome, W.T. 1990. Cortical microstimulation influences perceptual judgements of motion direction. *Nature, 346,* 174-177.

Searle, J.R. 1980. Minds, brains, and programs. *Behavioral and Brain Sciences, 3,* 417-457.

Searle, J.R. 1990. Consciousness, explanatory inversion and cognitive science. *Behavioral and Brain Sciences,* to appear.

Tenenbaum, J.M. & Barrow, H.G. 1976. Experiments in interpretation-guided segmentation. *Stanford Research Institute Tech. Note 123.*

Ullman, S. 1979. *The Interpretation of Visual Motion.* Cambridge, MA: MIT Press.

Ullman, S. 1980. Against direct perception. *The Behavioral and Brain Sciences, 3,* 373-415.

Zipser, D. & Andersen, R.A. 1988. A back-propagation programmed network that simulates response properties of a subset of posterior parietal neurons. *Nature, 331,* 679-684.

Vision tells you more than "what is where"

H. B. Barlow

Physiological Laboratory, Cambridge CB2 3EG, U.K.

David Marr started his book with the following question and answer:- "What does it mean, to see? The plain man's answer (and Aristotle's too) would be, to know what is where by looking" (Marr 1982). This mentions "knowing what", but most of the rest of the book was devoted to the problem of computing a representation, since he assumed that this was a necessary preliminary to the operations concerned with object recognition. This led him, and after him the rest of us, to appreciate what a very difficult computational task it is to go from the 2-D image, to a representation of the visible surface, and from there to a representation of the 3-D world with 3-D objects in it. I think it is fair to say that a large fraction of the effort in vision research over the past decade has been devoted to the problems he pointed to in that book, so this *tacit assumption that representation precedes recognition* has been very influential. But it is misleading.

I shall argue in this talk that representation and recognition proceed concurrently. The emphasis on representation has led to the neglect of something that is very important throughout image processing in the visual system, even in the very earliest stages. This is the assessment of the probabilistic and associative structure of sensory messages. "What unusual events have just happened?" "What unexpected combinations of events are occurring?" "Have such combinations happened often before?" It is necessary to take this sort of information into account in order to learn efficiently, and I shall argue that it is also essential for elementary object recognition, and even for selecting the primitive features to be used for representing visual scenes.

NEED FOR PRIOR PROBABILITIES IN LEARNING

To demonstrate the importance of paying attention to prior probabilities, it is easiest to start by taking an example at a high level, when objects and events have already been classified and recognised. At this level one can introspect and see the importance of associative structure and unexpectedness intuitively.

Consider Pavlov's dog. It is in a complex environment with all sorts of things going on that it does not understand; then from time to time something pleasant occurs in the form of a tasty morsel in its mouth, or something unpleasant such as a shock to its foot. The dog's brain discovers those features in the complex environment that reliably predict these reinforcements, and your own brain would do the same. Is this just a matter of the strengthening of a pathway from the conditional stimulus to the response by the repeated sequence of a feature followed by the unconditional stimulus, as a naive form of Hebb's thesis might suggest? There is more to it than that, for it is not just the high joint probability of conditional and unconditional stimulus that identifies a valid predictor, but a value approaching unity for the conditional probability of reinforcement, given the conditional stimulus. But P(U|C) = P(U.C)/P(C), so to get the conditional probability from the joint probability one must divide by the probability of the conditional stimulus. For efficient learning the prior probability of the conditional stimulus is a necessity.

To see this intuitively put yourself in the dog's position. It is obvious that the appropriate feature has to be identified among a host of alternative possible features, all of which will be stimulating the senses in the complex environment, many of them doing so in the right temporal relation to the reinforcing stimulus. But you need not consider, as potential predictors of reinforcement, those features of the environment that are there almost all the time; it's safe to assume that they will be there whether or not the reinforcing stimulus is given, so they have no predictive value. One should seek the conditional stimuli among the unusual features – those that are about as unusual as the reinforcement – but in order to select such features and reject those that are constantly present one obviously needs to know, not just that a particular feature is present, but also how often it occurs in the current environment.

Knowledge of the prior probability, or unexpectedness, of features in the environment actually helps in four ways. First, it is necessary in order to calculate the conditional probability, which has predictive value, from the joint probability, which means little by itself. Second, it helps one select candidate features that could be strongly correlated with reinforcement: if two things have different rates of occurrence they cannot be strongly correlated. Third, it points to the features for which the "pay-off" would be high if they were found to be correlated, and figure 1 shows the entropy reduction that can be effected if one knows about the correlation – not a very intuitive way of expressing the pay-off, but at least it's quantitative. Fourth, it helps to avoid concluding that a feature is predictive when in fact the correlation is simply accidental; for example if C and U each occur on 10% of all possible occasions, it is not in the least surprising if C precedes U on 1% of trials, because that is the predicted rate if they were statistically independent.

These four ways that knowledge of prior probabilities helps are just different aspects of the obvious fact that, to detect associations reliably, one has to do statistics, but I hope they show that, for learning at least, *simple representation is not enough: you also need to know the prior probability and unexpectedness of the sensory messages.*

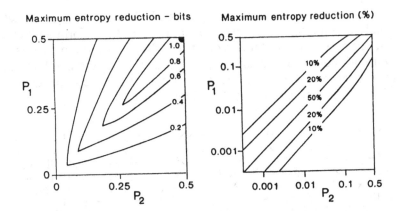

Figure 1. When an association is discovered one can recode to reduce entropy. This shows the maximum reduction of entropy possible if an association is discovered between two binary variables having probabilities P_1 and P_2. At left reduction is in bits, and the scales are linear; at right percentage reduction is shown, and scales are logarithmic. The greatest benefit is obtained when the two probabilities are nearly equal.

THE NEED FOR PRIOR PROBABILITIES IN OBJECT RECOGNITION

In the above discussion it was assumed that objects were represented in the parts of the brain where conditioning occurs, but this assumption begs a lot of questions (and is incidentally another example of a "misleading tacit assumption"). I want to suggest here that knowledge of the prior probabilities of the features that signal objects is likely to be needed for object recognition itself.

Object recognition is partly a matter of attaching a word such as "chair", "apple" or "pencil", to a subset of sensory messages that signal the features of the object, but there is another step that must be completed first. This is the task of segregation – identifying subsets of sensory messages that signal objects. I've got nothing to say about the naming process, but I think segregation is a statistical task that is similar in some ways to that of conditioning, and prior probabilities are likely to be equally important.

Conditioning depends on detecting the association between a conditional and an unconditional stimulus; segregating subsets of messages caused by the features of objects depends upon detecting that these messages frequently occur in association with each other – they form a "cliché" in fact.

A natural definition of an object might be "something that behaves as a unit or single entity, something that stays in one piece and that moves or can be moved about the world in one piece". But if the object stays in one piece, then it follows that it will tend to excite the same collection of sensory messages. There will of course be much variation from changes of perspective, illumination etc., but the constancy of the physical characteristics of the object is bound to cause the primitives it excites to co-occur at a rate greater than that expected by chance. Hence objects must give rise to sensory clichés.

In figure 2 there is an array of a large number of little figures, each of which is made up of a random combination of lines. In some cases these form an "object", ie a recognisable letter or another striking geometrical shape, but in most cases they do not. In the figure, all combinations are equally probable, but the combinations that form letters have been experienced more often than by chance expectation in the past, and I think it is this fact that makes them into "objects".

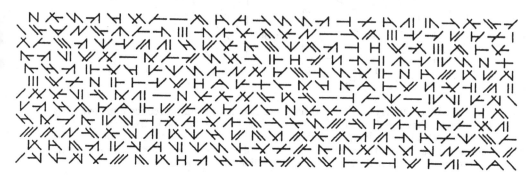

Figure 2. Each small figure is a random selection of three lines at successive positions. In most cases the figure is meaningless, but in some cases they are recognisable letters, and these stand out prominently as recognisable "objects". It is suggested that this is because this particular combination has been often experienced before, and therefore constitutes a "cliché" – a combination occurring with probability above chance expectation.

One can see at once that segregating the subsets of sensory messages that objects cause is similar to the conditioning task, but in this case there is no special class of sensory messages corresponding to the unconditional stimulus or reinforcement; one segregates sensory messages that have previously been associated with each other, rather than with the messages caused by an unconditional stimulus. This obviously makes the task more difficult simply because the number of possible associations is larger when there is no longer a restricted number of reinforcing stimuli having a special status. Furthermore there seems no simple definition of a conditional probability which "predicts" an object in the way that a properly selected high conditional probability can predict reinforcement. The way that evidence is combined to diagnose the presence of objects is complicated and largely unknown, but it is a safe bet that the prior probabilities of sensory messages are important for the efficient performance of the task.

PRIOR PROBABILITIES AND THE SELECTION OF PRIMITIVE FEATURES

Above I was following the traditional assumption that objects are signalled in the brain by combinatioins of certain primitive features such as edges, colour, texture, common direction of motion and common disparity – the properties the Gestaltists pointed to as causing grouping and figure/ground segregation, and the properties which neurophysiologists have subsequently discovered to be the features that neurons in the primary visual cortex respond to selectively. I think the relation between these primitives and the physical stimuli from natural images is similar to the relation between objects and features. Just as objects are the clichés in the world of feature signals, so the features themselves are the physical clichés in natural images. It is its ability to capture and exploit the associative structure in images of the world that causes a primitive to be effective.

It is often suggested that edges, colour, common motion and common disparity are used as features because they signal objects, but that is a dangerously circular argument. If these are the primitives we use, how else could an object be signalled? In contrast I am suggesting that their virtue results from properties of physical images, and that object segregation is a second stage. This captures the clichés in the physiological messages that carry signals of these primitives, so it is a repetition of the first stage. I am not suggesting that feature detectors are discovered anew by each brain, though there is a possible role for experience in developmental mechanisms (see below). No doubt evolutionary selection is mainly responsible for the primitives we use, but for this to have happened there must have been a selective advantage, and I am saying this is due to their relationship with properties of natural images. It should, then, be possible to point to the associative structure that is captured by each of them.

A patch of uniform colour, texture or disparity constitutes a cliché because there are strong auto-correlations for these properties over considerable distances in natural images, just as there are for luminance itself. This means that messages signalling the same colour, or disparity or texture, from neighbouring image regions will occur together more often than they would by chance, assuming independence, and that is the reason for using them as primitives. Edges and motion are a little different.

An edge is a region in the image that has translational symmetry in a particular direction; as you move along the edge, the profile of luminance at right angles to its direction stays the same. There is a family of edges of different orientations and also of different luminance profiles across the edge, but each member of the family is a cliché in the sense used here, because these translationally symmetric regions of the image occur in images more often than chance expectation. How do we know this is true?

Images that have been filtered so that their spatial frequency power spectra are as nearly as possible flat, and consequently have reduced autocorrelations, have edges as their most prominent remaining structural feature. This is not true of all images (for instance a random dot pattern) filtered in this way, but it is generally true of images of the world around us. It follows that the patches of translational symmetry that we call edges are commonly present in images and are properly dubbed clichés.

Motion is also different. Very often two successive images are related to each other by a simple displacement, so a single 2-vector allows the second to be completely predicted from the first. In almost all cases there are far fewer different motion vectors than there are individual objects. And at other times, provided that the 3-D information is available, movements over the whole image can be predicted from perspective transformations and a single displacement vector. Such arguments have always been the stock-in-trade of empiricist philosophers, but knowledge of the prior probabilities of movements is necessary in order to be able to conclude that specific combinations from different parts of the visual field occur together more often than they would by chance. Again, this knowledge is essential.

THE HIERARCHICAL USE OF PRIOR PROBABILITIES

I have discussed above three stages in the use of visual images, namely feature extraction, object recognition, and identifying conditional stimuli. At each stage (there are probably many more than three) it was proposed that something akin to statistical inference is done, namely detecting that a subset of the inputs occurs together more often than expected by chance. This subset then became the basis of a primitive feature, an object, or an identified conditional stimulus. These successive stages, and some of the requirements and consequences of such an arrangement, are shown in table 1.

The concept of representation in the visual system that such a scheme implies is radically different from that suggested by the tacit assumption that a process of simple representation, albeit in 3-dimensions, precedes recognition. In the new system representation and recognition proceed concurrently, and the representation becomes enormously richer, for it includes the stored knowledge of the prior probabilities and associative structure of the world. That is a heavy demand, but some such stored knowledge is surely required to explain the fact that every experience is seen in the light of previous experiences, not as something written for the first time on a clean slate. However there is an important aspect, complementary to the inference-like detection of associated subsets or clichés, that has not yet been described.

Table 1. Three inference-like steps in perception. To find that $P(C.U) > P(C) \times P(U)$, you must always know $P(C)$, where $P(C)$ is the prior probability of the physical stimulus, the primitive, or the object. This requires that unexpectedness (prior probability) must be signalled between levels. Note also that a representation formed on these principles is very different from a simple representatiion and includes stored knowledge of the associative structure of the world.

Primitives	**Objects**	**Conditional stimuli**
Suspicious coincidences of physical stimuli	Suspicious coincidences of primitives	Suspicious coincidences of object (C) with reinforcement (U)

IMPORTANCE OF INDEPENDENCE

One is forced to a hierarchical view of the organisation not only by the immensity of the problem of determining the associative structure of sensory messages from the world, but also by looking at the neuroanatomy: a neuron in the primary visual cortex can only be influenced directly by a small fraction of the whole visual image, so associations over larger regions must be handled by bringing together outputs from small subsections of the image. Each element at each level has the task of detecting a cliché – a subset of its inputs that is active together more often than by chance – but in each subsection there are many elements working in parallel on much the same input from the retina. This creates a problem: there will be a strong tendency for all the elements to identify the same cliché. If this happened it would be ruinous, for it would cause there to be associations passed on to higher levels that did not result from associations in the input, but from failure of the lower level units to detect different clichés in their inputs.

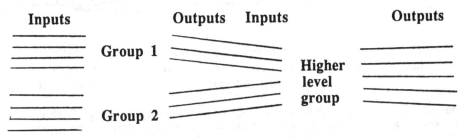

Figure 3 Inputs at the left go to two groups of elements, within which each element handles very much the same subset of inputs. They would therefore tend to detect the same clichés, but if they did the outputs from each group would be strongly associated with each other, and these associations would make it very difficult for elements at a higher level to detect new associations <u>between</u> outputs from group 1 and group 2. Outputs within a group must therefore be decorrelated.

To avoid this, one must postulate a mechanism for ensuring that the elements being fed from strongly overlapping sets of inputs adjust their outputs so that they are uncorrelated. One can regard this as a mechanism to justify the *a priori* assumption of independence by higher level elements, and it is necessary because it is only the assumption of independence that makes it possible to estimate the a priori probability of occurrence of a group of inputs from the lower levels. If these inputs were contaminated by strong associations resulting from overlap to the units at an earlier level, this would make it very difficult to detect new associations, which is the whole purpose of recombining the inputs.

A TENTATIVE MODEL

How might all this be done? Foldiak and I have a tentative model (Barlow and Földiák 1989; Földiák 1990) that is sketched in fig. 4. On the left, inputs converge on to a unit through Hebb-like synapses. Because these synapses are only strengthened when the output neuron fires, the synapses from a subset of the inputs that occur together will be strengthened, so such a unit will eventually tend to respond to the co-occurring inputs or clichés. In the middle are shown a set of such units with lateral interconnections that have anti-Hebbian properties. These can be thought of as inhibitory connections that increase in strength when a pair of units fire together, or alternatively as excitatory interconnections that decrease in strength when both units are active; in the latter case there would have to be a general backgound of inhibition (not shown) to prevent explosive effects from positive feedback. The effect of this anti-Hebbian mutual interaction is to decorrelate the outputs.

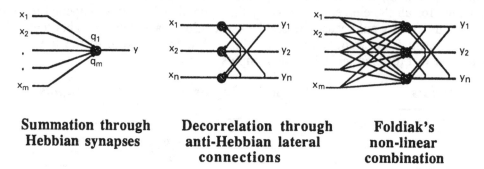

| Summation through Hebbian synapses | Decorrelation through anti-Hebbian lateral connections | Foldiak's non-linear combination |

Figure 4. The unit on the left forms the weighted sum of its inputs through Hebb-type synapses; the adjustments of the weights causes it to detect commonly occurring combinations or clichés. The lateral interconnections between the units in the centre have anti-Hebbian properties, which makes the outputs uncorrelated. At the right the two are combined, and the output is given a threshold. One then has a sytem that detects independent clichés, as required. The threshold can be adjusted locally to ensure a given probability of firing, thus fulfilling the requirement that prior probabilities are propagated through the network.

On the right the combined network is shown. It is not immediately obvious how nicely the two mechanisms co-operate. Anti-Hebbian decorrelation prevents a unit from firing unless it does so with statistical independence from its neighbours, but it is the firing of a neuron that determines when the Hebbian strengthening of a synapse from the input occurs. Hence it will make the different units respond to different subsets of co-occurring inputs, that is to different clichés.

The above account does not describe how a unit takes account of prior probabilities. The threshold for output of a unit can be set locally so that it has a predetermined probability of firing. Thus a unit at a later stage is in a position to "know" that, when a unit from the lower level fires, the event signalled has this definite prior probability. In a slightly more complex scheme, the rate of firing might be graded according to the unexpectedness of the event signalled. This does not say how the very first unit knows the prior probabilities of its inputs, but it would work for all later stages.

Elements with Hebbian forward connections and anti-Hebbian lateral connections, and with locally set thresholds that ensure particular prior probabilities for the outputs, seem very suitable for building up a hierarchichal network, but it must be added that much more is needed to make a working system. The detailed patterns of connection that is wired in, and the time intervals over which probabilities are estimated, are two of the more obvious gaps in the above account.

POSSIBLE NEUROPHYSIOLOGICAL CORRELATES

The idea of a Hebbian mechanism developing the sets of synaptic weights that detect clusters or clichés of co-active inputs provides an attractive explanation for many of the phenomena associated with visual deprivation during the sensitive or critical period (Hubel and Wiesel 1970, Rauschecker and Singer 1981, Singer 1990). It does not, however, explain why, under normal, undeprived, conditions a set of feature detectors develops that covers the whole array of orientations and ocular dominances; if the Hebbian mechanism was the only one at work, what is to prevent most of the units responding selectively to whatever orientation or eye happened to be stimulated most strongly or most often?

This is where the anti-Hebbian lateral interactions are required, and there are a range of quite different facts (reviewed in Barlow 1990) that support the existence of this mechanism. Briefly, many of the phenomena of pattern adaptation that are commonly explained in terms of fatigue of pattern selective neurons can be explained equally well by anti-Hebbian lateral interaction, and this mechanism alots an important functional role to the phenomenon. The connection between the network of fig 4 and some of the facts about plasticity in the visual system are summarised in table 2.

A network of this sort would also offer an explanation for one of the most surprising results to emerge from multi-unit recordings from the visual cortex, and it makes a testable prediction. It is often pointed out that, when one records from a single neuron in the visual cortex (or anywhere else for that matter) one is paying no attention to the behaviour of the large numbers of other neurons that are being activated by one's stimuli. This is true, and there is the possibility that a great deal of information is carried by the joint activity of sets of neurons. The difficult techniques for multi-unit recording and analysis have now been worked out and are being succesfully applied, but one of the interesting initial result seems to be that the great majority of pairs of neurons behave with a high degree of independence, even when they are picking up from the same area in the visual field. This is a remarkable result in view of the shared input that such neurons must have, but it is exactly what one would expect if there are anti-Hebbian lateral interactions between neurons that cause them to be decorrelated under ordinary conditions of excitation.

The prediction of the Foldiak-type network is that the uncorrelated activity of neighbouring neurons would be one of the first things to be affected by deprivation, and could perhaps be modified by even quite short periods of pattern adaptation, since we believe this causes modifications of the anti-Hebbian interactions. This prediction should be very easy to test.

Table 2 Two types of neural plasticity mediated by Hebbian forward connections and anti-Hebbian lateral connections, showing the period of plasticity, how it is demonstrated experimentally, and its functional role.

	Forward connections	Lateral connections
Synapse type	Hebbian	Anti-Hebbian
When operative	Post-natal period only	Throughout life
Experimental demonstration	Deprivation in sensitive period	Pattern adaptation
Functional role	Detecting clichés or coincidences	Decorrelation – making units detect different coincidences

CONCLUSIONS

1) Representation does not precede object recognition – they occur concurrently.

2) Learning, object recognition, and selection of primitives, all depend on an inference-like process of detecting clichés or suspicious coincidences – conjunctions that occur more often than chance expectation.

3) This requires that the prior probability or unexpectedness of messages be signalled from level to level.

4) There are two steps in using unexpectedness: detecting the cliché's and making sure they are different from each other.

5) The first requires a Hebb-like process, possibly exemplified by the effects of experience during the sensitive period.

6) The second requires an anti-Hebbian process, possibly exemplified by pattern-specific or contingent adaptation, and by many other effects proving that perception takes associative structure into account.

7) These two operations can be repeated hierarchically in a network, each level recombining messages in different ways.

8) A representation formed in this way stores knowledge of the associative structure of the normal environment, and would be an appropriately challenging task for cerebral cortex to perform.

REFERENCES

Barlow, H. B. (1990): A theory about the functional role and synaptic mechanism of visual after-effects. In C.B.Blakemore (Ed.), *Vision: coding and efficiency* Cambridge, UK.: Cambridge University Press.

Barlow, H. B., & Földiák, P. (1989): Adaptation and decorrelation in the cortex. In R. Durbin, C. Miall, & G. Mitchison (Ed.), *The Computing Neuron* (pp. 54-72). Wokingham, England: Addison-Wesley.

Földiák, P. (1990): Forming sparse representations by local anti-Hebbian learning. *Biological Cybernetics*, (In press)

Hubel, D. H., & Wiesel, T. N. (1970): The period of susceptibility to the physio logical effects of unilateral eye closure in kittens. *Journal of Physiology (London)*, *206*, 419-436

Marr, D. (1982): *Vision*. San Francisco: W H Freeman.

Rauschecker, J. P., & Singer, W. (1981): The effect of early visual experience on the cat's visual cortex and their possible explanation by Hebb synapses. *Journal of Physiology (London)*, *310*, 215-239

Singer, W. (1990): Search for coherence: a basic principle of cortical self-organisation. *Concepts in Neuroscience, 1*, 1-26.

Some strategic questions in visual perception

Bela Julesz

Laboratory of Vision Research, Rutgers University, New Brunswick, N.J.
and Biology Division, California Institute of Technology, Pasadena, CA

"Problems arise in a variety of ways, and it is often worthwhile to list the forms that they may take. Thus we can distinguish the following:

1. The classical problem, which has had much effort expended upon it, but without any acceptable solution.
2. The premature problem, which often is poorly formulated, or is not susceptible to attack.
3. The strategic problem, which seeks data on which a choice may be made between two or more basic assumptions or principles.
4. The stimulating problem, which may lead to a reexamination of accepted principles and may open up new areas for exploration.
5. The statistical question, which may be only a survey of possibilities.
6. The unimportant problem, which is easy to formulate and easy to solve.
7. The embarrassing question, commonly arising at meetings in the discussion of a paper, and rarely serving any useful purpose.
8. The pseudo problem, usually the consequence of different definitions or methods of approach. Another form of pseudo problem is a statement made in the form of a question. It is often the result of discussions in meetings.

It is frequently helpful to attempt to place a given problem in this array of possibilities, for such a classification may provide a hint as to the problem's significance, the difficulties involved in its attack, and the sort of solution that may be expected" (von Békésy, 1960).

I. On metascientific problems.

During the Spring and Summer of 1990, while I prepared this manuscript for *The Tacit Assumptions in Vision Research Symposium*, intended for my psychologists colleagues, I worked on and finished three other manuscripts: one written for physicists, the second for neurophysiologists and the third for philosophers interested in visual perception. The first, intended for physicists, is an elaborate review entitled *Early Vision and Focal Attention*, and was written at the request of the Editors of *Review of Modern Physics*. The second article of mine is an open peer review in which I answer, among others, the philosopher John Searle on his recent idea that the brain cannot have unconscious processes. This paper will be published sometime next year in the journal *Behavioral and Brain Research* and I will quote from my answer at a few places here. My third article, intended for neurophysiologists, entitled "Early vision is bottom-up except for focal attention," will appear in *Symposium #55: The Brain* celebrating the 100th anniversary of the Cold Spring Harbor Laboratory. These three articles, together with this one, span a large audience in different disciplines and while working on them permitted me to ponder over the state of psychobiology from four different perspectives.

Briefly I return to the article written to the physicist. There I expressed my opinion about the state of brain research, particularly about my own specialty, visual perception, claiming that it is in the same state as physics was prior to Galileo or biochemistry was prior to the discovery of the DNA double-helix by Watson and Crick. Obviously, the physicists and chemists have to be reminded that the human visual cortex, the most complex structure in the known Universe, is many orders of magnitude more complex than the atoms and molecules they usually study and it is no surprise that psychobiology is in such premature state. Of course, my colleagues in experimental psychology and neurophysiology know the fragmentation of our field in the absence of some fundamental insights yet to come. One of the basic and unanswerable questions is whether there is a really strategic problem in brain research - similar to the DNA double helix - or did brains evolve in ad hoc ways, and as result do operate by using "a bag of tricks" as Ramachandran (1985) suggests. Since I regard this question as a metascientific question, that is, beyond our present scientific understanding, I include this with some other metascientific questions that I list below and skip, and leave them to some more adventurous colleagues to ponder on.

So, I regard the following questions - perhaps among the most interesting problems of mankind - unsolvable at present even though some famous scientists think otherwise:

A) Why do higher organisms sleep?
B) Why do most sleeping organisms dream?
C) What are the mechanisms of short and long term memory?
D) Is there free will?
E) What is the essence of the state called consciousness?

Obviously these metascientific questions are not equally complex. Perhaps the riddle of short and long term memory will be solved in the not too distant future. However, it could be that "consciousness" belongs to a Gödel-like problem, that might exclude neural nets to inspect (solve) certain complex states of their own. ["Whether the human brain is equivalent or more powerful than a Universal Turing Machine (UTM)?" is another metascientific problem, although Penrose (1989) thinks he can discuss it scientifically. The fact that the relatively simple "halting problem" cannot be solved (computed) by a UTM should caution us about ever having the mental power to answer the really tough metascientific problems of our mental states.]

There is a metascientific question that interests me much more than consciousness. This is the problem of "sensations" (also called "qualia"). Wherever the neurophysiologists probe the brain with their microelectrodes, they seem to record similar histograms of neural spike activity, regardless of whether the corresponding sensations are brightness, color, pitch, itch, temperature, pain, pleasure, anxiety, hunger, or contentment. Johannes Müller (1844) in his doctrine of "specific nerve energies" already was aware of the problem of sensations, and stated that *specific sensations arise in specific brain areas*. While this doctrine might seem somewhat vacuous, I challenge any philosopher, or brain researcher to say more with certainty at present, than Johannes Müller said 150 years ago. [The interested reader might be referred to a debate between John Searle and the author (Julesz, 1991a, in press).]

The problem of sensations raises several related metascientific question that I list here with indices:

F1) Are the known laws of physics adequate to understand the emergence of "sensations (qualia)" in functioning brains, or are there some unknown forces (laws) of physics acting that are necessary to understand these mental phenomena?

F2) Is the mind-brain problem identical to F1) or different?

F3) Is the brain equivalent or more powerful than a Universal Turing Machine? [See, first section.]

Physics has similar metascientific, (in this case metaphysical) problems. "What was the Universe like (and its physical laws) before the Big-bang?"; "Why are elementary particles obeying probabilistic laws?"; "Why should all forces of Nature be unified into one super theory?"; "Why should there be some ultimate particles that could not be further reduced ad infinitum?', and so on. At this point I wish to clarify that I do not regard metascientific problems as ridiculous or useless. Indeed, many metaphysical problems at my birth have become respectable even solved problems in the last sixty years. The origin and age of the Universe as revealed by the 3 deg. Kelvin background radiation, the real meaning of $E = mc^2$ as revealed by nuclear fusion and fission, unification theories from quarks to super-strings, and so on, have become scientific problems and according to the epistemologist Karl Popper can now be falsifiable.

Here I repeat some points I raised in a debate I had with the philosopher John Searle, as I already mentioned earlier (Julesz, 1991, in press): "In every scientific field there are tacit assumptions and "tabu" problems that workers in the field tend to avoid. Most of these problems are avoided not because they are messy and not well formulated, but rather because our ignorance of these issues does not seem to impede progress. Indeed, these "luxury problems", whether solved or not, do not effect the daily routine of a scientific enterprise. For instance, "are mathematical structures and theorems (e.g. the Mandelbrot-set with its striking self-similarities, or analytical functions with their amazing properties, or twin-prime numbers and their distributions) discovered or invented?" is such a luxury question, usually asked by the layman while the professional mathematician goes ahead with creating new structures and proving theorems. I doubt that a creative mathematician ever pondered while solving a tough problem, whether he followed some logical steps ingrained in his brain, or has stumbled on some eternal truth. Of course, the philosopher could argue that the structure of the human brain evolved under the evolutionary pressures of the environment, that in turn obeys the laws of physics, so it is no wonder that the structure of the mathematical mind is reflecting some of the laws of the Universe. Again, whether such a statement is true or not, has no effect on the practicing mathematician! I am not regarding this problem trivial. After all, there is a qualitative difference between asking whether mankind might have stumbled eventually on, say, the four-color problem in any culture, or whether without a J.S. Bach the Six Brandenburg Concertos could have been ever created in the history of mankind. Therefore this problem might be rather deep in order to understand the difference between scientific and artistic activities, nevertheless, for the active mathematician it is a luxury problem. I could quote similar luxury problems in almost any scientific discipline, but these are so well known to the specialists, that I do not dwell on them further. While I admit that these luxury problems can be quite interesting, amusing, and thought provoking on their own merit, nevertheless, they are usually orthogonal to the strategic problems of the field they address and therefore I regard them metascientific.

While luxury problems belong to metascientific problems, I hasten to add, that not all metascientific problems are luxury problems. For instance, Darwin's theory of evolution, perhaps the greatest scientific insight ever made, in his time was a metascientific theory, since the carriers of evolutionary change - recombination and mutation of genes - were unknown (even though some of Mendel's experiments could have been known, but were not). In spite of that, the best scientists of that period immediately understood the significance of Darwin's theory. So, the criterion whether a theory or problem is metascientific and important, or metascientific and of the luxury-kind, depends on how the scientists receive them. Thus a luxury problem (theory) is a problem that good scientists almost by instinct disregard, because it does not help them in their research."

2. On Hilbert-like questions.

After this somewhat pessimistic introduction, I switch to a more optimistic tone. After all I was eyewitness to the monumental discoveries in psychobiology that took place in the last thirty years. There are now a gamut of exciting questions in vision research that can be attacked by existing scientific tools. In the following sections I will skip all the metascientific problems and leave them to Nobel laureates (but not to students in search for a PhD thesis topic), and turn to the more tractable ones that I regard strategic, in visual perception. I call them Hilbert-like problems, and I assure the reader that this is not immodesty on my part. Indeed, the German mathematician David Hilbert posed his famous problems in the early thirties as a scientific testament, and ironically, already during his lifetime he was proven wrong. Among his "strategic problems " the most celebrated ones were concerned with proving the consistency of several sub-fields of mathematics whose very foundations were shaken by the paradoxes that showed up in set theory. Unexpectedly, Kurt Gödel, among others, proved his epoch making theorem (known as the incompleteness theorem) that there exist mathematical problems that are undecidable. So, the most of the fundamental problems of Hilbert were preempted by Gödel, Alonzo Church, and Allen Turing, who proved that some rather simple mathematical problems could not be proved to be right or wrong, or in algorithmic form could not be computed. It is in this spirit that I pose my Hilbert-like problems in vision research, and I would be the happiest if some unexpected insights might preempt this list during my life-time!

Here is my list of strategic questions in vision research, with emphasis on *early vision*. Let me note that these problems are given in random order as they came into my mind, without any hierarchical order of importance or inherent structure, but in the next section I will single out my favorite ones.

PSYCHOLOGICAL QUESTIONS IN VISUAL PERCEPTION:

1) How, when, and where (in time and processing stages) do retinotopic coordinates convert into environmental (or ego-centric) ones?

2) Is the cortical-magnification-factor already compensated internally?

3) How is motion blur avoided without a "shutter"? Could it be that cortical shifter circuits are used?

4) Why do certain feature-conjunctions (i.e. color and orientation; color and motion) require scrutiny, while others (i.e. color and stereopsis; stereopsis and motion) do pop-out?

5) Is the preattentive/attentive dichotomy real, or it merely constitutes the extremes of a continuous scale?

6) How is fast search (i.e. scanning by scrutiny) at a 30-60 msec/item rate possible, when cuing locks focal attention to a place over 100 msec duration?

7) Is focal attention preventing inhibition by lateral elements within its aperture?

8) Under what conditions is "where" parallel while "what" is serial (as Sagi & Julesz (1985) found in textures)?

9) What is the role of figure-ground separation in texture segmentation?

10) Is asymmetry for same/different judgments the same for a few items transiently presented, as is texture segregation for many items (forming texture-like aggregates) when presented in a stationary way (due to lateral inhibition of nearby items)?

11) Can non-linear spatial filters explain filling-in effects, such as subjective-contours, and can they explain gap-completion effects too?

12) Is there only one search-light (or, two, or perhaps, seven)?

13) Are there pre-cues of attention that can be (voluntarily) ignored in perception.

14) Does aperture size of focal attention scale (similarly to eccentricity) with other factors (i.e. element size)?

15) What is the mechanism and where is the locus of focal attention?

16) Is there learning in early vision (e.g. slowly getting better in discriminating an array of T-s from L-s), is this learning retinotopic (as found by Karni and Sagi , 1990), and where is its locus?

17) Are 3-D and shape-from-shading cues texton-like?

18) What kind of feature differences (texton-gradients) can be detected preattentively, that is when focal attention is engaged in some identification task?

19) Are there features that can be recognized (identified) preattentively?

20) Is binocular hysteresis (Fender & Julesz, 1967) due to fusional or binocular-matching (label-preserving) mechanisms?

21) Is color utilized by the magnocellular system (e.g. for motion or stereopsis) at iso-luminance, and can such neurophysiological problems be decided by psychological methods?

22) Are there processes in early vision that are not bottom-up, but depend on top-down (semantic memory-like) processes (e.g. the object-superiority effect)?

23) At what stage and how do color, form (orientation), movement, and stereopsis begin to interact?

24) In spite of a 2 Diopter difference between blue and red color resolution due to the poor optics of the human eye (caused by the high chromatic aberration of the lens), why do we see both colors sharp when simultaneously presented?

25) Can one attend to scale (coarse versus fine detail, or low- versus high-spatial frequencies) the same way as to spatial location?

26) If a brain has N working neurons (and provided each neuron has 2 states), then the brain can be in $M=2^N$ states. Thus additional (M-N) retrieving neurons are needed to "sense" each state of the brain. These retrieving neurons can be regarded as the "grandmother cells". Those who do not accept such cells, what other retrieval principle could they propose instead?

27) How can seemingly similar spike histograms of aggregates of neurons give rise to unique sensations, such as color, pitch, itch, pain, etc.? Is it, that some important neural parameter of activity has been overlooked, or is it, that the brain areas that are stimulated determine the qualities of sensations as Johannes Müller assumed in his doctrine of "specific nerve energies"? Does any theory exist that would be more specific than Johannes Müller's concept formulated generations ago?

28) If the searchlight of attention is limited to about 30-60 msec/item scanning rate, how can observers perceive a large number of items in a complex image presented, say, for only 150 msec (as reported by Biederman, 1985)?

29) When is strabismus the cause of stereo-blindness, and when is binocular neural disfunction cause strabismus?

30) Are the "maturational windows (critical periods)" in early vision merely serving the need for fine-tuning to the environment, or are there some sophisticated perceptual skills that can be acquired only within the critical period (similar to language acquisition)?

31) Does the McCollough-effect belong to early vision or to some higher level cortical stages?

32) Elucidate the difference between perceptual mechanisms that process "projections onto the retinae" and those that "project out in space"?

MATHEMATICAL QUESTIONS IN VISUAL PERCEPTION:

1) Is it possible to extend the group concept such that it would only describe within an interval, but not outside of it? (E.g. that could explain why a rotated face appears familiar within a certain angle, but cannot be recognized outside.)

2) The aperture problem of motion perception: What mathematical constraints would disambiguate the orientation of perceived motion in an aperture?

3) Under what conditions can the 3-D shape of objects be determined from their 2-D projections (e.g. shade from shading)?

4) Extrema of curvature of 2-D projections of 3-D objects seems useful in perceptual segmentation (Hoffman & Richards, 1984). Can one establish a relationship between the extrema of curvatures of 3-D objects and the extrema (or some other geometrical singularity) of their 2-D projections?

5) Further research on iso-second and iso-third-order texture pairs. For instance, can one create iso-fourth-order texture pairs that perceptually segregate?

6) What kind of perturbations of objects can be perceived as one manipulates an object's shadow?

3. Some unlisted problems that became paradigms.

One of these questions that I did not list, since I found it so strategic and devoted much of my scientific life to it is: " *"How can the visual system reconstruct from two 2-D retinal projections a best possible 3-D representation of the environment?"* Similar variants of this question, for example, of *"How can the visual system reconstruct from a single 2-D retinal projection containing the reflectance distributions of Lambertian surfaces from a specified light source a 3-D relief?"* have been successfully attacked by Horn and Brooks (1985), Pentland (1986), Bülthoff and Mallot (1990), among others. The first question led to the introduction of random-dot stereograms (RDS) and random-dot cinematograms (RDC) into psychology (Julesz, 1960) and their impact on brain research is reviewed in the 25th Jubilee Issue of *Vision Research* (Julesz, 1986), and in the article in *Reviews of Modern Physics* (Julesz, 1991b, in press). In essence, computer-generated RDS and RDC showed that - contrary to common belief - stereopsis and motion perception are *global* processes, based on some cross-correlation-kind of operation, and do not require the enigmatic cues of semantics and Gestalt. This insight permitted to link global stereopsis and motion perception (of the short-range) to be linked to the neurophysiology of the input stages to the cortex. That is the reason why I call nowadays "cyclopean perception" rather "early visual perception.

The second strategic problem that I did not list for similar reasons was my 1962 question: What statistical parameters underlie preattentive texture discrimination?   This led to many mathematical inventions that produced stochastic texture-pairs with identical second- and third-order statistics.   The surprising outcome of this paradigm was the realization that - contrary to common belief - preattentive human texture discrimination was not governed by global (statistical) parameters, but was based on the density of *quasi-local* features, I called textons. Recently Fogel and Sagi (1989) and, independently, Malik and Perona (1990) developed texture-segmentation-algorithms based on local spatial filters (oriented Gabor-filters) followed by a quasi-local non-linear operation (simple squaring by Fogel and Sagi and some inhibition between neighboring elements by Malik and Perona) and a second spatial filter for final segmentation.   It was most impressive that it emulated human texture discrimination performance as measured by Kröse (1987), but still could not account for the asymmetry effects.   Therefore it is of great significance that Rubenstein and Sagi (1989) extended their model by determining the variances of the local texture elements' distributions after the non-linear stage, and found these variances asymmetric, particularly when the orientation of the elements was jittered, mimicking human performance.   Their model could account for the textural asymmetries reported by Gurnsey and Browse (1987) and probably will be able to handle some other asymmetries.   We also showed (Williams and Julesz, 1991, in press) that some of the textural asymmetries between gaped circles in closed circles and vice versa can be explained by the phenomenon of subjective contours that may close the gaps.   Since we know from the work of von der Heydt et al. (1988) that subjective contours of the Schumann (1904)- and Kanizsa (1976)- kind have neurophysiological explanation in V2 of the monkey cortex, it is gratifying that most phenomena of texture segmentation are in the realm of early vision.

For the author, who spent much of his scientific career in search for the elusive textons, it is anti-climactic, yet most satisfying, that quasi local spatial filters can extract texton gradients without having to specify complex concatenation rules between adjacent textons.   The reader familiar with speech research will recognize the familiarity between "phonemes" and "textons".   While phonemes were never well specified, and complex computer algorithms are now used to cope with the many ad hoc rules at their various concatenations in order to segment speech, nevertheless, the rudely defined phonemes permitted the development of phonetic writing, one of the great discoveries of human civilization.   Had the development of phonetic speech coincided with the invention of super-computers that could automatically segment speech and talk millennia ago, the skill of writing might have never developed.   Of course, the fact that our voice organs limit the number of phonemes to a few dozen helped their universal acceptance.   Similarly, the main insight from the texton theory was that of the infinite variety of 2-D textures, only a *limited number* of textons have perceptual significance and are evaluated *quasi-locally* in effortless texture discrimination.   And even though super-fast computers are needed for automatic texture segmentation, practitioners of visual skills - painters, designers of instrument panels, directors of movies, TV shows, and advertisements - can benefit from the texton theory by manipulating the viewer's eye.   Indeed, some of the great artists instinctively knew how to create a strong texton gradient to capture attention, or create a texton-equilibrium for which time-consuming scrutiny is needed to discover the hidden images.

In summary, I think that these two problems I raised in my youth were proven "strategic" and became paradigms as attested by my many colleagues who devote their efforts to contribute to them. Therefore, I turn now to some other, more neglected problems that are on my list.

4. Discussion of some of the listed questions.

A third problem that preoccupied me during the last decade is *focal attention* and the role it plays in early vision. In more general terms this problem is related to the *preattentive-attentive* dichotomy of vision. Although in 1962 I started to work on effortless texture discrimination, that is discrimination without scrutiny (Julesz, 1962), I was more interested in discrimination than in the process of scrutiny. Although I adopted Ulrich Neisser's term "preattentive" for "effortless," "instantaneous," and "automatic" I used before, it was only in 1979 when I became actively interested in focal attention. My first paper on textons and asymmetries in texture perception (Julesz, 1981) attests to this new focus. First with Peter Burt, followed by Jim Bergen we developed a technique of presenting a target element among an array of distractors followed my a masking array with variable delays (SOA). This backward masking proved quite effective, and correct responses as a function of SOA gave interesting insights in some properties of focal attention. Our work differs from the influential work of Anne Treisman basically in presenting targets and distractors as a dense array, so that they formed textures, and we have not much to say about sporadic targets and distractors that are widely scattered. Furthermore, even though masking is a tricky business, I trust more the direct SOA findings in the 30-120 msec range, than the indirect measurements of the slopes of reaction time (RT) curves, where the RTs are typically in the 1000 msec range.

Since work on focal attention is not as advanced as on the other two paradigms discussed above, I will take those questions from the list that are relevant, and will evaluate their present status. I start with Question 6). This question is a very profound one and concerns the paradox that it seems that *it takes considerable time to shift attention to a cue and even more time to disengage attention from this cue*, as shown by Nakayama and Mackeben (1989), among others, and therefore it is not clear *how can one shift attention rapidly from item to item.* Indeed these serial shifts of focal attention underlying scrutiny can take place rapidly without eye-movements as first observed by Helmholtz (1896), and elegantly studied by Posner (1978). [An English translation of this pioneering contribution by Helmholtz is provided by Nakayama and Mackeben (1989) from the third German edition after a century delay, because all existing English editions are based on the first German edition and Helmholtz added this important observation only to later editions.] While evidence is indirect, based on different models the experimental results are interpreted, the "searchlight of attention" scans about 30-60 msec/item (Sternberg 1966, Weichselgartner and Sperling 1987, Treisman and Gelade 1980, Bergen and Julesz 1983). Jukka Saarinen and I just measured directly the scanning speed of focal attention by briefly presenting (with masking) numbers at random locations, and though observers could not correctly report the order in the sequence, they could follow and identify as many as four consecutive numbers at 30 msec/item rates with few errors. Obviously this rate depends on the visibility of the texture gradients, and some parallel mechanism seems to facilitate serial search (Kröse and Julesz 1989, Wolfe and Cave 1990). These serial shifts of focal attention underlying scrutiny can take place rapidly without eye-movements as first observed by Helmholtz (1896), and elegantly studied by Posner (1978). The reason why Saarinen and I could obtain scanning rates in excess of 30 msec/item, while Sperling and his collaborators (Weichselgartner and Sperling, 1987) could only achieve at most 80 msec/item, was due to our modification of their paradigm. They presented characters in temporal sequence in a *single* spatial window that in our opinion slowed down the scanning rate. In contrast, we presented our items (numbers) at *different* random locations (on a circle around a fixation point) such that subsequent numbers were placed far from each other to avoid lateral inhibition that we observed years earlier (Sagi and Julesz, 1986). So it seems that if items are presented at some distance to avoid lateral interactions one can achieve the rapid scanning rates observed both by RT slopes and masking techniques. The question of why such presentations avoid the locking of attention to a given position is not answered. It seems that a new remote item suddenly presented at some distance acts as a stronger cue than just presenting some marker in the vicinity of an already presented item as practiced by Nakayama and Mackeben (1989) and Kröse and Julesz (1989). At the same time that nearby items inhibit each other, it is interesting that when observers identify a target by attending on them, this mental operation "lights up" the vicinity of the attended item and the detection of a dim test flash is much enhanced in this vicinity (Sagi and Julesz, 1986). As we see, many problems of the rapid scanning of attention are not solved. Particularly interesting is Question 12) that is related to focal versus divided attention. Is there only one search light, or are there more, and if there are more, are these discrete searchlights equivalent to the resource metaphor of workers in divided attention. So the question of which metaphors of attention, searchlight, divided resources, bottleneck, filters, and so on, are the best descriptions is another important unsolved question.

Here I turn to the related Questions 5) and 18): whether the preattentive-attentive dichotomy is real as posed by Treisman and her collaborators (Treisman and Gelade, 1980), and by us (Julesz and Bergen (1983), or just a metaphor pointing to the two extrema of a continuous scale of attention strength as we found in experiments where the positional accuracy of the texton-gradients was manipulated (Bergen and Julesz, 1983). Interestingly, the problem of preattentive-attentive dichotomy can be further refined. The first step in this direction was the finding by Sagi and Julesz (1985) that detecting the position of a texton-gradient could be accomplished in parallel (independent of the number of distractors), while identifying the actual orientation or color of an element at the location of the texton-gradient required element-by-element scrutiny (hence monotonically increased with the number of distractors). Unfortunately, in this study the question of detection versus identification is blurred. After all the detection of a very fine positional variation is similar to pattern recognition, and the course identification of a blurred pattern might permit the estimation of its position. Therefore a recent refinement of an experimental procedure by Braun and Sagi (1990) deserves special mention. They made the "two-visual system" concept even stronger by showing that while loading attention (by asking the observer to identify a letter) it was possible to carry out simultaneously the detection of texture gradients without interfering with the identification task, but, surprisingly, perceptual grouping affected identification. Now Jochen Braun and I joined forces in a systematic search for perceptual tasks that can be performed without interfering with the identification task carried on the same time. Of course, it is crucial that focal attention is engaged during the entire identification task, and it is avoided that focal attention quickly performs identification and quickly inspects the surround for further information collection.   That is one reason why in our masking paradigm it is so crucial to determine the fastest search speeds.   Regardless of the outcome of these studies, the existence of two visual systems that operate simultaneously, and one can perceive certain features in the environment without attention is a most interesting possible insight in brain research.

This renaissance of attentional studies by cognitive and perceptual psychologists is paralleled by the neurophysiologists Robert Desimone and his collaborators (Moran and Desimone 1985, Desimone and Ungerleider 1989), who found neurons in V4 whose firing for certain trigger features changes in accordance to the focal attention of the monkey.   I highly recommend to my psychologist colleagues to study carefully these important papers. Obviously, no speculation can answer Question 15) and only the neurophysiologist can work on this question, but has to use some of the sophisticated techniques of the psychophysicist.   Whether some of the novel non-invasive techniques, such as PET-scans, might have adequate spatio-temporal resolution to permit psychologists to work on locus questions, without the help of physiologists, is an important technical question that only time can answer.   [As the reader might have noticed I did not list any of the crucial technical breakthroughs in the making by phrasing them in question form, such as, "would it be possible to perform micro-surgery by ablating specific brain tissues with monoclonal antibodies" or "could the direct optical inspection of brain activity method, as pioneered by Grinvald and his collaborators (Ts'o et al.,1990) be further improved such that brains with many sulci can be studied in vivo?"]

The interesting Question 17) stems from the work of Enns and his collaborators, who made up arrays of little Necker-cubes with targets (cubes) favoring one kind of 3-D depth organization amidst cubes biased in the dual 3-D organization. This depth from 3-D perspective cues yields preattentive texture segmentation ("pop-out"), with the implication that in addition to the textons of brightness, color, orientation and aspect ratio of elongated blobs, flicker, motion, and stereopsis, even perceived depth in 2-D perspective drawings might act as a texton.   Accordingly some primitive 3-D processing of depth based on mechanisms processing shape-(depth)- from-shading and occlusion, might belong to early vision. Of course, the second possibility is that early vision cannot be solely bottom-up but must include (besides focal attention) some top-down processes too! This is Question 22), a fundamental problem in brain research. Indeed if we can demonstrate that most of the perceptual processes in global stereopsis, motion perception, and texture discrimination are essentially bottom-up, their linking to present neurophysiological results obtained in early cortical areas of V1, V2, V3, V4 or MT, is now possible. This is also in agreement with David Marr's view of computational vision being basically bottom-up. However, in a recent short monograph *Visual processing: Computational, Psychophysical, and Cognitive Research* Roger Watt (1988) argues for algorithms in early vision that are under the control of high-level processes and memory. This is an unsettled problem.   For example Jih Jie Chang and Julesz (1990) have shown that depth from shape-from-shading processes yields a monocularly strong percept, this depth is completely dominated by global stereopsis. This would argue again that global stereopsis is a bottom-up process that overrides many top-down processes or operates independently of them.   On the other hand there are many perceptual phenomena that depend on high-level processes, including semantic memory. A well-known example in cognition is the word-superiority effect, that denotes the fact that the recognition of certain letters is superior when contained in an English word than in a non-sense word. This makes a lot of sense, since recognition of letters and words is surely a high-level semantic process. However, as Naomi Weisstein and Charlie Harris (1974) have shown the same phenomenon exists in visual perception that they named the object-superiority effect. The detection of a line segment of certain orientation was greatly improved if the segment belonged to a line drawing that portrayed a 3-D object, deteriorated if the segment belonged to a random line-drawing, and was the worst if the line segment was shown in isolation.

Of course, object and form recognition are enigmatic, high-level processes in which semantics and Gestalt organization play a prominent role. Therefore the experiments by Gorea and Julesz (1990) are of special interest, who converted the object-superiority effect from an identification paradigm into a detection paradigm as follows. We presented an array of oblique line segments into which three horizontal and one vertical line segments were inserted. These four non-oblique line segments were clumped either in random fashion or representing a primitive human face (two vertical lines representing the eyes, the vertical line segment between the eyes representing the nose, and the bottom horizontal line segment portraying the mouth). Observers where not aware that occasionally a face was presented, and were only asked to detect any line segment that was not oblique. Surprisingly, observers detected the horizontal and vertical line segments significantly better if they belonged to the face, than when they belonged to the random clump (or to a symmetric four line segment symmetric pattern) that was not a face. I always have assumed that the detection of a line segment in a texture [based on a texture(texton)-gradient between adjacent orientation differences] was a simple parallel bottom-up process. And here is a case that even such a simple perceptual task might depend on top-down processing! I say "might", because the effect is very small (though statistically significant) and only four observers were tried. Because of the importance of this experiment, I would like if others would repeat this study of ours with more observers and perhaps inventing some other experimental design!

Next I take Question 28) from the list. This is related to the interesting experiments by Biederman (1985) who would flash a complex image briefly and report that his observers could correctly identify many objects and scenes contained in the image. This finding always puzzled me. However in the light of the Saarinen and Julesz experiments (discussed above) observers can identify numbers at 30 msec/item rates or even faster, which means that in an image flashed for about 150 msec it is possible to capture 5-7 items. If these items are objects in a real-life scene then these items are meaningful chunks with well-known relations between them. Now we know from the work of Chase and Simon (1973) that a grand-master of chess can reconstruct a briefly presented chess board from 5-7 chunks, but only if the board configuration resulted from a real game. If the board configuration was random, grand-masters' recall was not better than that of beginners. Since real-life scenes are rich in thousands of semantic cues that relate objects to each other, and many subset of objects that relate to each other form a chunk, it is not so surprising anymore that Biederman's observers do well with briefly presented TV frames. After all every human who is not legally blind is a grand-master of visual perception.

Time does not permit me to dwell on each of the many questions I listed. Most of them are self-explanatory, at least to the experts, and this article is written to the experts. Furthermore, I hope that for my   colleagues each question evokes rather different associations and I do not want to interfere with this creative process.   I am also convinced that many of my colleagues would add other questions to my list and remove several existing ones. Indeed, it is this lack of consensus in psychobiology that reminds me of the fragmentation of molecular biologists before the discovery of the structure and role of DNA. As a matter of fact, each time I read my own list some new questions come to my mind, and I am surprised that I did not think of them before. I would be grateful if colleagues, particularly those believe in my own paradigms would add to this list their strategic questions.

Just to illustrate what I mean by free associations about these problems, let me discuss Question 1) which I regard fundamental, but did not work on it directly (except for an interesting collaborative effort with Bruno Breitmeyer that was not conclusive, because of technical difficulties with the after-glow of phosphors in video-monitors, see Breitmeyer et al., 1982). If a worker in AI ponders on this question, the 3-level hierarchy of "early-sketch" - "2.5-sketch" - "stick figures" by David Marr (1982) will come to his mind. My own first thought is related to Emmert's law, and my first demonstrations of it to my introductory psychology classes.   I "burn" a bright spot on the retina of my students with a flash whose afterimage can last for several second and ask them to   "project" this afterimage on actual surfaces in the class-room at different distances from them. Since the size of the afterimage perceptually zooms in inverse fashion (to achieve size-constancy) with physical distance, one can appreciate that the problem of retinal coordinates and environmental coordinates are intimately coupled in early vision. I have many other associations in mind and leave the rest to my readers.

For illustration, let me elaborate on the Mathematical Question 2), also called the "aperture problem" by Wallach and O'Connell (1953). When a grating or a single line is moving behind an aperture such that the terminators of the line are hidden, the direction of the motion is ambiguous. Without additional cues usually one perceives motion at right angles to the moving gratings. Adelson and Movshon (1982), and Hildereth (1984) have studied this problem in detail. However, it was Reichardt and Egelhaaf (1988) who pointed out mathematically that the ambiguity of motion in an aperture (called "the aperture problem"), contrary to common belief, can be locally solved by two correlation-kind of motion detectors. Recently, Ramachandran (1990) made contributions to this problem by showing with drifting plaid patterns the aperture problem could be solved as long as the plaids were not transparent.   However, if one grating of the plaid pattern appeared transparent, so that the other grating was perceived as being behind the first, the two gratings would drift independently from each other in ambiguous directions. This demonstration serves to illustrate Ramachandran's view of the many bags of tricks the visual system uses in trying to disambiguate ill-posed problems. On the other hand, Werner Reichardt and his collaborators look for some general mathematical principles that are necessary and sufficient to solve the aperture problem. I think both approaches are useful in vision research.

It is not by error that I included Question 27) in my list in spite of calling it a metascientific problem in the Introduction. Somehow it is my feeling that with luck we can add a little more to Johannes Müller's doctrine of specific nerve energies.   One thing that comes to my mind is optokinetic movement perception.   During my sabbatical rear in 1976 in Günther Baumgartner's neurological laboratory at the University of Zürich I watch some experiments by the neurophysiologist Volker Henn and his collaborators. They would probe with a single microelectrode some vestibular neurons in the monkey cortex, while the chair of the monkey was rotated or an optical cylinder with vertical stripes was rotated around the monkey.   Similarly to humans both stimuli induced the percept of self-movement (similarly to the ambiguity of not knowing in a train whether one's train started to move or a nearby train started to leave the platform). The novelty was in Henn's experiments that he was recording from a neuron one synapse away from the vestibular organ, and this neuron responded the same way for mechanical acceleration of the chair, or optical acceleration of the grating.   Although Johannes Müller (1844) emphasized in his Handbook of Human Physiology (that I read in German recently and translated into English some relevant paragraphs, see Julesz, 1991, in press) that the same visual percepts (sensations) can be evoked either by visual stimulation or by producing mechanical or electrical phosphenes (e.g. by pressing slightly against the eye-ball), the situation here is slightly different. If, for instance, the vestibular neurons that are tuned to mechanical acceleration, or the visual neurons that are tuned to temporal acceleration of spatial gratings are very close in the brain to the neurons that respond equally to both, than there are two possibilities: Either this neuron is merely combining the two modalities and transmitting them to a further stage where the "specific nerve energy for perceiving self-motion" resides, or this neuron belongs to the neural structure that elicits the percept of circular vection. It might take decades to probe the brain in order to find out whether there are some higher centers that give rise to circular vection too, in addition to the ones already found.   However, if one is convinced that no higher center exists (a rather difficult claim to make), then there is some hope to elucidate the mysteries of qualia. Of course, similar arguments can be made for V4 as the earliest brain structure that gives rise to the sensation of color, but I believe that the vestibular neurons belong to a simpler system that is more appropriate for probing the doctrine of specific nerve energies.

Finally, I assume that the listed questions can be understood as stated, even though some are clearer than others.   For instance, the psychological Question 2) might not be obvious to some. Many colleagues believe that the periphery's poorer resolution can be compensated by presenting, say, letters with increased sizes at larger and larger eccentricites.   However, this is nor always the case. In a classic paper, Sperling and Melchner (1978) showed, that when such stimuli are shown (i.e. stimuli compensated by the cortical magnification factor) these cannot be focally attended as well as arrays of letters with equal sizes (provided the most peripheral letters are still resolvable.

Similarly, my last question: Question 32) might require further ellucidation. We often take for granted that 3-D stimuli that cast their 2-D projections on our retinae are perceived as being actually "projected out" in the environment at some x,y,z position.   That the latter is not an ability to be taken for granted is clearly demonstrated by Békésy (1960) who would stimulate two distant areas of the skin and ask of the perceived position in between, and even try whether one could learn to sense the position projected outside from the skin surface.

4. Conclusion.

In this paper I listed some questions that I regard strategic for vision research. Most of my questions are concerned about early vision, a sub-field of visual perception that is devoted mainly to the study of bottom-up processes except for focal attention, which is a clearly top-down process. This emphasis on early vision is brought about the present state of brain research, where recent advances in neurophysiology opened up the study of early processing stages in the monkey cortex. I hinted at some questions that require higher mental functions to understand form recognition and Gestalt Organization. I touched upon some "taboo" questions, such as speculations about the nature of "consciousness", since it is my belief that psychology without consciousness is as boring as mathematics would be if the concept of infinity were censored out. Nevertheless, the definition of a good scientist (according to a friend and former mentor of mine, John R. Pierce) is someone who does not work on problems he would like to solve, but rather on problems that he thinks can be solved, but are so complex that their solution requires utmost concentration and devotion. It is in this spirit that I pose my list of questions, and hope that these questions will inspire some colleagues in search for a topic. I hope that many other questions will be raised by my colleagues to expand my list in order to enhance progress of vision research.

REFERENCES

Adelson, E.H. and J.A. Movshon,1982, "The phenomenal coherence of moving patterns," *Nature,* **300**, 523-525.

Bergen, J.R. and B. Julesz, 1983, "Parallel versus serial processing in rapid pattern discrimination," *Nature* **303**, 696-698.

Biederman, I., 1985, "Recognition by components.: A theory of object recognition," *Computer Vision Graphics and Image Processing,* **32**, 29-73.

Breitmeyer, B.G., W. Kropfl, and B. Julesz, 1982, "The existence and role of retinotopic and spatiotopic forms of visual persistence," *Acta Psychologica* **52**, 175-196.

Braun, J. and D. Sagi, 1990, "Vision outside the focus of attention," *Perception and Psychophys.* **48**, 45-58.

Bülthoff, H.H. and H.A. Mallot, 1990, "Integration of stereo, shading, and texture," in *AI and the eye,* edited by A. Blake, and T. Troscianco, (J. Wiley, New York), p. 119-146.

Chang, J.J. and B. Julesz, 1990, "Low-level processing of disparity-tuned binocular neurons takes precedence of shape from shading," *Invest. Ophthalm. and Visual Sci.* **31**(4), 525.

Chase, W.G. and H.A. Simon, 1973, "The mind's eye in chess," in *Visual information processing,* edited by W.G. Chase (Academic Press, New York), pp. 218-281.

Desimone, R. and L.G. Ungerleider, 1989, "Neural mechanisms of visual processing in monkeys," in *Handbook of Neurophysiology. Vol.2,* edited by F. Boller, and J. Grafman (Elsevier Science Publ.), 267-269.

Fender, D. and B. Julesz, 1967, "Extension of Panum's fusional area in binocularly stabilized vision," *Jour. Opt. Soc. Am.* **57**(6), 819-30.

Fogel, I. and D. Sagi, 1989, Gabor filters as texture discriminators. *Biol. Cybern.* **61**, 103-113.

Gorea, A. and B. Julesz, 1990, "Context superiority in a detection task with line-element stimuli: a low-level effect," *Perception,* **19**, 5-16.

Gurnsey, R. and R. Browse, 1987, "Micropattern properties and presentation conditions influencing visual texture discrimination," *Perception and Psychophys.* **41**:239-252..

Helmholtz, H. von, 1896, *Handbuch der Physiologischen Optik. Dritter Abschnitt, Zweite Auflage,* Voss, Hamburg.

Hildereth, E., 1984, "The measurement of visual motion," The MIT Press, Cambridge, MA.

Hoffman, D.D. and W. Richards, 1984, "Parts of recognition," *Cognition* **18**, 65-96.

Horn, B.K.P. and M.J. Brooks, 1985, "The variational approach to shape from shading," *MIT Artificial Intelligence Lab. Memo.* No. **813**, 1-32.

Julesz, B., 1960, "Binocular depth perception of computer-generated patterns," *Bell Syst. Tech. Jour.* **39**, 1125-62.

Julesz, B., 1962, "Visual pattern discrimination," *IRE Trans. Info. Theory,* **IT-8**, 84-92.

Julesz, B., 1981, "Textons, the elements of texture perception and their interactions," *Nature* **290**, 91-97.

Julesz, B., 1986, "Stereoscopic vision," *Vision Res.* **26**(9), 1601-1612.

Julesz, B., 1990a, (in press), "Early vision is bottom-up, except for focal attention," in *Symposium 55: The Brain,* (Cold Spring Harbor Laboratory, New York).

Julesz, B., 1990b, "AI and early vision," in *AI and the eye,* edited by A. Blake, and T. Troscianco, (J. Wiley, New York), pp. 9-20.

Julesz, B., 1991a (in press), "Consciousness and focal attention," *Behavioral and Brain Sciences.*

Julesz, B., 1991b (in press), "Early vision and focal attention," *Rev. Mod. Phys.*

Kanizsa, G., 1976, "Subjective contours," *Scientific American,* **234**, 48-52.

Karni, A. and D. Sagi, 1990, "Where practice makes perfect in texture discrimination,"*The Weizmann Ins. of Science Technical Report CS90-02,* Jan.1990.

Kröse, B.J.A., 1987, Local structure analysers as determinants of preattentive pattern discrimination," *Biol. Cybern.* **55**, 289-298.

Kröse, B.J.A. and B. Julesz, 1989, "The control and speed of shifts of attention," *Vision Res.* **29**(11), 1607-1619.

Malik, J.M.A. and P. Perona, 1990, Preattentive texture discrimination with early vision mechanisms," *J. Opt. Soc. Am.* **A 7**(5), 923-932.

Marr, D., 1982, *Vision,* (Freeman, San Francisco,CA).

Moran, T. and R. Desimone, 1985, "Selective attention gates visual processing in the extrastriate cortex," *Science* **229**, 782-784.

Müller, J., 1844, *Handbuch der Physiologie des Menschen, 4th Edition,* Coblenz, Verlag von J. Hölscher.

Nakayama, K. and M. Mackeben, 1989, "Sustained and transient components of focal visual attention," *Vision Res,* **29,** 1631-164.

Penrose, R., 1989, *The emperor's new mind,* (Oxford U. Press, Oxford).

Pentland, A.P., 1986 "Shading into texture," *Artificial Intelligence,* **29,** 147 -170.

Posner, M.I., 1978, "Orienting of attention," *Quart. J. of Experimental Psychology,* **32,** 3-25.

Ramachandran, V.S., 1988, "Perception of shape from shading. *Nature* **331,** 163-166.

Ramachandran, V.S., 1990, "Visual perception in people and machines," In *AI and the eye,* edited by A. Blake, and T. Troscianco, (J. Wiley, New York), pp. 21-77.

Reichardt, W. and M. Egelhaaf, 1988, *Naturwissenschaften* **75,** 313-315.

Rubenstein, B. and D. Sagi, 1989, "Texture variability across the orientation spectrum can yield asymmetry in texture discrimination," *Perception* **18,** 517.

Saarinen, J. and B. Julesz, to be submitted, "The speed of attentional shifts in the visual field."

Sagi, D. and B. Julesz, 1985, "'Where' and 'what' in vision," *Science* **228,** 1217-1219.

Sagi, D. and B. Julesz, 1986, "Enhanced detection in the aperture of focal attention during simple discrimination tasks," *Nature* **321,** 693-695.

Schumann, F., 1904, "Einige Beobachtungen über die Zusammenfassung von Gesichtseindrucken zu Einheiten," *Psychol. Stud.* **1,** 1-32.

Sternberg, S., 1966, "High speed scanning in human memory. *Science* **153,** 652-654.

Treisman, A. and G. Gelade, 1980, "A feature-integration theory of attention," *Cognitive Psychology* **12,** 97-136.

Ts'o, D.Y., R.D. Frostig, E.E. Lieke, A. Grinvald, 1990, "Functional organization of primate visual cortex revealed by high resolution optical imaging," *Science* **249,** 417-420.

von Békésy, Georg, 1960, *Experiments in Hearing,* translated and edited by E.G. Wever, (McGraw-Hill, New York), p. 5.

von der Heydt, R., E. Peterhans, and G. Baumgartner, 1988, "Illusory contours and cortical neuron responses," *Science* **224,** 1260-1262.

Wallach, H. and D.N. O'Connell, 1953, "The kinetic depth effect," *J. of Exp. Psychol.* **45,** 205-217.

Watt, R., 1988, *Visual processing: Computational, psychophysical, and cognitive research*, (Lawrence Erlbaum Associates, Hillside, NJ).

Weichselgartner, E. and G. Sperling, 1987, "Dynamics of automatic and controlled visual attention," *Science* **238**, 778-779.

Weisstein, N. and C.S. Harris, 1974, "Visual detection of line segments: An object-superiority effect," *Science*, **186**, 752-755.

Williams, D.W. and B. Julesz, 1991, (in press), "Filters versus textons in human and machine texture discrimination," in *Neural networks for human and machine perception*, edited by H. Wechsler, (Academic Press, Orlando, FL).

Wolfe, J.M. and K.R. Cave, 1990, "Deploying visual attention: The guided search model," in *AI and the eye*, edited by A. Blake and T. Troscianco (J. Wiley, New York), p. 79-103.